Theoretical Chemistry and Computational Modelling

Modern Chemistry is unthinkable without the achievements of Theoretical and Computational Chemistry. As a matter of fact, these disciplines are now a mandatory tool for the molecular sciences and they will undoubtedly mark the new era that lies ahead of us. To this end, in 2005, experts from several European universities joined forces under the coordination of the Universidad Autónoma de Madrid, to launch the *European Masters Course on Theoretical Chemistry and Computational Modeling* (TCCM). The aim of this course is to develop scientists who are able to address a wide range of problems in modern chemical, physical, and biological sciences via a combination of theoretical and computational tools. The book series, *Theoretical Chemistry and Computational Modeling*, has been designed by the editorial board to further facilitate the training and formation of new generations of computational and theoretical chemists.

Prof. Manuel Alcami
Departamento de Química
Facultad de Ciencias, Módulo 13
Universidad Autónoma de Madrid
28049 Madrid, Spain

Prof. Ria Broer
Theoretical Chemistry
Zernike Institute for Advanced Materials
Rijksuniversiteit Groningen
Nijenborgh 4
9747 AG Groningen, The Netherlands

Dr. Monica Calatayud
Laboratoire de Chimie Théorique
Université Pierre et Marie Curie, Paris 06
4 place Jussieu
75252 Paris Cedex 05, France

Prof. Arnout Ceulemans
Departement Scheikunde
Katholieke Universiteit Leuven
Celestijnenlaan 200F
3001 Leuven, Belgium

Prof. Antonio Laganà
Dipartimento di Chimica
Università degli Studi di Perugia
via Elce di Sotto 8
06123 Perugia, Italy

Prof. Colin Marsden
Laboratoire de Chimie
et Physique Quantiques
Université Paul Sabatier, Toulouse 3
118 route de Narbonne
31062 Toulouse Cedex 09, France

Prof. Otilia Mo
Departamento de Química
Facultad de Ciencias, Módulo 13
Universidad Autónoma de Madrid
28049 Madrid, Spain

Prof. Ignacio Nebot
Institut de Ciència Molecular
Parc Científic de la Universitat de València
Catedrático José Beltrán Martínez, no. 2
46980 Paterna (Valencia), Spain

Prof. Minh Tho Nguyen
Departement Scheikunde
Katholieke Universiteit Leuven
Celestijnenlaan 200F
3001 Leuven, Belgium

Prof. Maurizio Persico
Dipartimento di Chimica e Chimica
Industriale
Università di Pisa
Via Risorgimento 35
56126 Pisa, Italy

Prof. Maria Joao Ramos
Chemistry Department
Universidade do Porto
Rua do Campo Alegre, 687
4169-007 Porto, Portugal

Prof. Manuel Yáñez
Departamento de Química
Facultad de Ciencias, Módulo 13
Universidad Autónoma de Madrid
28049 Madrid, Spain

More information about this series at http://www.springer.com/series/10635

Coen de Graaf · Ria Broer

Magnetic Interactions in Molecules and Solids

Coen de Graaf
Department of Physical and Inorganic
  Chemistry
Universitat Rovira i Virgili / ICREA
Tarragona
Spain

Ria Broer
Zernike Institute for Advanced Materials
University of Groningen
Groningen
The Netherlands

ISSN 2214-4714   ISSN 2214-4722   (electronic)
Theoretical Chemistry and Computational Modelling
ISBN 978-3-319-22950-8   ISBN 978-3-319-22951-5   (eBook)
DOI 10.1007/978-3-319-22951-5

Library of Congress Control Number: 2015947103

Springer Cham Heidelberg New York Dordrecht London

Printed on acid-free paper

Springer International Publishing AG Switzerland is part of Springer Science+Business Media
(www.springer.com)

To W.C. Nieuwpoort

Preface

Magnetic interactions are not only fascinating from an academic viewpoint, they also play an increasingly important role in chemistry, especially in the chemistry that is aimed at designing materials with predefined properties. Many of these materials are magnetic, either in their ground states or by external perturbation and have found their way into real-world applications as molecular switches, sensors or memories. Although magnetic interactions are commonly orders of magnitude weaker than other interactions like covalent bonding, due to these interactions small changes in composition or external conditions may have huge consequences for the properties. Think for example of perovskite-type manganese oxides, where chemical doping affects the interplay between magnetic and electric properties, leading to *giant* or *collossal* magnetic resistance. An obvious example dealing with molecular (non-bulk) moieties can be found in the design of single-molecule magnets. Obtaining systems with tailor-made properties heavily depends on our knowledge of the interactions between local magnetic sites.

This textbook aims to explain the theoretical basis of magnetic interactions at a level that will be useful for master's students in chemistry. Although it has been written as a volume in the series "Theoretical and Computational Chemistry", the book is intended to be also helpful for students of physical, inorganic and organic chemistry. Most chemistry textbooks give only a brief general introduction, whereas textbooks treating magnetic interactions at a more advanced level are mostly written from the perspective of solid-state physics, aiming at physics students.

This volume gives a treatment of magnetic interactions in terms of the phenomenological spin Hamiltonians that have been such powerful tools in chemistry and physics in the past half century. On the other hand, it also explains the magnetic properties using many-electron quantum mechanical models, first at a simple level and then working towards more and more advanced and accurate treatments. Connecting the two perspectives is an essential aspect of the book. It makes clear that in many cases one can derive magnetic coupling parameters not only from experiment, but also, independently, from accurate *ab initio* calculations. Combining the two approaches leads, in addition, to a deeper understanding of the relation between physical phenomena and basic properties and how we can influence these. Think for example of magnetic anisotropy and spin-orbit coupling.

Throughout the book the text is interlarded with exercises, stimulating the students to not only read but also verify the assertions and perform (parts of) derivations by themselves. In addition, each chapter ends with a number of problems that can be used to check whether the material has been understood.

The first chapter of this volume introduces a number of basic concepts and tools necessary for the development of the theories and methods treated in the following chapters. It explains various ways to generate many-electron spin-adapted functions, gives an introduction to perturbation theories and to effective Hamiltonian theory. Chapter 2 treats atoms with and without an external magnetic field. This is followed by a chapter on systems containing more than one magnetic center. In this chapter the phenomenological Hamiltonians are introduced, beginning with the Heisenberg and the Ising Hamiltonian and ending with Hamiltonians that include biquadratic, cyclic or anisotropic exchange. Chapter 4 explains how quantum chemical methods, reaching from simple mean field methods to accurate models, can help to understand the magnetic properties. The simple models can give a qualitative understanding of the phenomena. The more accurate models, such as post Hartree-Fock models like DDCI, CASPT2 and NEVPT2 or broken symmetry models based on density functional theory, are able to produce accurate predictions of the energies and wave functions of the relevant states. Making accurate computations is one thing, mapping the results back onto the intuitive models yielding parameters that can be compared with the ones deduced from experiments is another. Effective Hamiltonian theory is a powerful tool to make these connections, as shown in Chap. 5. The last chapter explains how the magnetic interactions in solid-state compounds can be treated, with embedded cluster models and with periodic approaches. It gives an account of the double exchange mechanism in mixed valence systems, explaining the Goodenough-Kanamori rules. Finally, an account is given of spin wave theory for (anti-)ferromagnets.

The book covers a full Master's course, but a shorter course can be distilled from it in many ways. One of them includes Chap. 2, the first two sections of Chap. 3 and optionally one of the subsections of 3.4 to get acquainted with the spin Hamiltonian formalism. After that, Sects. 4.1.1 and 4.1.2 combined with Sects. 4.3.1, 4.3.2 and 4.3.4 can be studied to connect the quantitative and qualitative computational viewpoints of magnetic interactions. From Chap. 5, we recommend to include Sects. 5.1.1 and 5.3, which provide us with the basic tools for analysis. If time permits, one can close the short course with a brief account on some issues related to the solid state: Sects. 6.3 and 6.5 provide some basic notions on this topic.

We end by noting that the outstanding book by the late Prof. Olivier Kahn, O. Kahn, Molecular Magnetism, VCH Publishers, 1993, has been an inspiration for the entire book.

Tarragona   Coen de Graaf
Groningen   Ria Broer
July 2015

Acknowledgments

This book can be considered to a large extent as a product of sharing knowledge in the so-called *Jujols* community over the past 25 years. Without the continual interactions during conferences, visits and intense collaborations on many aspects of the theoretical description of magnetic phenomena in molecules and solids, this book would never have reached the degree of completeness and clarity that we hope to have reached. Special thanks are due to Jean-Paul Malrieu, Nathalie Guihéry, Carmen Calzado, Rosa Caballol, Rémi Maurice, Celestino Angeli, Nicolas Ferré, Eliseo Ruiz, Joan Cano, Remco Havenith, Alex Domingo and Gerjan Lof for inspiration, clarifications, resolving doubts, affirmations, corrections, proofreading, etc., during the process of writing the book. We are grateful to the late Olivier Kahn, for sharing his broad and deep knowledge with us, in person, but also through the great legacy of his book on molecular magnetism.

Contents

1 Basic Concepts 1
  1.1 Slater Determinants and Slater–Condon Rules 1
  1.2 Generation of Many Electron Spin Functions 5
      1.2.1 Many Electron Spin Functions by Projection 9
      1.2.2 Spin Functions by Diagonalization 10
      1.2.3 Genealogical Approach 12
      1.2.4 Final Remarks 20
  1.3 Perturbation Theory 21
      1.3.1 Rayleigh–Schrödinger Perturbation Theory 21
      1.3.2 Møller–Plesset Perturbation Theory 25
      1.3.3 Quasi-Degenerate Perturbation Theory 27
  1.4 Effective Hamiltonian Theory 28
  Problems 31
  References 32

2 One Magnetic Center 33
  2.1 Atomic Magnetic Moments 33
  2.2 The Eigenstates of Many-Electron Atoms 34
  2.3 Further Removal of the Degeneracy of the N-electron States 39
      2.3.1 Zero Field Splitting 39
      2.3.2 Splitting in an External Magnetic Field 43
      2.3.3 Combining ZFS and the External Magnetic Field 52
  Problems 54
  References 57

3 Two (or More) Magnetic Centers 59
  3.1 Localized Versus Delocalized Description
      of the Two-Electron/Two-Orbital Problem 59
  3.2 Model Spin Hamiltonians for Isotropic Interactions 68
      3.2.1 Heisenberg Hamiltonian 69
      3.2.2 Ising Hamiltonian 74

	3.2.3	Comparing the Heisenberg and Ising Hamiltonians.	76
3.3		From Micro to Macro: The Bottom-Up Approach	77
	3.3.1	Monte Carlo Simulations, Renormalization Group Theory	81
3.4		Complex Interactions	87
	3.4.1	Biquadratic Exchange	88
	3.4.2	Four-Center Interactions	90
	3.4.3	Anisotropic Exchange	95
Problems			101
References			102

4   From Orbital Models to Accurate Predictions 105
4.1		Qualitative Valence-Only Models	105
	4.1.1	The Kahn–Briat Model	105
	4.1.2	The Hay–Thibeault–Hoffmann Model	108
	4.1.3	McConnell's Model	110
4.2		Magnetostructural Correlations	113
4.3		Accurate Computational Models	120
	4.3.1	The Reference Wave Function and Excited Determinants	121
	4.3.2	Difference Dedicated Configuration Interaction	123
	4.3.3	Multireference Perturbation Theory	127
	4.3.4	Spin Unrestricted Methods	131
	4.3.5	Alternatives to the Broken Symmetry Approach	136
Problems			138
References			139

5   Towards a Quantitative Understanding 141
5.1		Decomposition of the Magnetic Coupling	141
	5.1.1	Valence Mechanisms	142
	5.1.2	Beyond the Valence Space	148
	5.1.3	Decomposition with MRPT2	151
5.2		Mapping Back on a Valence-Only Model	152
5.3		Analysis with Single Determinant Methods	157
5.4		Analysis of Complex Interactions	159
	5.4.1	Decomposition of the Biquadratic Exchange	159
	5.4.2	Decomposition of the Four-Center Interactions	166
	5.4.3	Complex Interactions with Single Determinant Approaches	168
Problems			174
References			175

6   Magnetism and Conduction 177
   6.1   Electron Hopping................................. 177
   6.2   Double Exchange................................. 182
   6.3   A Quantum Chemical Approach to Magnetic Interactions
         in the Solid State 189
         6.3.1   Embedded Cluster Approach.................... 190
         6.3.2   Periodic Calculations 193
   6.4   Goodenough–Kanamori Rules 197
   6.5   Spin Waves for Ferromagnets 204
   Problems 211
   References 212

**Appendix A: Effect of the $\hat{l}$ Operator and the Matrix Elements
      of the p and d Orbitals** 213

**Appendix B: Effect of the $\hat{S}$ Operator and the Matrix Elements
      for $\frac{1}{2} \leq S \leq \frac{5}{2}$** 215

**Appendix C: Matrix Representation of the ZFS Model
      Hamiltonian** 217

Appendix D: Analytical Expressions for $\chi(T)$ 219

Appendix E: Solutions 223

Index 243

Acronyms

AF	Antiferromagnetic
BS	Broken Symmetry
CAS (n, m)	Complete Active Space with n electrons and m orbitals
CASPT2	Complete Active Space second-order Perturbation Theory
CASSCF	Complete Active Space Self-Consistent Field
CI	Configuration Interaction
CISD	Configuration Interaction of Singles and Doubles
CSF	Configuration State Function
DDCI	Difference Dedicated Configuration Interaction
DFT	Density Functional Theory
F	Ferromagnetic
GK	Goodenough-Kanamori
HS	High Spin
HTH	Hay-Thibeault-Hoffmann
IR	Irreducible representation
KS	Kohn-Sham
LDA	Local Density Approximation
LMCT	Ligand-to-Metal Charge Transfer
LS	Low Spin
MLCT	Metal-to-Ligand Charge Transfer
MO	Molecular Orbital
MR	Multideterminantal/Multiconfigurational reference; Multireference
NEVPT2	N-Electron Valence state second-order Perturbation Theory
NH	Non Hund
QDPT	Quasi Degenerate Perturbation Theory
REKS	Restricted Ensemble Kohn-Sham
RHF	Restricted Hartree Fock
ROKS	Restricted Open-shell Kohn-Sham
VB	Valence Bond
WF	Wave function

ZFS	Zero-field splitting
ψ, φ, ϕ	one-electron functions, orbitals
Ψ	N-electron wave function
Φ	Slater determinant
$\widetilde{\psi}$	Projection of Ψ on a model space
$\widetilde{\psi}'$	Normalized projection of Ψ on a model space
$\widetilde{\psi}^{\perp}$	Orthonormalized projection of Ψ on a model space
$\widetilde{\psi}^{\dagger}$	Biorthogonal projection of Ψ on a model space
$\widetilde{\psi}'^{\dagger}$	Normalized biorthogonal projection of Ψ on a model space
$E^{(n)}$	n-th order correction to the energy
$\psi^{(n)}$	n-th order correction to the wave function

Chapter 1
Basic Concepts

Abstract In this chapter we examine some basic concepts of quantum chemistry to give a solid foundation for the other chapters. We do not pretend to review all the basics of quantum mechanics but rather focus on some specific topics that are central in the theoretical description of magnetic phenomena in molecules and extended systems. First, we will shortly review the Slater–Condon rules for the matrix elements between Slater determinants, then we will extensively discuss the generation of spin functions. Perturbation theory and effective Hamiltonians are fundamental tools for understanding and to capture the complex physics of open shell systems in simpler concepts. Therefore, the last three sections of this introductory chapter are dedicated to standard Rayleigh–Schrödinger perturbation theory, quasi-degenerate perturbation theory and the construction of effective Hamiltonians.

1.1 Slater Determinants and Slater–Condon Rules

The Slater determinant is the central entity in molecular orbital theory. The exact N-electron wave function of a stationary molecule in the Born-Oppenheimer approximation is a $4N$-dimensional object that depends on the three spatial coordinates and a spin coordinate of the N electrons in the system. This object is of course too complicated for any practical application and is, in first approximation, replaced by a product of N orthonormal 4-dimensional functions that each depend on the coordinates of only one of the electrons in the system.

$$\Psi(x_1, y_1, z_1, \sigma_1, x_2, y_2, z_2, \sigma_2, \ldots, x_N, y_N, z_N, \sigma_N)$$
$$= \phi_a(x_1, y_1, z_1, \sigma_1)\phi_b(x_2, y_2, z_2, \sigma_2)\ldots\phi_\omega(x_N, y_N, z_N, \sigma_N) \qquad (1.1)$$

These one-electron functions are commonly referred to as spin orbitals and the product is known as the Hartree product Π. Obviously, the product suffers from important deficiencies with respect to the foundations of Quantum Mechanics. The wave function is not antisymmetric with respect to the permutation of any two electrons, and

© Springer International Publishing Switzerland 2016
C. Graaf and R. Broer, *Magnetic Interactions in Molecules and Solids*,
Theoretical Chemistry and Computational Modelling,
DOI 10.1007/978-3-319-22951-5_1

hence, does not fulfill the Pauli principle. However, by replacing the product by a determinant

$$\Psi(1, 2, \ldots N) = \frac{1}{\sqrt{N!}} \begin{vmatrix} \phi_a(1) & \phi_b(1) & \cdots & \phi_\omega(1) \\ \phi_a(2) & \phi_b(2) & \cdots & \phi_\omega(2) \\ \vdots & & & \vdots \\ \phi_a(N) & \phi_b(N) & \cdots & \phi_\omega(N) \end{vmatrix} \quad (1.2)$$

this requirement is automatically fulfilled. Shorthand notations for this Slater determinant are

$$\Psi(1, 2, \ldots N) = |\phi_a(1)\phi_b(2) \ldots \phi_\omega(N)| = |\phi_a \phi_b \ldots \phi_\omega| \quad (1.3)$$

where only the diagonal elements of the determinant are shown, the four coordinates are compacted in one index, and the normalization factor is implicit. The one-electron functions are ordered by columns *(from left to right)* and the electrons by rows *(from top to bottom)*. An alternative, more explicit way of writing the wave function is obtained by defining an operator that antisymmetrizes the Hartree product Π

$$\Psi = \hat{A}\Pi = \hat{A}[\phi_a(1)\phi_b(2) \ldots \phi_\omega(N)] \quad (1.4)$$

with

$$\hat{A} = \frac{1}{\sqrt{N!}} \sum_{\gamma=0}^{N-1} (-1)^\gamma \hat{P}_\gamma = \frac{1}{\sqrt{N!}} \left(1 - \sum_{i<j} \hat{P}_{ij} + \sum_{i<j<k} \hat{P}_{ijk} - \ldots \right) \quad (1.5)$$

where $\hat{P}_{ij}$ permutes the electron labels i and j in the Hartree product, $\hat{P}_{ijk}$ replaces the electron labels ijk by jki and kij.

1.1 Write out explicitly the wave function $\Psi(1, 2, 3) = |\phi_a(1)\phi_b(2)\phi_c(3)|$ and show that $\Psi(2, 1, 3) = -\Psi(1, 2, 3)$. What happens to the wave function when two electrons are described by the same one-electron function?

A serious deficiency is that neither a Hartree product nor a Slater determinant can be an eigenfunction of the N-electron Hamilton operator. Therefore Ψ cannot be a solution of the time-independent electronic Schrödinger equation. The reason is that the N-electron Hamiltonian cannot be written as a sum of N one-electron Hamiltonians, due to the repulsive Coulomb interactions between the electrons. Nevertheless, in practice it turns out that we can work rather well with an approximate wave function consisting of only one Slater determinant if we choose that particular Slater determinant Ψ that yields the lowest energy expectation value $\langle \Psi | \hat{H} | \Psi \rangle$. In other words, we must vary the spin orbitals in Ψ until we have reached the lowest value of $\langle \Psi | \hat{H} | \Psi \rangle$. The variation theorem tells us that this lowest value is still above the exact ground state energy E.

This variational procedure leads to a set of equations

$$\hat{f} \phi_i = \varepsilon_i \phi_i \tag{1.6}$$

called the Hartree–Fock equations, which determine the spin orbitals in Ψ. The set of equations (1.6) can be seen as effective one-electron Schrödinger equations, whose eigenvalues ε are called one-electron energies or orbital energies. There is an operator $\hat{f}$, called Fock operator, for each electron in the molecule, and they are all identical. Much can be said about the Hartree–Fock equations, their eigenvalues ε and their eigenfunctions, the spin orbitals ϕ but here we restrict ourselves to a few aspects that are relevant later in this chapter. Firstly, $\hat{f}$ depends on the spin orbitals to be found, which has the consequence that the equations have to be solved iteratively and secondly, the energy expectation value E is not equal to the sum of the one electron energies. Summing the N individual Fock operators for the electrons of the molecule gives an N-electron Hamiltonian, $\hat{H}^{(0)}$, that is not equal to the true N-electron Hamiltonian, but that we will use later as zeroth order Hamiltonian in a perturbation expansion. All Slater determinants Φ_k, $k = 1, 2, \ldots$ that can be built from the spin orbitals of Eq. 1.6 are eigenfunctions of $\hat{H}^{(0)}$, with eigenvalues $E_k^{(0)}$ equal to the sum of the orbital energies of the spin orbitals used in Ψ_k.

The calculation of the energy of a Slater determinant and the interaction between two different Slater determinants may seem a rather complicated task given the large number of terms ($N!$) when the determinant is written in its explicit form. However, the Slater–Condon rules given in Table 1.1 establish a few simple relations to calculate matrix elements between two Slater determinants.

Table 1.1 Slater–Condon rules for the matrix elements between two Slater determinants

Matrix element	Differences	One-electron term	Two-electron term
$\langle \Phi_K | \hat{H} | \Phi_K \rangle$	0	$\sum\limits_{m}^{N} \langle \phi_m | \hat{h} | \phi_m \rangle$	$\sum\limits_{m<n}^{N} \langle \phi_m \phi_n | \frac{1-\hat{P}_{12}}{r_{12}} | \phi_m \phi_n \rangle$
$\langle \Phi_K | \hat{H} | \Phi_L \rangle$	1	$\langle \phi_m | \hat{h} | \phi_p \rangle$	$\sum\limits_{n}^{N} \langle \phi_m \phi_n | \frac{1-\hat{P}_{12}}{r_{12}} | \phi_p \phi_n \rangle$
$\langle \Phi_K | \hat{H} | \Phi_M \rangle$	2	0	$\langle \phi_m \phi_n | \frac{1-\hat{P}_{12}}{r_{12}} | \phi_p \phi_q \rangle$
$\langle \Phi_K | \hat{H} | \Phi_N \rangle$	3 or more	0	0

The entry 'differences' indicates the number of different spin orbitals in the determinants of the *bra* and *ket*

$\Phi_K = |\phi_a \phi_b \ldots \phi_m \phi_n \phi_o \ldots \phi_\omega|$
$\Phi_L = |\phi_a \phi_b \ldots \phi_p \phi_n \phi_o \ldots \phi_\omega|$
$\Phi_M = |\phi_a \phi_b \ldots \phi_p \phi_q \phi_o \ldots \phi_\omega|$
$\Phi_N = |\phi_a \phi_b \ldots \phi_p \phi_q \phi_r \ldots \phi_\omega|$

$\hat{P}_{12}$ is the permutation operator that interchanges the coordinates of electron 1 and 2

To derive these rules it is convenient to introduce two formal properties of the antisymmetrizer $\hat{A}$

$$\hat{A}\hat{A} = \sqrt{N!}\hat{A} \qquad \hat{A}\hat{H} = \hat{H}\hat{A} \tag{1.7}$$

where $\hat{H}$ is the many-electron Hamiltonian.

1.2 Write down the anti-symmetrization operator $\hat{A}$ for a two-particle wave function. Show that $\hat{A}\hat{A}$ applied on the Hartree product $\varphi_1\varphi_2$ gives the same result as applying $\sqrt{N!}\hat{A}$.

Then the energy of the determinant Φ_K can be written as

$$E = \langle \Phi_K|\hat{H}|\Phi_K\rangle = \langle \hat{A}\Pi|\hat{H}|\hat{A}\Pi\rangle = \sqrt{N!}\langle \Pi|\hat{H}|\hat{A}\Pi\rangle = \sum_{\gamma=0}^{N-1}(-1)^\gamma \langle \Pi|\hat{H}|\hat{P}_\gamma\Pi\rangle \tag{1.8}$$

and instead of working with determinants, the energy can be calculated from the Hartree products. In the first place, we take a closer look on the one-electron part of the Hamiltonian. For $\gamma = 0$ and $\hat{h}(1)$ we obtain

$$\langle \phi_a(1)\phi_b(2)\ldots\phi_\omega(N)|\hat{h}(1)|\phi_a(1)\phi_b(2)\ldots\phi_\omega(N)\rangle$$
$$= \langle \phi_a(1)|\hat{h}(1)|\phi_a(1)\rangle\langle \phi_b(2)\ldots\phi_\omega(N)|\phi_b(2)\ldots\phi_\omega(N)\rangle = \langle \phi_a|\hat{h}|\phi_a\rangle = h_a \tag{1.9}$$

Using $\hat{h}(2)$ leads to h_b and all other electron coordinates give similar results. On the contrary, the evaluation of the matrix elements with $\gamma = 1$, that is one permutation in Π, leads to zero due to the orthogonality of the orbitals. For example, the action of $\hat{P}_{12}$ gives

$$- \langle \phi_a(1)\phi_b(2)\ldots\phi_\omega(N)|\hat{h}(1)|\phi_b(1)\phi_a(2)\ldots\phi_\omega(N)\rangle$$
$$= -\langle \phi_a(1)|\hat{h}(1)|\phi_b(1)\rangle\langle \phi_b(2)\ldots\phi_\omega(N)|\phi_a(2)\ldots\phi_\omega(N)\rangle = 0 \tag{1.10}$$

where the minus sign arises from the $(-1)^\gamma$ factor in the energy expression. The two-electron part can be determined with a similar reasoning. First we focus on the $\gamma = 0$ case with the coordinates of electron 1 and 2.

$$\langle \phi_a(1)\phi_b(2)\phi_c(3)\ldots\phi_\omega(N)|\frac{1}{r_{12}}|\phi_a(1)\phi_b(2)\phi_c(3)\ldots\phi_\omega(N)\rangle$$
$$= \langle \phi_a(1)\phi_b(2)|\frac{1}{r_{12}}|\phi_a(1)\phi_b(2)\rangle\langle \phi_c(3)\ldots\phi_\omega(N)|\phi_c(3)\ldots\phi_\omega(N)\rangle$$
$$= \langle \phi_a\phi_b|\frac{1}{r_{12}}|\phi_a\phi_b\rangle = J_{ab} \tag{1.11}$$

Similar Coulomb integrals J are obtained for other combinations of electron coordinates. The next step is to see what integrals are obtained for $\gamma = 1$ interchanging electrons 1 and 2:

$$- \langle \phi_a(1)\phi_b(2)\phi_c(3)\ldots\phi_\omega(N)|\frac{1}{r_{12}}|\phi_b(1)\phi_a(2)\phi_c(3)\ldots\phi_\omega(N)\rangle$$

$$= -\langle \phi_a(1)\phi_b(2)|\frac{1}{r_{12}}|\phi_b(1)\phi_a(2)\rangle\langle\phi_c(3)\ldots\phi_\omega(N)|\phi_c(3)\ldots\phi_\omega(N)\rangle$$

$$= -\langle \phi_a\phi_b|\frac{1}{r_{12}}|\phi_b\phi_a\rangle = -K_{ab} \tag{1.12}$$

This integral is known as the exchange integral and usually written as K_{ab}. Other combinations of permutations and electron coordinates lead to similar $K's$ but higher-order permutations will always result in zero contributions due to the orthogonality. Hence, the terms can be collected and the expression given in the top row of Table 1.1 emerges.

The evaluation of the interaction matrix elements between Slater determinants with different occupations follows the same mechanics and can be derived as a useful exercise by the reader.

1.2 Generation of Many Electron Spin Functions

In a non-relativistic setting the N-electron wave function Ψ can be chosen to be also an eigenfunction of $\hat{S}^2$ and one of its components, we denote this component $\hat{S}_z$.

$$\hat{S}^2\Psi_{S,M_S} = S(S+1)\Psi_{S,M_S} \tag{1.13a}$$

$$\hat{S}_z\Psi_{S,M_S} = M_S\Psi_{S,M_S} \tag{1.13b}$$

with S the total spin quantum number and the magnetic spin quantum number M_S running from $-S$ to S in steps of 1.

1.3 Give the degeneracy of Ψ_{S,M_S} in terms of S assuming that spin-orbit coupling (see Sect. 2.1) can be neglected.

Before looking in more detail to the N-electron wave functions, we will first shortly summarize the most important aspects of the spin part of a one-electron wave function. We will follow the common practice to use lower case symbols when dealing with one-particle wave functions and uppercase for many-particle systems.

The one-electron spin functions to be considered have the quantum numbers $s = \frac{1}{2}$ and $m_s = \pm\frac{1}{2}$ and can be written in different formats:

$$|s, m_s\rangle = [|^1/_2, ^1/_2\rangle, |^1/_2, -^1/_2\rangle] = [\alpha, \beta] = [\uparrow, \downarrow] \tag{1.14}$$

where [...] denotes the set of functions. When the spatial part of the wave function is explicitly written, a similar notation can be used for spin orbitals:

$$|s, m_s\rangle = [\varphi_1, \overline{\varphi}_2] \tag{1.15}$$

where the barred orbital carries the electron with $m_s = -^1/_2$. The notations by α, β and $\varphi_1, \overline{\varphi}_2$ are most frequently used and will also be followed here. The corresponding eigenvalues of the total spin operator $\hat{s}^2$ and the z-component of it ($\hat{s}_z$) are

$$\hat{s}^2\alpha = ^1/_2 \, (^1/_2 + 1)\,\alpha = ^3/_4\alpha \qquad\qquad \hat{s}_z\alpha = ^1/_2\alpha \tag{1.16a}$$
$$\hat{s}^2\beta = -^1/_2 \, (-^1/_2 + 1)\,\beta = ^3/_4\beta \qquad\qquad \hat{s}_z\beta = -^1/_2\beta \tag{1.16b}$$

The ladder operators $\hat{s}^\pm = \hat{s}_x \pm i\hat{s}_y$ change the m_s quantum number of the spin functions by the following action

$$\hat{s}^+|s, m_s\rangle = \sqrt{s(s + 1) - m_s(m_s + 1)}|s, m_s + 1\rangle \tag{1.17}$$
$$\hat{s}^-|s, m_s\rangle = \sqrt{s(s + 1) - m_s(m_s - 1)}|s, m_s - 1\rangle$$

This leads to the following simple relations when applied to the one-electron spin functions α and β:

$$\hat{s}^+\alpha = 0 \qquad\qquad\qquad \hat{s}^-\alpha = \beta \tag{1.18a}$$
$$\hat{s}^+\beta = \alpha \qquad\qquad\qquad \hat{s}^-\beta = 0 \tag{1.18b}$$

The substitution of $\hat{s}_x = \frac{1}{2}(\hat{s}^+ + \hat{s}^-)$ and $\hat{s}_y = \frac{1}{2i}(\hat{s}^+ - \hat{s}^-)$ in the expression of the total spin operator $\hat{s}^2 = \hat{s}_x^2 + \hat{s}_y^2 + \hat{s}_z^2$ gives a simple working equation to evaluate the expectation value of $\hat{s}^2$ for spin functions:

$$\hat{s}^2 = \hat{s}^+\hat{s}^- - \hat{s}_z + \hat{s}_z^2 \tag{1.19}$$

For completeness, we also give the results of operating with $\hat{s}_x$ and $\hat{s}_y$ on α and β

$$\hat{s}_x\alpha = \frac{1}{2}\beta \qquad\qquad\qquad \hat{s}_y\alpha = -\frac{1}{2i}\beta \tag{1.20a}$$
$$\hat{s}_x\beta = \frac{1}{2}\alpha \qquad\qquad\qquad \hat{s}_y\beta = \frac{1}{2i}\alpha \tag{1.20b}$$

1.4 (a) Demonstrate that the normalization factor is one for the application of $\hat{s}^+$ to β and zero for α. (b) Derive the expression of the total spin operator $\hat{s}^2$ in terms of $\hat{s}^+$, $\hat{s}^-$ and $\hat{s}_z$. Remember that $[\hat{s}^+, \hat{s}^-] = 2\hat{s}_z$. (c) Calculate the expectation value of $\hat{s}^2$ of α and β using Eq. 1.19.

In the case of N-electron systems, the spin operators have to be applied on Slater determinants or linear combinations of these. The action of the N-electron operator $\hat{S}^2$ is most conveniently evaluated in the N-electron version of Eq. 1.19 with $\hat{S}_z$, $\hat{S}^+$ and $\hat{S}^-$ defined as the sum of the corresponding one-electron operators.

$$\hat{S}^2 = \hat{S}^+\hat{S}^- - \hat{S}_z + \hat{S}_z^2 \tag{1.21}$$

with

$$\hat{S}_z = \sum_{i=1}^{N} \hat{s}_z(i) \qquad \hat{S}^+ = \sum_{i=1}^{N} \hat{s}^+(i) \qquad \hat{S}^- = \sum_{i=1}^{N} \hat{s}^-(i) \tag{1.22}$$

The multi-electron version of Eq. 1.17 is

$$\hat{S}^+|S, M_S\rangle = \sqrt{S(S+1) - M_S(M_S+1)}|S, M_S + 1\rangle$$
$$\hat{S}^-|S, M_S\rangle = \sqrt{S(S+1) - M_S(M_S-1)}|S, M_S - 1\rangle \tag{1.23}$$

whereas many-electron functions consisting of one Slater determinant are always eigenfunctions of $\hat{S}_z$ with an eigenvalue given by the difference of the number of α and β electrons multiplied by one half, this is in general not the case for $\hat{S}^2$. To illustrate this, we apply the two operators on the Slater determinants $|\varphi_1\varphi_2|$ and $|\varphi_1\overline{\varphi}_2|$.

$$\hat{S}_z|\varphi_1\varphi_2| = \hat{S}_z\frac{(\varphi_1\varphi_2 - \varphi_2\varphi_1)}{\sqrt{2}} = \frac{\varphi_1\varphi_2 - \varphi_2\varphi_1}{\sqrt{2}}(\hat{s}_z(1) + \hat{s}_z(2))\alpha\alpha$$
$$= \frac{\varphi_1\varphi_2 - \varphi_2\varphi_1}{\sqrt{2}}\left(\frac{1}{2}\alpha\alpha + \frac{1}{2}\alpha\alpha\right) = 1 \cdot \frac{\varphi_1\varphi_2 - \varphi_2\varphi_1}{\sqrt{2}} = 1 \cdot |\varphi_1\varphi_2| \tag{1.24}$$

$$\hat{S}_z|\varphi_1\overline{\varphi}_2| = \hat{S}_z\frac{(\varphi_1\overline{\varphi}_2 - \overline{\varphi}_2\varphi_1)}{\sqrt{2}}$$
$$= \frac{\varphi_1\varphi_2(\hat{s}_z(1) + \hat{s}_z(2))\alpha\beta - \varphi_2\varphi_1(\hat{s}_z(1) + \hat{s}_z(2))\beta\alpha}{\sqrt{2}}$$
$$= \frac{\varphi_1\varphi_2(1/2\,\alpha\beta - 1/2\,\alpha\beta) - \varphi_2\varphi_1(-1/2\,\beta\alpha + 1/2\,\beta\alpha)}{\sqrt{2}}$$
$$= 0 \cdot \frac{\varphi_1\overline{\varphi}_2 - \overline{\varphi}_2\varphi_1}{\sqrt{2}} = 0 \cdot |\varphi_1\overline{\varphi}_2| \tag{1.25}$$

This shows that the Slater determinants are eigenfunctions of $\hat{S}_z$ and the corresponding eigenvalues M_S are equal to 1 and 0, respectively. The result of applying $\hat{S}^2$ is rather straightforward using Eq. 1.21 with the notion that $(-\hat{S}_z + \hat{S}_z^2)$ gives zero when applied on the spin functions $\alpha\alpha$, $\alpha\beta$ and $\beta\alpha$. Only remains to determine the action of the two ladder operators to check whether the determinants are eigenfunctions of $\hat{S}^2$

$$\hat{S}^+\hat{S}^-|\varphi_1\varphi_2| = \frac{\varphi_1\varphi_2 - \varphi_2\varphi_1}{\sqrt{2}}(\hat{s}^+(1) + \hat{s}^+(2))(\hat{s}^-(1) + \hat{s}^-(2))(\alpha\alpha)$$

$$= \frac{\varphi_1\varphi_2 - \varphi_2\varphi_1}{\sqrt{2}}(\hat{s}^+(1) + \hat{s}^+(2))(\beta\alpha + \alpha\beta)$$

$$= \frac{\varphi_1\varphi_2 - \varphi_2\varphi_1}{\sqrt{2}}(\alpha\alpha + \alpha\alpha) = 2 \cdot \frac{\varphi_1\varphi_2 - \varphi_2\varphi_1}{\sqrt{2}} = 2|\varphi_1\varphi_2| \quad (1.26)$$

$$\hat{S}^+\hat{S}^-|\varphi_1\overline{\varphi}_2| = \frac{\varphi_1\varphi_2}{\sqrt{2}}(\hat{s}^+(1) + \hat{s}^+(2))(\hat{s}^-(1) + \hat{s}^-(2))(\alpha\beta)$$

$$- \frac{\varphi_2\varphi_1}{\sqrt{2}}(\hat{s}^+(1) + \hat{s}^+(2))(\hat{s}^-(1) + \hat{s}^-(2))(\beta\alpha)$$

$$= \frac{\varphi_1\varphi_2}{\sqrt{2}}(\hat{s}^+(1) + \hat{s}^+(2))(\beta\beta) - \frac{\varphi_2\varphi_1}{\sqrt{2}}(\hat{s}^+(1) + \hat{s}^+(2))(\beta\beta)$$

$$= \frac{\varphi_1\varphi_2}{\sqrt{2}}(\alpha\beta + \beta\alpha) - \frac{\varphi_2\varphi_1}{\sqrt{2}}(\alpha\beta + \beta\alpha) = |\varphi_1\overline{\varphi}_2| + |\overline{\varphi}_1\varphi_2|$$

$$\neq S(S + 1)|\varphi_1\overline{\varphi}_2| \quad (1.27)$$

Hence, the single Slater determinant $|\varphi_1\varphi_2|$ is a proper spin eigenfunction, while $|\varphi_1\overline{\varphi}_2|$ is not. In general, linear combinations of Slater determinants are necessary to ensure that the wave function is an eigenfunction of $\hat{S}^2$.

In the following, three strategies will be illustrated to construct spin eigenfunctions from scratch based on (i) projection techniques to eliminate the contributions of unwanted spin eigenfunctions, (ii) diagonalization of the matrix representation of $\hat{S}^2$, and (iii) the genealogical construction of spin functions in which spins are added one-by-one.

In the above demonstrations we have first developed the Slater determinants and then applied the spin operators. This strategy becomes of course very laborious for functions with more than two electrons. It should however be noted that it is not necessary to work with the fully expanded determinants, one gets the same results when working with the product of the diagonal elements.

1.5 Apply the total spin operator on $\Phi_1 = |\varphi_1\overline{\varphi}_1|$ and $\Phi_2 = \{|\varphi_1\overline{\varphi}_2| + |\overline{\varphi}_1\varphi_2|\}/\sqrt{2}$ and check that the same result is obtained when the determinants are fully expanded.

1.2.1 Many Electron Spin Functions by Projection

In general the spin eigenfunction of a $2S + 1$ spin multiplet is a linear combination of N-electron Slater determinants that all have the same M_S quantum number

$$^{2S+1}\Psi = \sum_L c_L \Phi_L \tag{1.28}$$

with $\hat{S}_z \Phi_L = M_S \Phi_L$ for all L. If $\{\Psi_i\}$ is a complete set of M eigenfunctions of $\hat{S}^2$ with the same M_S, i.e. $\hat{S}^2 \Psi_i = S_i(S_i + 1)\Psi_i$ and $\hat{S}_z \Psi_i = M_S \Psi_i$ for $i = 1, M$, then any of the determinants Φ_L can be written as a linear combination of these spin eigenfunctions. In other words, any determinant Φ_L can be seen as a linear combination of different spin eigenfunctions Ψ_i and to obtain the expression of a proper spin eigenfunction one should eliminate all the undesired terms from the sum. A natural way to proceed is to apply projection techniques. Since the spin eigenvalue of Ψ_i is equal to $S_i(S_i + 1)$, the operator

$$\hat{P}_L^e = \left[\hat{S}^2 - S_L(S_L + 1)\right] \tag{1.29}$$

eliminates the L-component from the determinant Φ. Hence the subsequent application of $\hat{P}_i^e$, $\hat{P}_j^e$, ... $\hat{P}_M^e$ (except $\hat{P}_k^e$) preserves the k-component and leads to $^{2S_k+1}\Psi_k$. This procedure is illustrated in Fig. 1.1 for a trivial example of a vector with two components. After projecting the vector on the x-axis, one subtracts this projection from the total vector to obtain the y-component.

The general expression of the operator to obtain spin eigenfunction Ψ_k from a determinant Φ_L is

$$\hat{P}_k = \prod_{l \neq k}^M \hat{P}_l^e = \prod_{l \neq k}^M \left[\hat{S}^2 - S_l(S_l + 1)\right] \tag{1.30}$$

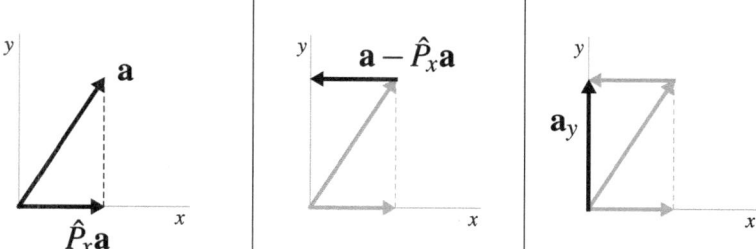

Fig. 1.1 Illustration of the projection method to eliminate undesired components of a vector. *Left* **a** is projected on the x-axis; *Middle* the projection ($\hat{P}_x\mathbf{a}$) is subtracted from **a**; *Right* The result of the operation is the y-component of **a**

These expressions produce projections that are not necessarily normalized to one, but this can easily be done at the end of the process. The procedure is most conveniently illustrated by deriving the singlet and triplet open-shell spin eigenfunctions with $M_S = 0$ for a two-electron in two-orbitals case. In the notation of Eq. 1.15 the two determinants are

$$\Phi_1 = |\varphi_1 \overline{\varphi}_1 \varphi_2 \overline{\varphi}_2 \ldots \varphi_a \overline{\varphi}_b| = |a\overline{b}|$$
$$\Phi_2 = |\varphi_1 \overline{\varphi}_1 \varphi_2 \overline{\varphi}_2 \ldots \overline{\varphi}_a \varphi_b| = |\overline{a}b| \tag{1.31}$$

There are two possible spin eigenfunctions, singlet and triplet, with S equal to 0 and 1, respectively. The projection operators are directly obtained from Eq. 1.30

$$\hat{P}_0 = \hat{S}^2 - 2 \qquad \hat{P}_1 = \hat{S}^2 - 0 \tag{1.32}$$

The result of applying $\hat{S}^2$ on Φ_1 is given in Eq. 1.27, and hence, the projection operators give

$$\hat{P}_0|a\overline{b}| = (\hat{S}^2 - 2)|a\overline{b}| = |a\overline{b}| + |\overline{a}b| - 2|a\overline{b}| = |\overline{a}b| - |a\overline{b}| \tag{1.33a}$$
$$\hat{P}_1|a\overline{b}| = (\hat{S}^2 - 0)|a\overline{b}| = |a\overline{b}| + |\overline{a}b| \tag{1.33b}$$

The functions have to be multiplied by $\frac{1}{\sqrt{2}}$ to obtain the properly normalized expressions.

1.6 (a) Find the other two components of the triplet spin eigenfunctions by applying the ladder operators on the $M_S = 0$ component of the triplet function. (b) Derive the singlet and triplet spin eigenfunctions by projection using Φ_2 of Eq. 1.31.

1.2.2 Spin Functions by Diagonalization

One way to find the eigenvalues and eigenvectors of an operator is to diagonalize the matrix representation of the operator in a complete basis. Therefore, a natural alternative to the projection method is the process of diagonalizing the matrix representation of the $\hat{S}^2$ operator. The basis of the matrix representation is formed by the individual determinants. The resulting eigenvectors are the spin eigenfunctions (linear combinations of these basis functions, the determinants) and the corresponding eigenvalues indicate the spin of the eigenfunction. The method is straightforward in its application but can require a substantial amount of analytical work since all matrix elements of $\hat{S}^2$ are needed, which can become rather cumbersome for systems with an elevated number of unpaired electrons. The method is illustrated for a system with three electrons in three distinct orbitals. The basis set spanned in the $M_S = \frac{1}{2}$ space contains three determinants

$$\Phi_1 = |ab\bar{c}| \qquad \Phi_2 = |a\bar{b}c| \qquad \Phi_3 = |\bar{a}bc| \tag{1.34}$$

The matrix representation of $\hat{S}^2$ can be constructed by analyzing the effect of $\hat{S}^+\hat{S}^-$, $\hat{S}_z$ and $\hat{S}_z^2$ (cf. Eq. 1.21) on the three basis functions.

$$\hat{S}^+\hat{S}^-|ab\bar{c}| = \hat{S}^+\left(|\bar{a}b\bar{c}| + |a\bar{b}\bar{c}| + 0\right) = |ab\bar{c}| + |\bar{a}bc| + |ab\bar{c}| + |a\bar{b}c| \tag{1.35a}$$

$$\hat{S}_z|ab\bar{c}| = \left(\frac{1}{2} + \frac{1}{2} - \frac{1}{2}\right)|ab\bar{c}| = \frac{1}{2}|ab\bar{c}| \tag{1.35b}$$

$$\hat{S}_z^2|ab\bar{c}| = \frac{1}{4}|ab\bar{c}| \tag{1.35c}$$

The other two determinants give analogous results and from this we evaluate the action of $\hat{S}^2$ on the three basis functions:

$$\hat{S}^2|ab\bar{c}| = {}^7\!/_4|ab\bar{c}| + |a\bar{b}c| + |\bar{a}bc| \tag{1.36a}$$

$$\hat{S}^2|a\bar{b}c| = |ab\bar{c}| + {}^7\!/_4|a\bar{b}c| + |\bar{a}bc| \tag{1.36b}$$

$$\hat{S}^2|\bar{a}bc| = |ab\bar{c}| + |a\bar{b}c| + {}^7\!/_4|\bar{a}bc| \tag{1.36c}$$

which leads to the following matrix representation of $\hat{S}^2$

$$
\begin{array}{c|ccc}
 & |ab\bar{c}\rangle & |a\bar{b}c\rangle & |\bar{a}bc\rangle \\
\hline
\langle ab\bar{c}| & \frac{7}{4} & 1 & 1 \\
\langle a\bar{b}c| & 1 & \frac{7}{4} & 1 \\
\langle \bar{a}bc| & 1 & 1 & \frac{7}{4}
\end{array}
\tag{1.37}
$$

As can be seen, the matrix has non-zero off diagonal matrix elements showing that the basis set of determinants is not a basis of eigenfunctions of the $\hat{S}^2$ operator. From here, the search for spin eigenfunctions follows standard diagonalization schemes. First, the eigenvalues are determined by finding the x-values for which the secular determinant is zero

$$\begin{vmatrix} \frac{7}{4} - x & 1 & 1 \\ 1 & \frac{7}{4} - x & 1 \\ 1 & 1 & \frac{7}{4} - x \end{vmatrix} = 0 \tag{1.38}$$

This gives $x_1, x_2 = \frac{3}{4}$ and $x_3 = \frac{15}{4}$, corresponding to two doublet functions $(\frac{1}{2}(\frac{1}{2} + 1) = \frac{3}{4})$ and one quartet function $(\frac{3}{2}(\frac{3}{2} + 1) = \frac{15}{4})$. The corresponding eigenvectors are determined by substituting the respective x-values in the secular equations.

$$
\begin{pmatrix} \frac{7}{4} & 1 & 1 \\ 1 & \frac{7}{4} & 1 \\ 1 & 1 & \frac{7}{4} \end{pmatrix} \begin{pmatrix} c_1 \\ c_2 \\ c_3 \end{pmatrix} = x \begin{pmatrix} c_1 \\ c_2 \\ c_3 \end{pmatrix} \tag{1.39}
$$

This gives $c_1 = c_2 = c_3 = \frac{1}{\sqrt{3}}$ for $x = \frac{15}{4}$, where the normalization condition is used to determine the numerical value. The quartet spin function with $M_S = \frac{1}{2}$ is given by

$$
{}^4\Psi = \frac{1}{\sqrt{3}} \left(|ab\bar{c}| + |a\bar{b}c| + |\bar{a}bc| \right) \tag{1.40}
$$

1.7 Find the $M_S = -\frac{1}{2}, \pm\frac{3}{2}$ components of the quartet function with the ladder operators.

The situation for the doublet functions is more complicated. The resulting equations for the coefficients are linear dependent ($c_2 + c_3 = -c_1$; $c_1 + c_3 = -c_2$; $c_1 + c_2 = -c_3$) and no unique solution can be determined. This is expected since the two functions have the same eigenvalues of $\hat{S}^2$ and any linear combination of the two doublet functions is also an eigenfunction. In some cases the spatial symmetry of the system imposes extra restrictions on the coefficients such that a unique solution emerges. For instance, in a system with inversion symmetry and center b located on the inversion center, c_1 must be equal to $\pm c_3$ and the following two doublet functions fulfil spatial and spin symmetry conditions.

$$
{}^2\Psi_A = \frac{1}{\sqrt{2}} \left(|ab\bar{c}| - |\bar{a}bc| \right) \tag{1.41a}
$$

$$
{}^2\Psi_B = \frac{1}{\sqrt{6}} \left(2|a\bar{b}c| - |ab\bar{c}| - |\bar{a}bc| \right) \tag{1.41b}
$$

1.8 (a) Check that the two doublets are orthogonal. (b) Check that the expectation value of $\hat{S}^2$ for ${}^2\Psi_A$ is 3/4.

1.2.3 Genealogical Approach

The third method to obtain spin eigenfunctions is based on a stepwise generation of the N-electron spin eigenfunction through a one-by-one addition of one-electron spin functions to a known spin eigenfunction. This genealogical way of constructing

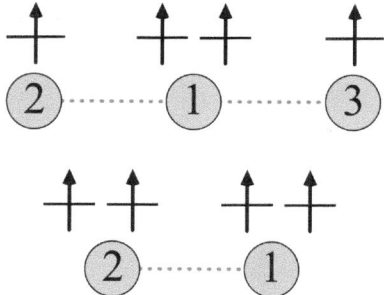

Fig. 1.2 Model system with four unpaired electrons on three centers (*top*). The two electrons on center 1 are coupled to a triplet state as stated by Hund's rule. Singlet coupling on center 1 leads to states that are much higher in energy and not directly relevant for the magnetic interactions. The *lower part* shows the system with four electrons on two centers

spin eigenfunctions is described in great detail by Pauncz [1, 2] and we refer to these books for further reading. Here, we will describe the main characteristics of the method and illustrate it with a system with four unpaired electrons localized on two or three magnetic centers as shown in Fig. 1.2.

The starting point of the method is the one electron spin function α with $S = \frac{1}{2}$ to which a second electron spin can be *added* to give an $S = 1$ spin function or *subtracted*, resulting in an $S = 0$ function. Subsequently more electron spins can be added or subtracted until the desired number of spins are described in the spin eigenfunctions. The use of Clebsch–Gordon coefficients ensures that linear combinations of determinants are produced that are eigenfunctions of $\hat{S}^2$ at each stage of the procedure. An advantage of this method is that one can specifically construct a certain spin eigenfunction among all possible with the required spin couplings between the electrons. This is best illustrated in the branching diagram shown in Fig. 1.3, which represents the different routes that can be taken to construct a spin function with a given S-value (on the y-axis) for a certain number of electrons (on the x-axis). The way up along the branching diagram represents *adding* an electron spin (increasing S by $\frac{1}{2}$) and going downwards indicates that an electron spin is *subtracted*, that is, S is diminished from S to $S - \frac{1}{2}$. The number in the circles gives the number of different routes that can be taken to arrive at that point. For instance, there are two ways to construct a singlet spin function with four electrons, three different triplet functions and one quintet. The branching diagrams allows us to choose one specific path to reach the desired spin function. This can be very useful, for example, to impose high spin coupling between unpaired electrons on the same magnetic center to fulfill Hund's rule.

The formulas for adding and subtracting an electron spin look somewhat awkward but are rather simple in their application. Moreover, the method is very well suited for translation into a computer program. *Adding* a spin is done with

Fig. 1.3 Branching diagram

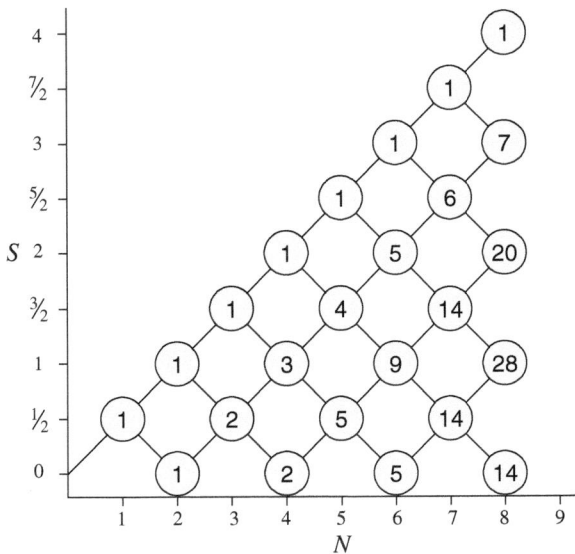

$$\Psi(N, S, M_S) = \left[(S + M_S)^{\frac{1}{2}}\Psi(N - 1, S - \frac{1}{2}, M_S - \frac{1}{2})\alpha(N)\right.$$

$$\left. + (S - M_S)^{\frac{1}{2}}\Psi(N - 1, S - \frac{1}{2}, M_S + \frac{1}{2})\beta(N)\right]$$

$$\times (2S)^{-\frac{1}{2}} \tag{1.42}$$

and the *subtraction* of a spin requires

$$\Psi(N, S, M_S) = \left[-(S - M_S + 1)^{\frac{1}{2}}\Psi(N - 1, S + \frac{1}{2}, M_S - \frac{1}{2})\alpha(N)\right.$$

$$\left. + (S + M_S + 1)^{\frac{1}{2}}\Psi(N - 1, S + \frac{1}{2}, M_S + \frac{1}{2})\beta(N)\right]$$

$$\times (2S + 2)^{-\frac{1}{2}} \tag{1.43}$$

The first example that will be discussed concerns the generation of the spin eigenfunctions relevant for the magnetic interactions in a system with three magnetic centers and four unpaired electrons, see Fig. 1.2 (top). Two of these four electrons are localized on the same center and coupled to a local triplet state. Hund's rule tells us that the singlet coupling of these two electrons gives rise to an electronic configuration that is much higher in energy and not directly relevant for the magnetic interactions as will be discussed in Chap. 2. Starting by assigning α to electron 1, the pathway marked in the first diagram of Fig. 1.4 shows that adding a second electron spin provides us the required triplet spin eigenfunction for the two electrons on the

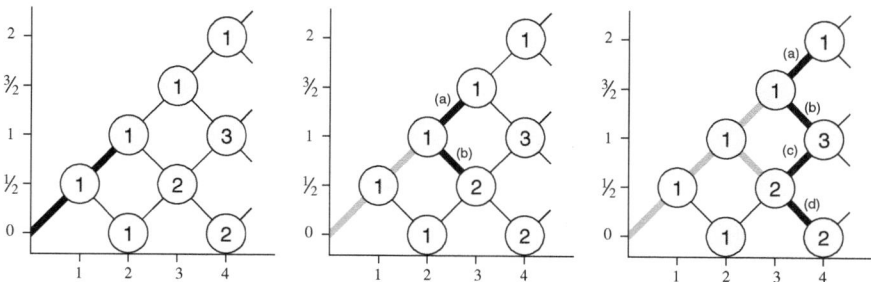

Fig. 1.4 Paths to generate spin eigenfunctions with 2 (*left*), 3 (*middle*) and 4 (*right*) electrons under the restriction of triplet coupling of electrons 1 and 2

same magnetic center. The *downwards* path leads to singlet coupling and is ruled out for this example. Equation 1.42 is applied with $N = 2$, $S = 1$ and $M_S = 1$.

$$\Psi(2, 1, 1) = \left[(1+1)^{\frac{1}{2}} \alpha(1)\alpha(2) + (1-1)^{\frac{1}{2}} \cdot 0 \cdot \beta(2) \right] (2 \cdot 1)^{-\frac{1}{2}} = \alpha\alpha \quad (1.44)$$

The third electron spin, localized on the second magnetic center can be coupled parallel or anti-parallel to this triplet, giving a quartet ($S = \frac{3}{2}$) or doublet ($S = \frac{1}{2}$) function, as shown in the middle diagram of Fig. 1.4. The quartet function is obtained from Eq. 1.42 with $N = 3$, $S = \frac{3}{2}$ and $M_S = \frac{3}{2}$.

$$\Psi(3, \tfrac{3}{2}, \tfrac{3}{2}) = \left[(\tfrac{3}{2} + \tfrac{3}{2})^{1/2} \alpha(1)\alpha(2)\alpha(3) + (\tfrac{3}{2} - \tfrac{3}{2})^{1/2} \cdot 0 \cdot \beta(3) \right]$$
$$\times (2 \cdot \tfrac{3}{2})^{-\frac{1}{2}} = \alpha\alpha\alpha \quad (1.45)$$

On the other hand, the doublet spin function is generated with Eq. 1.43 with $N = 3$, $S = \frac{1}{2}$ and $M_S = \frac{1}{2}$; and $\Psi(N - 1, S + 1/2, M_S - 1/2)$ is obtained by applying the $\hat{S}^-$ operator to $\Psi(2, 1, 1)$ given in Eq. 1.44:

$$\Psi(3, \tfrac{1}{2}, \tfrac{1}{2}) = \left[-(\tfrac{1}{2} - \tfrac{1}{2} + 1)^{1/2} \frac{1}{\sqrt{2}} [\alpha(1)\beta(2) + \beta(1)\alpha(2)]\alpha(3) \right.$$
$$\left. + (\tfrac{1}{2} + \tfrac{1}{2} + 1)^{1/2} \alpha(1)\alpha(2)\beta(3) \right] (2 \cdot \tfrac{1}{2} + 2)^{-1/2}$$
$$= \frac{1}{\sqrt{6}} (2\alpha\alpha\beta - \alpha\beta\alpha - \beta\alpha\alpha) \quad (1.46)$$

The incorporation of the fourth electron spin can be done in four different ways. $\Psi(3, \tfrac{3}{2}, \tfrac{3}{2})$ creates a quintet and a triplet state, while $\Psi(3, \tfrac{1}{2}, \tfrac{1}{2})$ leads to a second triplet and a singlet state, as illustrated in the right diagram of Fig. 1.4.

$$\Psi(4, 2, 2) = \left[(2+2)^{\frac{1}{2}}\alpha(1)\alpha(2)\alpha(3)\alpha(4) + (2-2)^{\frac{1}{2}} \cdot 0 \cdot \beta(4)\right](2 \cdot 2)^{-\frac{1}{2}}$$

$$= \alpha\alpha\alpha\alpha \tag{1.47}$$

To generate the triplet function by subtraction, we need the expression of $\Psi(3, {}^3/_2, {}^1/_2)$, which can be obtained by operating on $\Psi(3, {}^3/_2, {}^3/_2)$ with the $\hat{S}^-$ operator.

$$\Psi(4, 1, 1) = \left[-(2 - 2 + 1)^{\frac{1}{2}}\frac{1}{\sqrt{3}}\,[\beta(1)\alpha(2)\alpha(3) + \alpha(1)\beta(2)\alpha(3)\right.$$

$$\left. + \alpha(1)\alpha(2)\beta(3)]\,\alpha(4) + (1 + 1 + 1)^{\frac{1}{2}}\alpha(1)\alpha(2)\alpha(3)\beta(4)\right](2 \cdot 1 + 2)^{-\frac{1}{2}}$$

$$= \frac{1}{2\sqrt{3}}(3\alpha\alpha\alpha\beta - \beta\alpha\alpha\alpha - \alpha\beta\alpha\alpha - \alpha\alpha\beta\alpha) \tag{1.48}$$

The generation of the second triplet function from the doublet state by addition gives

$$\Psi'(4, 1, 1) = \left[(1 + 1)^{\frac{1}{2}}\frac{1}{\sqrt{6}}[2\alpha(1)\alpha(2)\beta(3) - \alpha(1)\beta(2)\alpha(3)\right.$$

$$\left. - \beta(1)\alpha(2)\alpha(3)]\alpha(4) + (1 - 1)^{\frac{1}{2}} \cdot 0 \cdot \beta(4)\right](2 \cdot 1)^{-\frac{1}{2}}$$

$$= \frac{1}{\sqrt{6}}(2\alpha\alpha\beta\alpha - \alpha\beta\alpha\alpha - \beta\alpha\alpha\alpha) \tag{1.49}$$

Note that $\Psi(4, 1, 1)$ and $\Psi'(4, 1, 1)$ are degenerate with respect to the $\hat{S}^2$ operator, and therefore any linear combination of these two functions is equally valid. In analogy to the discussion for the doublet states in the previous section, the spatial symmetry can impose extra conditions on the values of the coefficients of the determinants. If the second and third magnetic center are symmetry equivalent, the interchange of the coordinates of electron 3 and 4 should leave the wave function unaltered, except for a possible sign change. This is obviously not the case for the here generated spin functions, but the linear combinations $\Psi(4, 1, 1) + \frac{2}{\sqrt{2}}\Psi'(4, 1, 1)$ and $\Psi(4, 1, 1) - \frac{1}{\sqrt{2}}\Psi'(4, 1, 1)$ give

$$\widetilde{\Psi}(4, 1, 1) = \frac{1}{\sqrt{2}}(\alpha\alpha\alpha\beta - \alpha\alpha\beta\alpha)$$

$$= \frac{1}{\sqrt{2}}[\alpha\alpha(\alpha\beta - \beta\alpha)] \tag{1.50a}$$

$$\widetilde{\Psi}'(4, 1, 1) = \frac{1}{2}(\alpha\alpha\alpha\beta + \alpha\alpha\beta\alpha - \alpha\beta\alpha\alpha - \beta\alpha\alpha\alpha)$$

$$= \frac{1}{2}[\alpha\alpha(\alpha\beta + \beta\alpha) - (\alpha\beta + \beta\alpha)\alpha\alpha] \tag{1.50b}$$

which are (anti-)symmetric under the permutation of electron 3 and 4. Furthermore, these two linear combinations clearly reveal the triplet coupling of electron 1 and 2, and the singlet ($\widetilde{\Psi}$) or triplet coupling ($\widetilde{\Psi}'$) for electron 3 and 4.

Remains to evaluate the function generated by the incorporation of the fourth electron spin by subtraction from the three-electron doublet function. This gives a singlet spin function characterized by the triplet coupling of electron 1 and 2 (following Hund's rule) and of electron 3 and 4. To apply Eq. 1.43, the $\Psi(3, \frac{1}{2}, -\frac{1}{2})$ function has to be generated by acting with $\hat{S}^-$ on $\Psi(3, \frac{1}{2}, \frac{1}{2})$.

$$
\begin{aligned}
\Psi(4, 0, 0) = \Big[&-(0-0+1)^{\frac{1}{2}} \frac{1}{\sqrt{6}}[-2\beta(1)\beta(2)\alpha(3) + \beta(1)\alpha(2)\beta(3) \\
&+ \alpha(1)\beta(2)\beta(3)]\alpha(4) + (0+0+1)^{\frac{1}{2}} \frac{1}{\sqrt{6}}[2\alpha(1)\alpha(2)\beta(3) \\
&- \alpha(1)\beta(2)\alpha(3) - \beta(1)\alpha(2)\alpha(3)]\beta(4)\Big](2\cdot 0 + 2)^{-\frac{1}{2}} \\
= &\frac{1}{2\sqrt{3}}[2(\alpha\alpha\beta\beta + \beta\beta\alpha\alpha) - \alpha\beta\alpha\beta - \alpha\beta\beta\alpha - \beta\alpha\alpha\beta - \beta\alpha\beta\alpha] \\
= &\frac{1}{2\sqrt{3}}[2(\alpha\alpha\beta\beta + \beta\beta\alpha\alpha) - (\alpha\beta + \beta\alpha)(\alpha\beta + \beta\alpha)] \qquad (1.51)
\end{aligned}
$$

1.9 (a) Construct a branching diagram and mark the path to generate the $N = 4$ triplet and singlet spin states with singlet coupling for electron 1 and 2. (b) Construct $\Psi(2, 0, 0)$ with the genealogical approach.

Two-by-two additions: The process of generating spin functions by the genealogical approach can be made a little less tedious by considering the incorporation of two electrons at the same time. An additional advantage of doing so is that one better controls the spin coupling of electron pairs. The triplet functions with four electrons, $\Psi(4, 1, 1)$ and $\Psi'(4, 1, 1)$ Eqs. 1.48 and 1.49 do have triplet coupling among electron 1 and 2, but turn out to be mixtures of singlet and triplet coupling for electrons 3 and 4. Only after taking the correct linear combination, spin functions could be constructed with clear-cut spin couplings of both electron pairs. This can be achieved directly with the Serber variant of the genealogical approach [3, 4] illustrated in the branching diagram of Fig. 1.5.

Starting with an $N-2$-electron spin function of spin S', singlet or triplet coupled two-electron functions are added to obtain $\Psi(N, S)$ with $S = S'+1$, S' or $S'-1$. The branching diagram shows that four different cases can be distinguished, for which the following formulas need to be considered:

- case 1: Singlet incorporation (gray solid lines); $S = S'$

$$
\Psi(N, S, M_S) = \Psi(N - 2, S, M_S)\Phi_a \qquad (1.52)
$$

Fig. 1.5 Branching diagram for two-by-two electron incorporations. *Gray* solid lines represent singlet coupled additions. The other lines represent triplet additions with $S = S' - 1$ (*black dashed*), $S = S'$ (*gray dashed*) and $S = S' + 1$ (*black solid*)

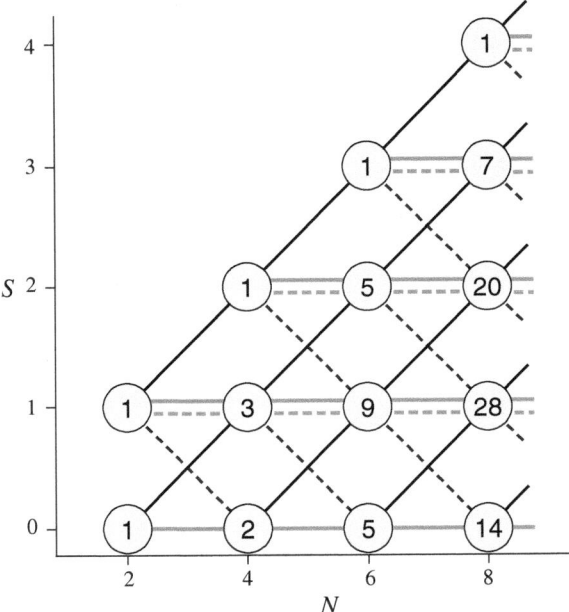

- case 2: Triplet incorporation (black solid lines); $S = S' + 1$

$$\Psi(N, S, M_S) = \Big[\{(S + M_S)(S + M_S - 1)\}^{\frac{1}{2}} \Psi(N - 2, S - 1, M_S - 1)\Phi_b$$
$$+ \{2(S + M_S)(S - M_S)\}^{\frac{1}{2}} \Psi(N - 2, S - 1, M_S)\Phi_c$$
$$+ \{(S - M_S)(S - M_S - 1)\}^{\frac{1}{2}} \Psi(N - 2, S - 1, M_S + 1)\Phi_d\Big]$$
$$\times [2S(2S - 1)]^{-\frac{1}{2}} \tag{1.53}$$

- case 3: Triplet incorporation (gray dashed lines); $S = S'$

$$\Psi(N, S, M_S) = \Big[-\{(S + M_S)(S - M_S + 1)\}^{\frac{1}{2}} \Psi(N - 2, S, M_S - 1)\Phi_b$$
$$+ \sqrt{2} M_S \Psi(N - 2, S, M_S)\Phi_c$$
$$+ \{(S - M_S)(S + M_S + 1)\}^{\frac{1}{2}} \Psi(N - 2, S, M_S + 1)\Phi_d\Big]$$
$$\times [2S(S + 1)]^{-\frac{1}{2}} \tag{1.54}$$

- case 4: Triplet incorporation (black dashed lines): $S = S' - 1$

$$\Psi(N, S, M_S) = \left[\{(S - M_S + 2)(S - M_S + 1)\}^{\frac{1}{2}}\Psi(N - 2, S + 1, M_S - 1)\Phi_b\right.$$
$$- \{2(S - M_S + 1)(S + M_S + 1)\}^{\frac{1}{2}}\Psi(N - 2, S + 1, M_S)\Phi_c$$
$$\left.+ \{(S + M_S + 1)(S + M_S + 2)\}^{\frac{1}{2}}\Psi(N - 2, S + 1, M_S + 1)\Phi_d\right]$$
$$\times [(2S + 2)(2S + 3)]^{-\frac{1}{2}} \tag{1.55}$$

with

$$\Phi_a = \frac{1}{\sqrt{2}}[\alpha(N - 1)\beta(N) - \beta(N - 1)\alpha(N)] \tag{1.56a}$$

$$\Phi_b = \alpha(N - 1)\alpha(N) \tag{1.56b}$$

$$\Phi_c = \frac{1}{\sqrt{2}}[\alpha(N - 1)\beta(N) + \beta(N - 1)\alpha(N)] \tag{1.56c}$$

$$\Phi_d = \beta(N - 1)\beta(N) \tag{1.56d}$$

The method is nicely illustrated for a system with two magnetic centers, both with two unpaired electrons as shown in Fig. 1.2 (bottom). Hund's rule dictates that the electrons on each magnetic center are preferably coupled to a local triplet. Hence, the starting point in the Serber diagram is $\Psi(2, 1, 1) = \alpha(1)\alpha(2)$ and depending on the route taken one obtains a quintet (black), a triplet (gray dashed) or a singlet (black dashed) state. Equations 1.53 and 1.55 lead to the same expressions for the quintet as singlet state as with the standard genealogical approach. In contrast, the triplet state obtained from Eq. 1.54 is directly the correct expression and not a linear combination of singlet and triplet coupling among electron 3 and 4 as before (see Eqs. 1.48–1.50).

$$\Psi(4, 1, 0) = \left[-\{(1 + 0)(1 - 0 + 1)\}^{\frac{1}{2}}\beta(1)\beta(2)\alpha(3)\alpha(4)\right.$$
$$+ \sqrt{2} \cdot 0 \cdot \frac{1}{\sqrt{2}}[\alpha(1)\beta(2) + \beta(1)\alpha(2)]\frac{1}{\sqrt{2}}[\alpha(3)\beta(4) + \beta(3)\alpha(4)]$$
$$\left.+ \{(1 + 0)(1 + 0 + 1)\}^{\frac{1}{2}}\alpha(1)\alpha(2)\beta(3)\beta(4)\right]\{2 \cdot 1(1 + 1)\}^{-\frac{1}{2}}$$
$$= \frac{1}{\sqrt{2}}[\alpha\alpha\beta\beta - \beta\beta\alpha\alpha] \tag{1.57}$$

1.10 (a) Check that the expressions in Eqs. 1.50b and 1.57 are two different M_S components of the same triplet. (b) Construct $\Psi(4, 1, 0)$ with singlet coupling of electron 1 and 2 and triplet coupling for 3 and 4. Use the Serber variant of the genealogical approach.

1.2.4 Final Remarks

The examples given in the precedent paragraphs cover many real-world cases. The two-electrons in two-orbitals case discussed in Sect. 1.2.1 is representative of all types of Cu^{II} binuclear complexes or organic biradicals. It is important to remind that the intuitive representation with an up and down spin often used to indicate the open-shell singlet state in these cases (see Fig. 1.6) is not the most rigorous representation of this quantum state and may lead to confusion. It corresponds to the $|\alpha(1)\beta(2)|$ determinant and is actually a mixture of the $M_S = 0$ components of singlet and triplet functions.

Similar considerations hold for the three unpaired electrons case of Sect. 1.2.2. While the three parallel electrons on the transition metal centers on the left side of Fig. 1.7 give a satisfactory representation of the high-spin situation, i.e. the quartet

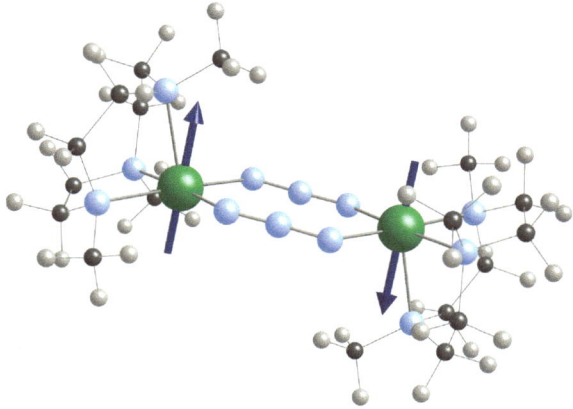

Fig. 1.6 Ball and stick representation of a $Cu_2(\mu\text{-}N_3)_2$ complex. The arrows on the Cu^{II} ions (*green*) are indications of the spin moment of the unpaired electron in the Cu-3d orbitals

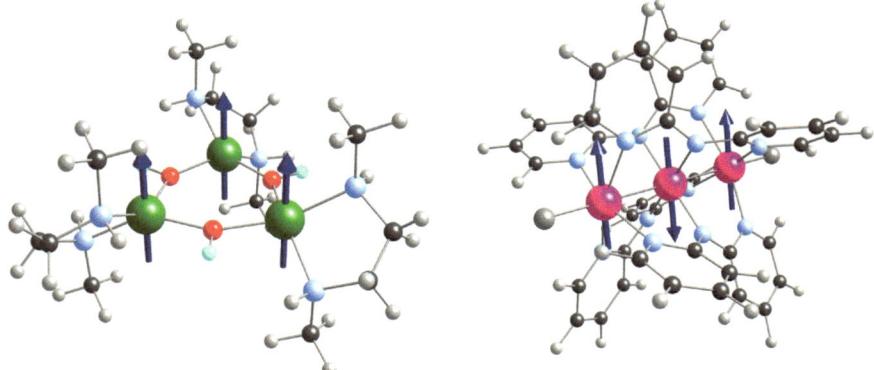

Fig. 1.7 Ball and stick representation of a $Cu_3(OH)_3$ complex (*left*) and an extended metal atom chain (EMAC) made of three Cr^{2+} ions hold together by four tridentate organic ligands

state, the up-down-up situation depicted on the right is not the doublet state but rather a superposition of the doublet and quartet spin functions given in Eqs. 1.40 and 1.41.

> **1.11** Calculate the overlap of the quartet and doublet spin functions given in Eqs. 1.40 and 1.41 with the $|\alpha\beta\alpha|$ determinant.

1.3 Perturbation Theory

Many body perturbation theory is one of the fundamental tools in Quantum Chemistry. It takes a central place both in the calculation of accurate energies and wave functions, and in the analysis of results for reaching a better understanding of the sometimes complicated physics contained in the system. There are basically two flavors of many-body perturbation theory. The first is what one calls the diagonalize-and-then-perturb method, and the second one inverts this order, it follows a perturb-and-then-diagonalize approach.

When one is only interested in a single state that is well separated from all the others, for example a non-degenerate ground state, the distinction is not very relevant. Using a proper zeroth-order wave function such as the one provided by the Hartree–Fock approach, the effect of electron correlation can be estimated with any standard perturbation scheme, being the Møller–Plesset implementation the most common one.

However, it becomes a little more subtle when one wants to describe a collection of states of a quantum system that are close in energy, or when states with a marked multiconfigurational character have to be described. The reference space is now spanned by several Slater determinants that define a collection of electronic states. Most approaches first diagonalize the reference space and then introduce the effect of the external determinants with perturbation theory. In contrast, quasi-degenerate perturbation theory (QDPT), first addresses the external determinants for all the matrix elements among the reference determinants and then diagonalizes the reference space to obtain the energies and wave functions of the states of interest.

1.3.1 Rayleigh–Schrödinger Perturbation Theory

There are only few systems for which the Schrödinger equation can be solved exactly. Therefore, many schemes have been developed to obtain as accurate as possible approximate solutions. The perturbative treatment is based on the partition of the full Hamiltonian of the system in two parts.

$$\hat{H} = \hat{H}^{(0)} + \lambda \hat{V} \tag{1.58}$$

The first term is the Hamiltonian of a model system with a complete set of known (normalized) solutions

$$\hat{H}^{(0)} \psi_i^{(0)} = E_i^{(0)} \psi_i^{(0)} \tag{1.59}$$

and $\hat{V}$ is the perturbation operator, which *perturbs* the model system. The parameter λ can be varied from zero (no perturbation) to one (complete Hamiltonian). In addition to this splitting of the Hamiltonian, the energy and the wave function are expanded in Taylor series writing the exact solutions as the sum of the model system solutions and corrections in the first, second, third, and higher order of the perturbation

$$\psi_0 = \psi_0^{(0)} + \lambda \psi_0^{(1)} + \lambda^2 \psi_0^{(2)} + \lambda^3 \psi_0^{(3)} + \cdots$$
$$E_0 = E_0^{(0)} + \lambda E_0^{(1)} + \lambda^2 E_0^{(2)} + \lambda^3 E_0^{(3)} + \cdots \tag{1.60}$$

where the subscript "0" makes reference to the ground state. The substitution of Eqs. 1.58 and 1.60 in the Schrödinger equation of the full system leads to

$$(\hat{H}^{(0)} + \lambda \hat{V}) \left(\psi_0^{(0)} + \lambda \psi_0^{(1)} + \lambda^2 \psi_0^{(2)} + \cdots \right)$$
$$= \left(E_0^{(0)} + \lambda E_0^{(1)} + \lambda^2 E_0^{(2)} + \cdots \right) \left(\psi_0^{(0)} + \lambda \psi_0^{(1)} + \lambda^2 \psi_0^{(2)} + \cdots \right) \tag{1.61}$$

Since λ can in principle take any value between 0 and 1, this equation only has a solution when the sum of the left-hand terms of a certain power of λ are equal to the sum of the right-hand terms of the same power of λ. This permits us to split the equation and group the terms by the power of λ

$$\lambda^0 : \quad \hat{H}^{(0)} \psi_0^{(0)} = E_0^{(0)} \psi_0^{(0)} \tag{1.62a}$$

$$\lambda^1 : \quad \hat{H}^{(0)} \psi_0^{(1)} + \hat{V} \psi_0^{(0)} = E_0^{(0)} \psi_0^{(1)} + E_0^{(1)} \psi_0^{(0)} \tag{1.62b}$$

$$\lambda^2 : \quad \hat{H}^{(0)} \psi_0^{(2)} + \hat{V} \psi_0^{(1)} = E_0^{(0)} \psi_0^{(2)} + E_0^{(1)} \psi_0^{(1)} + E_0^{(2)} \psi_0^{(0)} \tag{1.62c}$$

1.12 Write down the equation for the terms that are cubic in λ.

These equations can now be solved one-by-one to determine the different corrections to $E^{(0)}$ and $\psi^{(0)}$ in order to approximate the solutions of the full system. The equation that stems from the terms that are independent of λ defines the model system and does not provide new information. The first-order correction to the energy ($E_0^{(1)}$) can be determined from the equation with the linear λ terms. For that purpose we first multiply all terms with $\psi_0^{(0)*}$ and then integrate over the electron coordinates.

$$\langle \psi_0^{(0)}|\hat{H}^{(0)}|\psi_0^{(1)}\rangle + \langle \psi_0^{(0)}|\hat{V}|\psi_0^{(0)}\rangle = E_0^{(0)}\langle \psi_0^{(0)}|\psi_0^{(1)}\rangle + E_0^{(1)}\langle \psi_0^{(0)}|\psi_0^{(0)}\rangle \quad (1.63)$$

Since the zeroth-order wave function of the ground state is normalized, this equation can be rewritten to

$$E_0^{(1)} = \langle \psi_0^{(0)}|\hat{H}^{(0)}|\psi_0^{(1)}\rangle + \langle \psi_0^{(0)}|\hat{V}|\psi_0^{(0)}\rangle - E_0^{(0)}\langle \psi_0^{(0)}|\psi_0^{(1)}\rangle \quad (1.64)$$

The only unknown quantity on the right-hand-side of this equation is $\psi_0^{(1)}$. Therefore it is expanded as a linear combination of excited state wave functions of the unperturbed system.

$$\psi_0^{(1)} = \sum_{i\neq 0} a_i \psi_i^{(0)} \quad (1.65)$$

These wave functions of the excited states of the model system are all known and together with $\psi_0^{(0)}$ they form a complete set of functions. The orthogonality to $\psi_0^{(0)}$ is ensured by excluding this term from the linear combination. The substitution of the expansion in Eq. 1.64 leads to

$$E_0^{(1)} = \sum_{i\neq 0}\langle \psi_0^{(0)}|\hat{H}^{(0)}|\psi_i^{(0)}\rangle + \langle \psi_0^{(0)}|\hat{V}|\psi_0^{(0)}\rangle - \sum_{i\neq 0} E_0^{(0)}\langle \psi_0^{(0)}|\psi_i^{(0)}\rangle \quad (1.66)$$

By realizing that $\hat{H}^{(0)}|\psi_i^{(0)}\rangle = E_i^{(0)}|\psi_i^{(0)}\rangle$, the orthogonality of the different eigenfunctions of the zeroth-order model makes that all right-hand-side terms are zero, except the second one. Hence, the first-order correction to the energy is

$$E_0^{(1)} = \langle \psi_0^{(0)}|\hat{V}|\psi_0^{(0)}\rangle, \quad (1.67)$$

which corresponds to the expectation value of the perturbation operator for the unperturbed wave function. To determine the first-order corrected wave function, we need to find the values of the expansion coefficients a_i of Eq. 1.65. This can be done by substituting the expansion in the equation linear in λ Eq. 1.62b and after multiplying by $\psi_k^{(0)*}$ we integrate over the electron coordinates

$$\sum_{i\neq 0} a_i \langle \psi_k^{(0)}|\hat{H}^{(0)}|\psi_i^{(0)}\rangle + \langle \psi_k^{(0)}|\hat{V}|\psi_0^{(0)}\rangle$$

$$= \sum_{i\neq 0} a_i \langle \psi_k^{(0)}|\psi_i^{(0)}\rangle E_0^{(0)} + \langle \psi_k^{(0)}|\psi_0^{(0)}\rangle E_0^{(1)} \quad (1.68)$$

Taking into account the orthogonality of the zeroth-order eigenvectors, the only non-zero terms in the summation are those when $i = k$ and the equation simplifies to

$$a_k E_k^{(0)} + \langle \psi_k^{(0)} | \hat{V} | \psi_0^{(0)} \rangle = a_k E_0^{(0)} + 0 \tag{1.69}$$

from which a_k and $\psi^{(1)}$ follow immediately

$$a_k = \frac{\langle \psi_k^{(0)} | \hat{V} | \psi_0^{(0)} \rangle}{E_0^{(0)} - E_k^{(0)}} \qquad \psi^{(1)} = \sum_{i \neq 0} \frac{\langle \psi_i^{(0)} | \hat{V} | \psi_0^{(0)} \rangle}{E_0^{(0)} - E_i^{(0)}} \psi_i^{(0)} \tag{1.70}$$

The second-order correction to the energy is obtained from the quadratic equation in λ Eq. 1.62c in a similar fashion as the first-order correction. First, we multiply the equation by $\psi_0^{(0)*}$ and then we integrate over the electron coordinates

$$\langle \psi_0^{(0)} | \hat{H}^{(0)} | \psi_0^{(2)} \rangle + \langle \psi_0^{(0)} | \hat{V} | \psi_0^{(1)} \rangle$$
$$= E_0^{(0)} \langle \psi_0^{(0)} | \psi_0^{(2)} \rangle + E_0^{(1)} \langle \psi_0^{(0)} | \psi_0^{(1)} \rangle + E_0^{(2)} \langle \psi_0^{(0)} | \psi_0^{(0)} \rangle \tag{1.71}$$

Orthogonality causes the first and second term on the right-hand-side to be zero and the substitution of $\psi_0^{(2)}$ by a linear combination of zeroth-order eigenfunctions leads to the following equation

$$\sum_{j \neq 0} b_j \langle \psi_0^{(0)} | \hat{H}^{(0)} | \psi_j^{(0)} \rangle + \langle \psi_0^{(0)} | \hat{V} | \psi_0^{(1)} \rangle = E_0^{(2)} \tag{1.72}$$

The first left-hand-side term is zero and after substituting Eq. 1.70, the second order correction to the energy is obtained

$$E_0^{(2)} = \sum_{i \neq 0} \frac{\langle \psi_0^{(0)} | \hat{V} | \psi_i^{(0)} \rangle \langle \psi_i^{(0)} | \hat{V} | \psi_0^{(0)} \rangle}{E_0^{(0)} - E_i^{(0)}} \tag{1.73}$$

Higher order corrections can be derived in a similar way, but the expressions get more complicated rapidly.

When excited state energies of the model system ($E_i^{(0)}$) are close to $E_0^{(0)}$, the corresponding terms in the summations of $\psi^{(1)}$ and $E^{(2)}$ diverge, unless the matrix elements of these terms are zero. In case of a degenerate ground state of the model system, say $E_0^{(0)} = E_i^{(0)} = \cdots = E_k^{(0)}$, we can solve this problem by first diagonalizing the full $\hat{H}$ in the basis of $\psi_0^{(0)}, \psi_i^{(0)}, \ldots, \psi_k^{(0)}$. This yields linear combinations of $\psi_0^{(0)}, \psi_i^{(0)}, \ldots, \psi_k^{(0)}$ that are equally valid as zeroth-order wave functions while the divergence problem is avoided since the diagonalization process made all non-diagonal matrix elements equal to zero. Note that in case of near, but not strict degeneracy, this *diagonalize-and-then-perturb* procedure yields model wave functions that are no longer eigenfunctions of $\hat{H}^{(0)}$.

1.3.2 Møller–Plesset Perturbation Theory

As mentioned above, a common implementation of many-body perturbation theory in quantum chemistry is based on the zeroth-order Hamiltonian proposed by Møller and Plesset. When the Hartree–Fock wave function Φ_{HF} is known, the zeroth-order Hamiltonian can be defined as the sum of the Fock operators

$$\hat{H}^{(0)} = \sum_{i}^{N} \hat{f}(i) \tag{1.74}$$

with N is the number of electrons and

$$\hat{f}(i) = \hat{h}(i) + \sum_{k} \left(\hat{J}_k(i) - \hat{K}_k(i) \right) = \hat{h}(i) + \hat{g}(i) \tag{1.75}$$

The perturbation operator corresponds to the difference of the instantaneous electron–electron interaction operator and the mean-field electron–electron interaction of the Hartree–Fock description

$$\hat{V} = \hat{H} - \hat{H}^{(0)} = \sum_{i} \hat{h}(i) + \sum_{i} \sum_{j>i} \frac{1}{r_{ij}} - \sum_{i} \left(\hat{h}(i) + \hat{g}(i) \right) = \sum_{i} \sum_{j>i} \frac{1}{r_{ij}} - \sum_{i} \hat{g}(i) \tag{1.76}$$

The zeroth-order (known) solutions are defined by

$$\hat{H}^{(0)} \Psi_k^{(0)} = E_k^{(0)} \Psi_k^{(0)} \tag{1.77}$$

with

$$\Psi_0^{(0)} = \Phi_{HF} = |\phi_1 \phi_2 \ldots \phi_j \phi_k \ldots \phi_N| \qquad E_0^{(0)} = \sum_{m} \varepsilon_m \tag{1.78a}$$

$$\Psi_j^{(0)} = |\phi_1 \phi_2 \ldots \phi_k \ldots \phi_N \phi_a| = \Phi_j^a \qquad E_j^{(0)} = E_0^{(0)} - \varepsilon_j + \varepsilon_a \tag{1.78b}$$

$$\Psi_{jk}^{(0)} = |\phi_1 \phi_2 \ldots \ldots \phi_N \phi_a \phi_b| = \Phi_{jk}^{ab} \qquad E_{jk}^{(0)} = E_0^{(0)} - \varepsilon_j - \varepsilon_k + \varepsilon_a + \varepsilon_b \tag{1.78c}$$

The first order correction to the energy can be calculated with Eq. 1.67 and using the Slater–Condon rules one arrives at the following expression

$$E_0^{(1)} = \langle \Psi^{(0)} | \hat{V} | \Psi^{(0)} \rangle = \langle \Phi_{HF} | \sum_i \sum_{i>j} \frac{1}{r_{ij}} | \Phi_{HF} \rangle - \langle \Phi_{HF} | - \sum_i \hat{g}(i) | \Phi_{HF} \rangle$$

$$= \frac{1}{2} \sum_i \langle \phi_i | \sum_j \hat{J}_j - \hat{K}_j | \phi_i \rangle - \sum_i \langle \phi_i | \sum_j \hat{J}_j - \hat{K}_j | \phi_i \rangle$$

$$= -\frac{1}{2} \sum_i \langle \phi_i | \sum_j \hat{J}_j - \hat{K}_j | \phi_i \rangle \tag{1.79}$$

This leads the following expression for the energy at first-order

$$E_0 = E_0^{(0)} + E_0^{(1)} = \sum_i \varepsilon_i - \frac{1}{2} \sum_i \langle \phi_i | \sum_j \hat{J}_j - \hat{K}_j | \phi_i \rangle = E_{HF} \tag{1.80}$$

The first order correction to the wave function is

$$\Psi_0^{(1)} = \sum_{i \neq 0} \frac{\langle \Psi_i^{(0)} | \hat{V} | \Psi_0^{(0)} \rangle}{E_0^{(0)} - E_i^{(0)}} | \Psi_i^{(0)} \rangle \tag{1.81}$$

The numerator can be simplified by replacing $\hat{V}$ by $\hat{H} - \hat{H}^{(0)}$

$$\langle \Psi_i^{(0)} | \hat{V} | \Psi_0^{(0)} \rangle = \langle \Psi_i^{(0)} | \hat{H} | \Psi_0^{(0)} \rangle - \langle \Psi_i^{(0)} | \hat{H}^{(0)} | \Psi_0^{(0)} \rangle$$

$$= \langle \Psi_i^{(0)} | \hat{H} | \Psi_0^{(0)} \rangle - E_0^{(0)} \langle \Psi_i^{(0)} | \Psi_0^{(0)} \rangle = \langle \Psi_i^{(0)} | \hat{H} | \Psi_0^{(0)} \rangle \tag{1.82}$$

This last term is zero for determinants that arise from single excitations Eq. 1.78b because of Brillouin's theorem. It is also zero for determinants with more than two electron replacements, and hence, only the double excitations Eq. 1.78c need to be considered. This observation also serves to simplify the second-order correction to the energy

$$E_0^{(2)} = \sum_{i \neq 0} \frac{\langle \Psi_0^{(0)} | \hat{H} | \Psi_i^{(0)} \rangle \langle \Psi_i^{(0)} | \hat{H} | \Psi_0^{(0)} \rangle}{E_0^{(0)} - E_i^{(0)}} = \sum_{\substack{a<b \\ i<j}} \frac{\langle \Psi_0^{(0)} | \hat{H} | \Phi_{ij}^{ab} \rangle \langle \Phi_{ij}^{ab} | \hat{H} | \Psi_0^{(0)} \rangle}{\varepsilon_i + \varepsilon_j - \varepsilon_a - \varepsilon_b}$$

$$\tag{1.83}$$

Again, only the doubly excited determinants have to be considered to calculate the second-order correction to the energy. The expression for the third-order correction to the energy is slightly more complicated but as most salient feature introduces the effect of the interaction between excited determinants.

$$E_0^{(3)} = \sum_{\substack{a<b \\ i<j}} \sum_{\substack{c<d \\ k<l}} \frac{\langle \Psi_0^{(0)}|\hat{H}|\Phi_{ij}^{ab}\rangle \langle \Phi_{ij}^{ab}|\hat{H}|\Phi_{kl}^{cd}\rangle \langle \Phi_{kl}^{cd}|\hat{H}|\Psi_0^{(0)}\rangle}{(\varepsilon_i + \varepsilon_j - \varepsilon_a - \varepsilon_b)(\varepsilon_k + \varepsilon_l - \varepsilon_c - \varepsilon_d)}$$

$$- E_0^{(1)} \sum_{\substack{a<b \\ i<j}} \frac{|\langle \Psi_0^{(0)}|\hat{H}|\Phi_{ij}^{ab}\rangle|^2}{(\varepsilon_i + \varepsilon_j - \varepsilon_a - \varepsilon_b)^2} \qquad (1.84)$$

1.3.3 Quasi-Degenerate Perturbation Theory

The second-order correction to the energy given in Eq. 1.83 diverges when the denominator goes to zero, that is when the zeroth-order energy of excited Slater determinants becomes close to $E^{(0)}$. Moreover, in such situations one is usually interested not only in the lowest state, but in a number of low-lying *nearly-degenerate* states. In such cases one should go beyond the single determinant description of the zeroth-order problem and extend the reference with other low-energy determinants.

Let $\mathbb{S}$ be the collection of Slater determinants that span the Hilbert space of the full Hamiltonian of a system. The complete space is divided in a model space $\mathbb{S}_0$ and an external space $\mathbb{S}'$.

$$\mathbb{S} = \mathbb{S}_0 + \mathbb{S}' \qquad (1.85)$$

with $\mathbb{S}_0 = \{\Phi_I, \Phi_J, \ldots\}$ and $\mathbb{S}' = \{\Phi_R, \Phi_S, \ldots\}$. The model space contains all the determinants that significantly contribute to the (multiconfigurational) wave functions of the lowest, nearly degenerate electronic states. In ordinary many-body perturbation theory, one would first diagonalize the full Hamiltonian in the subspace $\mathbb{S}_0$ to construct the reference wave functions $\Psi^{(0)}$ and then include the effect of the determinants of $\mathbb{S}'$ through the expressions of the second- (or higher-) order perturbation theory in a state-by-state manner as schematically illustrated in Fig. 1.8. This is the *diagonalize-and-then-perturb* approach. On the contrary, quasi-degenerate perturbation theory first takes into account the effect of the external determinants on the interactions among the determinants of $\mathbb{S}_0$ and then diagonalizes the resulting matrix to obtain the N-electron wave functions and energies of the states of interest. This modification of the matrix elements of $\mathbb{S}_0$ is often called *dressing* or *screening* and leads to an effective Hamiltonian that not only describes the bare coupling between the determinants of the model space, but also the effects of electron correlation. The expression for the effective Hamiltonian at the second-order of perturbation is

$$\langle \Phi_I|\hat{H}^{eff}|\Phi_J\rangle = \langle \Phi_I|\hat{H}|\Phi_J\rangle + \sum_{R\in\mathbb{S}'} \frac{\langle \Phi_I|\hat{H}|\Phi_R\rangle \langle \Phi_R|\hat{H}|\Phi_J\rangle}{E_J^{(0)} - E_R^{(0)}} \qquad (1.86)$$

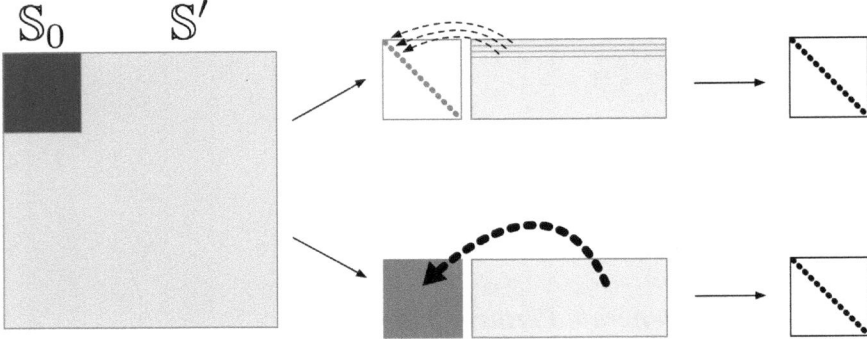

Fig. 1.8 Schematic representation of the diagonalize-then-perturb approach (*top*) and perturb-then-diagonalize (*bottom*) approaches. In the upper scheme, the model space is diagonalized and then the effect of the external determinants is included state-by-state. In the lower scheme, all matrix elements of the model space are perturbed and subsequently the model space is diagonalized

It is obvious from the denominator in the second term that the matrix will become non-Hermitian when the zeroth-order energies of the determinants in $\mathbb{S}_0$ are not the same. Therefore, this recipe only works for (nearly-)degenerate states. One advantage of this '*perturb-and-then-diagonalize*' approach is that the length of the wave function expansion remains of the dimension of the model space, and hence, especially suitable for analysis purposes. Multideterminantal perturbation schemes that follow the '*diagonalize-and-then-perturb*' approach are described in Sect. 4.3.3.

1.4 Effective Hamiltonian Theory

The exact N-electron wave function can be thought of to be an infinite linear combination of Slater determinants built from an infinitely large orbital set. While such a wave function is only a hypothetical object, lengthy wave function expansions can be considered to be good approximations to the exact solution. Hence, they will provide us with accurate energies and other observables of the system that can be extracted from the wave function by calculating the expectation value of the corresponding operator. However, such lengthy wave functions are often not easily *understood* and the extraction of simple models with predictive and interpretative power is not straightforward. Ideally, one would like to have a compact wave function with only a small number of Slater determinants, without loosing the accuracy of the nearly exact wave function.

Effective Hamiltonian theory establishes a connection between accuracy and interpretation. It is used in many fields of chemistry and physics in different variants and sometimes confused with model Hamiltonians. In the scope of this monograph, the latter term is used for simple Hamiltonians that find their origin in physical/chemical intuition, and hence, are phenomenological in nature. These model Hamiltonians have been put forward to interpret experimental measurements and capture the complex physics of a system in simpler concepts. The parameters of the model are usually determined by fitting the experimental data to analytical expressions derived from the model Hamiltonian.

In the significance used here the effective Hamiltonian maps a lengthy, highly-accurate wave function onto a much smaller subspace in such a way that the diagonalization of the subspace gives exactly the same eigenvalues as those of the nearly exact wave functions and the corresponding eigenvectors are projections of the original wave functions. The dimension of the subspace is typically the same as the dimension of the space spanned by some widely used model Hamiltonian. In this way, a one-to-one correspondence can be established between the *ab initio* calculations and the model Hamiltonian. Hence, the effective Hamiltonian theory provides a rigorous procedure to extract model parameters from accurate calculations.

Similar to what is done in QDPT, a model space $\mathbb{S}_0$ of dimension N is defined as a subspace of the full Hilbert space $\mathbb{S}$ of dimension M. Remember that QDPT is used to determine accurate wave functions and energies starting from a limited description of the system based on the model space. However, in the present case, the accurate energies and wave functions are already known and the action goes in the opposite direction; the lengthy wave function of length M is mapped on the smaller subspace $\mathbb{S}_0$ ensuring a minimum loss of the information contained in the full solution.

In the first place, the eigenfunctions of $\mathbb{S}$ (Ψ_k) have to be projected onto the model space by applying the projection operator

$$\hat{P}_{\mathbb{S}_0} = \sum_{i=1}^{N} |\Phi_i\rangle\langle\Phi_i| \tag{1.87}$$

where Φ_i is the basis of the model space. Among the projected vectors $\widetilde{\Psi}_k = \hat{P}_{\mathbb{S}_0}\Psi_k$, the N projections are selected that have the largest norm. These vectors are often defined as the basis of the so-called target space $\mathbb{S}_T$ and are used to construct the effective Hamiltonian. However, the vectors $\widetilde{\Psi}_k$ are in general not orthogonal. In the original formulation of Bloch [5] the projections are transformed to their biorthogonal form by

$$\widetilde{\Psi}_k^\dagger = S^{-1}\widetilde{\Psi}_k \tag{1.88}$$

with

$$\langle\widetilde{\Psi}_k|\widetilde{\Psi}_l^\dagger\rangle = \langle\widetilde{\Psi}_k^\dagger|\widetilde{\Psi}_l\rangle = \delta_{kl} \qquad \langle\widetilde{\Psi}_k|\widetilde{\Psi}_l\rangle = \langle\widetilde{\Psi}_k^\dagger|\widetilde{\Psi}_l^\dagger\rangle = S_{kl} \tag{1.89}$$

The effective Hamiltonian can now be expressed in its spectral decomposition

$$\hat{H}^{eff} = \sum_{k\in\mathbb{S}_T} |\widetilde{\Psi}_k\rangle E_k \langle\widetilde{\Psi}_k^\dagger| \tag{1.90}$$

However, this definition leads to a non-Hermitian Hamiltonian, which may not be the most optimal representation for interpretation. Therefore, one often adopts the orthogonalization of the projections proposed by des Cloizeaux, which involves the $S^{-1/2}$ overlap matrix [6]:

$$\widetilde{\Psi}_k^{\perp} = S^{-1/2} \, \widetilde{\Psi}_k \tag{1.91}$$

This procedure produces orthogonal vectors and all elements of the target space are affected in a similar degree. The effective Hamiltonian constructed with these vectors is hermitian and reads

$$\hat{H}^{eff} = \sum_{k \in \mathbb{S}_T} |\widetilde{\Psi}_k^{\perp}\rangle E_k \langle \widetilde{\Psi}_k^{\perp}| \tag{1.92}$$

The third possibility for processing the projected vectors is the Gram-Schmidt orthogonalization, in which the projections are sequentially orthogonalized. Starting with the normalization of $\widetilde{\Psi}_1$, the second vector is orthogonalized by projecting out the component of vector 1. Then $\widetilde{\Psi}_3$ is orthogonalized on $\widetilde{\Psi}_1$ and $\widetilde{\Psi}_2$, and so on. This means that the first vectors in the process are only slightly affected by the orthogonalization, while the last one is completely determined by the orthogonality condition. This loss of information—the coefficients of the projection of the last vector are not used—may be advantageous when some roots, i.e. computed (approximate) wavefunctions, in the target space are (nearly) degenerate with other roots in the external space. In such cases, the norm of the projection may be rather small and the information hold in the projections is not always well-founded, since strong mixing may have occurred with the states that are in the external space. The energy of these states can be considered as reliable, mixing among (nearly) degenerate states does not affect the energy.

In short, an effective Hamiltonian can be constructed from the following recipe.

- Choose a relevant model space of dimension N and write down the Slater determinants that constitute the basis of this space. It may be handy to work out all the matrix elements of the model Hamiltonian.
- Select the N eigenfunctions of the full Hilbert space (e.g. obtained in an *ab initio* calculation) with the largest projection onto the model space. (Bi-)orthonormalize the projections of these vectors and take the total energy of one of the roots as zero of energy.
- Calculate the matrix elements $\langle \Phi_I | \hat{H}^{eff} | \Phi_J \rangle$ of the effective Hamiltonian using the definition given in Eq. 1.90 or Eq. 1.92. One can check the procedure by diagonalizing the resulting matrix. This should give the same energies as found in the *ab initio* calculation and the corresponding eigenvectors have to be identical to the projections of these roots.

When the effective Hamiltonian is constructed from *ab initio* wave functions the resulting matrix is numerical in nature. This matrix can be used to determine the values of the parameters of a phenomenological model Hamiltonian, but also to check the validity of the model. In most cases the structure of the effective Hamiltonian matrix coincides with the structure of the model Hamiltonian, but when significant deviations are observed, it should not be discarded that important interactions are missing in the model. For instance, when non-zero matrix elements appear in the effective Hamiltonian at places where the model Hamiltonian is zero, one should revise the expression of the model Hamiltonian. This brings us to the second type of effective Hamiltonians that falls under the scope of this monograph; the analytical effective Hamiltonian. The above sketched procedure to derive a numerical Hamiltonian can also be used to map the analytical expressions of a precise, but complicated model Hamiltonian onto a simpler one. In this way one can rigorously derive new model Hamiltonians, for example when the comparison between a numerical effective Hamiltonian and a simple model Hamiltonian fails.

Problems

1.1 Ordering by spatial or spin part. In the notation of multideterminantal wave functions, one can either respect as much as possible the order of the spatial part in the different determinants, or strictly maintain the order of the spin part. Construct singlet and triplet functions for a two-electrons in two-orbitals case respecting (i) the order of the spatial part and (ii) the order of the spin part of the total wave function.

1.2 Coulomb, exchange or other. Classify the following two-electron integrals as Coulomb, exchange or other integral and assign a relative size (*large*, *medium*, or *small* to the integrals:

(a) $\langle \phi_a(1)\phi_b(2) | \dfrac{1}{r_{12}} | \phi_a(1)\phi_b(2) \rangle$

(b) $\displaystyle\int \dfrac{\phi_a(1)\phi_b(1)\phi_a(2)\phi_b(2)}{r_{12}} d\tau_1 d\tau_2$

(c) $\langle \phi_a(1)\overline{\phi}_c(2) | \dfrac{1 - \hat{P}_{12}}{r_{12}} | \phi_c(1)\overline{\phi}_a(2) \rangle$

(d) $\langle \phi_b(1)\phi_c(2) | \dfrac{1}{r_{12}} | \phi_c(1)\phi_d(2) \rangle$

(e) $\displaystyle\int \phi_a(1)\phi_c(2) \dfrac{1}{r_{12}} \phi_c(1)\phi_a(2) d\tau_1 d\tau_2$

(f) $\langle \phi_b(1)\phi_d(2) | \dfrac{\hat{P}_{12}}{r_{12}} | \phi_d(1)\phi_b(2) \rangle$

ϕ_a and ϕ_b are centered on site A, ϕ_c and ϕ_d on site B.

1.3 Perturbation theory. The prototype *particle in a box* problem is perturbed by a finite potential V_0 of width γ centered at $x = \frac{1}{2}L$. Calculate the first-order energy correction for the ground state, and the first and second excited states.

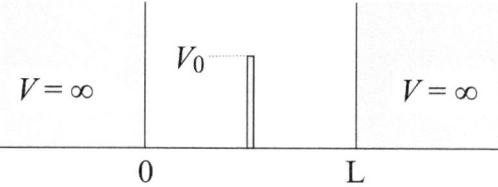

Reminder: $\psi_n^{(0)}(x) = \sqrt{\frac{2}{L}} \sin \frac{n\pi x}{L}$ and $E_n^{(0)} = \frac{hn^2}{8mL^2}$, Assume that γ is small enough to consider $\psi^{(0)}$ constant in the $\frac{1}{2}L - \frac{1}{2}\gamma \ldots \frac{1}{2}L + \frac{1}{2}\gamma$ interval.

1.4 Effective Hamiltonians: The following (hypothetical) model Hamiltonian is used to analyze a certain experimental observation

	$\lvert\Phi_1\rangle$	$\lvert\Phi_2\rangle$	$\lvert\Phi_3\rangle$
$\langle\Phi_1\rvert$	0		
$\langle\Phi_2\rvert$	μ	Δ_1	
$\langle\Phi_3\rvert$	γ	$(\gamma-4\mu)/2$	Δ_2

To get insight in the parameters of the model Hamiltonian an *ab initio* calculation was performed giving the following multideterminantal wave functions Ψ_k and energies E_k.

	Ψ_1	Ψ_2	Ψ_3	Ψ_4	Ψ_5
Φ_1	0.4804	0.8486	−0.0381	−0.2147	0.0387
Φ_2	0.3203	0.3990	−0.1391	−0.7732	0.3480
Φ_3	0.1601	0.0495	0.9468	0.0437	0.2714
Φ_4	0.8006	0.3397	−0.1109	0.4293	−0.2167
Φ_5	0.0000	0.0526	−0.2656	0.4122	0.8699
E	−0.50	−0.38	−0.40	−0.36	−0.20

a. Determine the norm of the projections of Ψ_k on the model space.
b. Select the three roots with the largest norm and orthogonalize the projections $\tilde{\Psi}_k$
c. Construct the 3×3 effective Hamiltonian and diagonalize the resulting matrix. Are the eigenvalues of $\hat{H}^{eff}$ equal to the eigenvalues of Ψ_k?
d. Determine the value of the model parameters. Is the model Hamiltonian consistent with the *ab initio* result?

References

1. R. Pauncz, *Spin Eigenfunctions* (Plenum Press, New York, 1979)
2. R. Pauncz, *The Construction of Spin Eigenfunctions: An Exercise Book* (Kluwer Academic/Plenum Publishers, New York, 2000)
3. R. Serber, Phys. Rev. **45**, 461 (1934)
4. R. Serber, J. Chem. Phys. **2**, 697 (1934)
5. C. Bloch, Nucl. Phys. **6**, 329 (1958)
6. J. des Cloizeaux, Nucl. Phys. **20**, 321 (1960)

Chapter 2
One Magnetic Center

Abstract This chapter discusses some of the magnetic phenomena that can be observed in systems with a single paramagnetic center. After shortly reviewing the basics of the magnetic moments of a free atom, we analyze the effect of spin-orbit coupling and an external magnetic field on the M_S levels of the ground state of larger systems. In a step-by-step procedure, we will first derive the model Hamiltonian to describe the magnetic anisotropy without external field, the so-called *zero-field splitting*. Secondly, the role of the external field is explored and a relation is established with the magnetic susceptibility, a macroscopic quantity. The chapter is closed by discussing the model Hamiltonian that combines the zero-field splitting and the anisotropy of the *g*-tensor to complete the description of the splitting of the M_S levels in systems with one, anisotropic, magnetic center.

2.1 Atomic Magnetic Moments

The two main sources for the magnetic moment of a free atom or molecule are the electronic spin moment and the angular moment. The motion of electrons relative to the nucleus in atoms with filled shells and in closed shell molecules leads to zero spin moment and zero angular moment. Therefore, such atoms and molecules can only have an induced magnetic moment when placed in an external magnetic field.

The simplest system with an intrinsic non-zero magnetic moment is an isolated one-electron atom or ion, treating both particles as point charges and neglecting the possible nuclear spin. The motion of the electron around the charged nucleus induces a microscopic current that produces a microscopic magnetic field. This leads to a so-called orbital magnetic moment which is proportional to the angular moment of the electron. Taking z as our quantization axis, its z-component equals

$$m_z = -\mu_B m_l \tag{2.1}$$

where m_l is the magnetic quantum number of the electron: $m_l = -l, -l+1 \ldots l-1, l$. The quantity $\mu_B = eh/4\pi m_e$ (1/2 in atomic units) is the elementary unit of magnetic moment, called the Bohr magneton. Its value is 9.27×10^{-24} JT^{-1}.

© Springer International Publishing Switzerland 2016

C. Graaf and R. Broer, *Magnetic Interactions in Molecules and Solids*,
Theoretical Chemistry and Computational Modelling,
DOI 10.1007/978-3-319-22951-5_2

There is another contribution to the total magnetic moment of the electron in a hydrogenic atom, this is due to the electron spin. From a *classical* viewpoint, this contribution (the spin magnetic moment) is due to the rotation of the charged electron around its axis. Its magnitude is given by

$$m_{sz} = -g_e \mu_B m_s \qquad (2.2)$$

with $m_s = \frac{1}{2}, -\frac{1}{2}$. The factor g_e turns out to be equal to 2.002319314. This value is slightly different from the value of 2 that might at first sight be expected from the analogy with the orbital magnetic moment, showing that this classical approach may be misleading. The difference with the value of 2 can be accounted for by the theory of quantum electron dynamics. The spin-orbit interaction is a relativistic effect, which appears in a natural way if we use Dirac's instead of Schrödinger's equation of motion. We can also describe it in an approximate sense by adding to the non-relativistic one-electron Hamiltonian a term that is proportional to the inner product of the vector operators $\hat{l}$ and $\hat{s}$

$$\hat{H}_{so} = \xi(r)\hat{l} \cdot \hat{s} \qquad (2.3)$$

The average of $\xi(r)$ over r is written $hc\zeta$ and ζ is called the spin-orbit constant. The spin-orbit constant of a hydrogenic atom turns out to be strongly dependent on Z and on the quantum numbers n and l of the electronic wave function

$$\zeta_{n,l} = \frac{\alpha^2 R Z^4}{n^3 l \left(l + \frac{1}{2}\right)(l + 1)} \qquad (2.4)$$

where α is the fine-structure constant ($\sim 1/137$) and R is the Rydberg constant. The non-relativistic one-electron Hamiltonian commutes with $\hat{l}^2$, $\hat{s}^2$ and (taking z as the quantization axes) with $\hat{l}_z$ and $\hat{s}_z$. Clearly, when we add $\hat{H}_{so}$ to the Hamiltonian, the Hamiltonian no longer commutes with these four operators and l, s, m_l and m_s are no longer "good" quantum numbers. The only remaining quantum numbers are j (with values $l + \frac{1}{2}$ and $l - \frac{1}{2}$) and $m_j = m_l + m_s$ (with values $j, j - 1, \ldots - j$).

2.1 Calculate $\zeta_{n,l}$ for the hydrogenic atoms H-$2p^1$, Ca^{19+}-$3p^1$, Ca^{19+}-$3d^1$, U^{91+}-$2p^1$, U^{91+}-$6d^1$ and U^{91+}-$5f^1$.

2.2 The Eigenstates of Many-Electron Atoms

In many-electron atoms we have an analogous situation, be it that the electron-electron interactions have to be included from the very beginning. We focus first on free many-electron atoms or ions, i.e. atoms or ions in a zero or uniform external field, for the time being neglecting any spin-orbit coupling. The total orbital angular moment operators $\hat{L}^2$, $\hat{L}_z$ and the total spin angular moment operators $\hat{S}^2$ and $\hat{S}_z$ commute with the Hamilton operator $\hat{H}$ where each electron moves in a field of spherical symmetry due to the nucleus and the field due the other electrons. Therefore, the eigenfunctions of $\hat{H}$ are in general also eigenfunctions of the other four operators. Only in the case of degenerate eigenstates of $\hat{H}$ one may choose (or find) eigenfunctions that are not simultaneously eigenfunctions of the other four operators. In that case, however, the eigenfunctions of $\hat{H}$ may always be rotated within the degenerate set to become also eigenfunctions of the other operators.

This implies that these eigenfunctions of $\hat{H}$ can be labelled using the quantum numbers S, M_S, L and M_L. The energy eigenvalues only depend on the eigenvalues of $\hat{L}^2$ and $\hat{S}^2$, and not on M_S and M_L. Therefore, the degenerate set of eigenfunctions of the free-atom Hamiltonian corresponding to one eigenvalue of $\hat{L}^2$ and $\hat{S}^2$ can be labelled by their values for L and S. It has become customary to use as labels not the spin moment S and orbital moment L but rather the spin multiplicity $2S + 1$ and L, in the notation ^{2S+1}L, where a *spectroscopic notation* $S, P, D, F, G, \ldots$ is used for $L = 0, 1, 2, 3, 4, \ldots$ The degenerate set of $(2S + 1) \times (2L + 1)$ eigenfunctions of ^{2S+1}L is commonly called an LS term. Examples of free atom (ion) LS terms are ^1S, ^2P, ^4F, $\ldots$

Since there is a one-to-one correspondence of the different L eigenvalues with the irreducible representations (IR) of the spherical symmetry group SO(3), the labels of the LS terms are simultaneously symmetry labels. Note that the angular moment operators $\hat{L}_x$, $\hat{L}_y$, $\hat{L}_z$ transform as the rotation operators $\hat{R}_x$, $\hat{R}_y$, $\hat{R}_z$, i.e. as P [1]. The behavior of atoms and ions, free and in compounds, depends for a large part on the ground term and the lowest excited terms. The symmetry of these lowest LS terms and the ordering of their energies can in general be well deduced using a simple one configuration model.

So far we have not considered any relativistic effects for the atoms. In particular, spin appears in the wave function but not in the non-relativistic Hamiltonian. Traditionally, spin is introduced *ad hoc* to explain the splitting of a beam of silver atoms into two parts in the famous experiment of Stern and Gerlach in 1922. A similar splitting was observed for a beam of hydrogen atoms in a later experiment. The splitting indicates the presence of an angular moment, but it cannot be an orbital angular moment since both atoms have an $L = 0$ ground state. Therefore, the spin property introduced to explain the splitting is considered to reflect an intrinsic angular moment s, which for electrons must be 1/2, and hence it is concluded that each electron has an additional quantum number $s = 1/2$, with a z-component s_z of either $+1/2$ or $-1/2$. The individual spin angular moments of the electrons in an atom can be coupled together to give a total spin angular moment S, analogous to the coupling of the individual orbital angular moments l to a total angular moment L. Since there is no spin operator in the Hamiltonian, there is also no coupling between spin and orbital angular moment.

The relativistic Hamilton operator for an electron can be derived, using the correspondence principle, from its relativistic classical Hamiltonian and this leads to the one-electron Dirac equation, which does contain spin operators. From the one-electron Dirac equation it seems trivial to define a many-electron relativistic equation, but the generalization to more electrons is less straightforward than in the non-relativistic case, because the electron-electron interaction is not unambiguously defined. The non-relativistic Coulomb interaction is often used as a reasonable first approximation. The relativistic treatment of atoms and molecules based on the many-electron Dirac equation leads to so-called *four-component* methods. The name stems from the fact that the electronic wave functions consist of four instead of two components. When the couplings between spin and orbital angular moment are comparable to the electron-electron interactions this is the preferred way to explain the electronic structure of the lowest states.

In most cases, however, the relativistic effects are rather weak and may be separated into spin-orbit coupling effects and scalar effects. The latter lead to compression and/or expansion of electron shells and can rather accurately be treated by modifying the one-electron part of the non-relativistic many-electron Hamiltonian. With this *scalar-relativistic* Hamiltonian the (modified) energies and wave functions are computed and subsequently an effective spin-orbit part $\hat{H}^{SO}$ is added to the Hamiltonian. The effects of the spin-orbit term on the energies and wave functions are commonly estimated using second-order perturbation theory. More information for the interested reader can be found in excellent textbooks on relativistic quantum chemistry [2, 3].

The standard way to include relativistic angular moment couplings in the notation of eigenvalues and eigenfunctions of the thus obtained energies and wave functions is the so-called Russell–Saunders coupling scheme. It is adequate if the spin-orbit coupling is considered to be weak compared to the electron-electron interactions. For a free atom or ion the Russell–Saunders scheme implies that the one-electron moments l and s are first coupled to a many-electron angular moment L and spin moment S, which are subsequently coupled to a total angular moment J. Due to the spin-orbit coupling the wave functions are no longer eigenfunctions of the $\hat{L}$ and $\hat{S}$ operators (L and S are no longer "good quantum numbers") but only of the $\hat{J}$ operator and the degeneracy of the states belonging to one LS term is partly removed. Only the states corresponding to a particular J eigenvalue are degenerate, but nevertheless the states of one LS term are close in energy. Such a set of nearly-degenerate states originating from one LS term is called a Russell–Saunders (RS) term and commonly denoted by a Russell-Saunders term symbol $^{2S+1}L_J$. For example, the lowest energy RS term of an atom with a single valence p-electron is $^2P_{1/2}$, with the RS term $^2P_{3/2}$ having a slightly higher energy. The $^2P_{1/2}$ term is two-fold degenerate ($M_J = 1/2, -1/2$) while the $^2P_{3/2}$ term is four-fold degenerate ($M_J = 3/2, 1/2, -1/2, -3/2$).

2.2 Give the Rusell–Saunders term symbol for the ground state of an atom with three electrons in the $2p$ orbitals.

In cases where the spin-orbit coupling is strong compared to the electron-electron interactions it is more reasonable to account for the spin-orbit coupling by using the so-called j-j coupling scheme. Here the orbital moment l and the spin s of each electron are coupled to give an individual angular moment for each electron. The individual j for each electron are then coupled to give a total angular moment J. Modern relativistic many-electron quantum mechanical computational treatments are able to treat the entire range of angular moment couplings, from negligible to dominant spin-orbit coupling. The results can be expressed in either terms of Russell-Saunders states or in j-j coupled states, whatever representation gives better insight. For core-excited states where the core spin-orbit coupling is much larger than the valence spin-orbit coupling, a mixed notation is sometimes used, in which the open core shell is j-j coupled and the open valence shell is Russell–Saunders coupled.

The many-electron states of an atom in a crystal field or a molecule can obviously not be labelled by the IRs of SO(3), since the Hamilton operator, the angular moment operator and therefore also the many-electron wave functions transform according to the IRs of a less symmetric point group. The lower symmetry may also remove the degeneracies of the LS terms. For example, the ^2P ground term of a boron atom becomes $^2T_{1u}$ in an octahedral crystal field so that the three fold degeneracy is retained, but splits into two LS terms of ^2E and $^2A_1$ symmetry when the crystal field symmetry is lowered to C_{3v}.

Orbital moment quenching: The eigenfunctions of the angular moment operator $\hat{l}^2$ are the spherical harmonics, characterized by the quantum numbers l and m.

$$\hat{l}^2 Y_{l,m} = l(l+1) Y_{l,m} \tag{2.5}$$

where $\hbar$ is put to 1. On the other hand, these functions cannot be eigenfunctions simultaneously of the three components $\hat{l}_{x,y,z}$, because these operators do not commute. Choosing z to be the quantization axis gives

$$\hat{l}_z Y_{l,m} = m Y_{l,m}$$

$$\hat{l}_x Y_{l,m} = \frac{1}{2}\sqrt{(l-m)(l+m+1)} Y_{l,m+1} + \frac{1}{2}\sqrt{(l-m+1)(l+m)} Y_{l,m-1}$$

$$\hat{l}_y Y_{l,m} = \frac{1}{2i}\sqrt{(l-m)(l+m+1)} Y_{l,m+1} - \frac{1}{2i}\sqrt{(l-m+1)(l+m)} Y_{l,m-1}$$

$$\tag{2.6}$$

In general the spherical harmonics $Y_{l,m}$ are complex functions, but linear combination can be made such that the eigenfunctions of $\hat{l}^2$ become real. For example for the spherical harmonics with $l = 1$ of a p^1 electronic configuration:

$$p_x = \frac{1}{\sqrt{2}}\left[-Y_{1,1} + Y_{1,-1}\right] \qquad p_y = \frac{+i}{\sqrt{2}}\left[Y_{1,1} + Y_{1,-1}\right] \qquad p_z = Y_{1,0} \qquad (2.7)$$

The first two functions are no longer eigenfunctions of $\hat{l}_z$, because linear combinations are made of functions with different m quantum number. The action of the $\hat{l}_{x,y,z}$ on these functions is worked out for one case and then summarized in Eq. 2.9.

$$\hat{l}_z p_x = \frac{1}{\sqrt{2}}\hat{l}_z\left[-Y_{1,1} + Y_{1,-1}\right] = \frac{1}{\sqrt{2}}\left[-Y_{1,1} - Y_{1,-1}\right] = ip_y \qquad (2.8)$$

ψ	$\hat{l}_x|\psi\rangle$	$\hat{l}_y|\psi\rangle$	$\hat{l}_z|\psi\rangle$	
p_x	0	$-ip_z$	ip_y	(2.9)
p_y	ip_z	0	$-ip_x$	
p_z	$-ip_y$	ip_x	0	

The matrix elements of $\hat{l} = \hat{l}_x + \hat{l}_y + \hat{l}_z$ in the basis of the $p_{x,y,z}$ functions are easily obtained

$\langle\hat{l}\rangle$	p_x	p_y	p_z	
p_x	0	$-i$	i	(2.10)
p_y	i	0	$-i$	
p_z	$-i$	i	0	

and the subsequent diagonalization results in two non-zero expectation values of $\langle\hat{l}\rangle$. In analogy to the spin operators the N-electron angular moment operators $\hat{L}$ are obtained by summing over the corresponding one-electron operators. The N-electron eigenfunctions are products of the one-electron spherical harmonics $Y_{l,m}$. Notice that the orbital moment of the p^5 electron configuration can be treated as if it were a one-electron system because of the hole-electron analogy.

2.3 Confirm that $\langle p_x|\hat{l}_z|p_y\rangle = -\langle p_y|\hat{l}_z|p_x\rangle = -i$.

The situation changes drastically when an external potential removes the degeneracy of the three functions. This can be caused by the crystal field exerted by the ions in an extended lattice for solid state compounds or by the ligands in the coordination sphere of an ion in a coordination complex. Although magnetic phenomena are more common for systems with incomplete d-shells, we will continue with our example concerning the p^1 configuration for simplicity. Imagine an external potential that makes, for example, the state with one electron in the p_x orbital lowest in energy and places the other two orbitals at slightly higher energy. Then, the orbital angular moment of the ground state is defined from the matrix element $\langle p_x|\hat{l}|p_x\rangle$ only, which is zero as can be seen in the matrix 2.10. In many physics textbooks, this effect is called *orbital angular moment quenching*. On the other hand, if the external potential destabilizes p_x with respect to $p_{y,z}$, the orbital angular moment of the p^1 electronic configuration is defined by the lower-right 2×2 sub-block of matrix 2.10 and results in non-zero expectation values. Now, the orbital angular moment is not quenched.

The general condition for non-zero orbital angular moment for a given Russell-Saunders term $^{2S+1}\Gamma$ is that the direct product $\Gamma \times \Gamma$ contains irreducible representations of the orbital moment operators $\hat{L}_{x,y,z}$. Since these operators have identical transformation properties as the rotation operator $\hat{R}_{x,y,z}$ (which is usually listed in the character tables of the symmetry point groups), it is easier to work with the rotation operator. For example, the ground state of the d^1 electronic configuration in an octahedral surrounding is $^2T_{2g}$. The rotation operator transforms as T_{1g} in the O_h point group. Since the direct product $T_{2g} \times T_{2g} = A_{1g} + E_g + T_{1g} + T_{2g}$ contains the irreducible representation of the rotation operator, one expects a non-zero orbital angular moment for this system. Note, however, that the d^1 electronic configuration is Jahn-Teller active and the geometry spontaneously distorts to a lower symmetry group accompanied by a (partial) *quenching* of the orbital angular moment.

2.4 Predict the (non-)existence of a net orbital angular moment for the high-spin d^2 electronic configuration in complexes with tetrahedral, octahedral and C_{2v} symmetry.

2.3 Further Removal of the Degeneracy of the N-electron States

The first two columns of Fig. 2.1 show how the free atom levels of a d^7 configuration are split by a distorted tetrahedral ligand-field. In this example, the states are labeled by the IR's of the D_{2d} subgroup of T_d and the ^4F (with a degeneracy of $(2S + 1) \times (2L + 1) = 4 \times 7 = 28$) is split in five energy levels. Based on the discussion in the previous section, one only expects a non-zero orbital moment for the ^4E states. The inset of the figure zooms in on the levels of the $^4B_1$ state and shows how the degeneracy is removed under the influence of spin-orbit coupling and when the system is placed in an external magnetic field. In the following two subsections we will discuss these two effects.

2.3.1 Zero Field Splitting

In $3d$ transition metal complexes, the splitting of the Russell–Saunders terms due to spin-orbit coupling is in general more important than the one caused by the external magnetic field typically used in EPR experiments. Therefore, the description of the

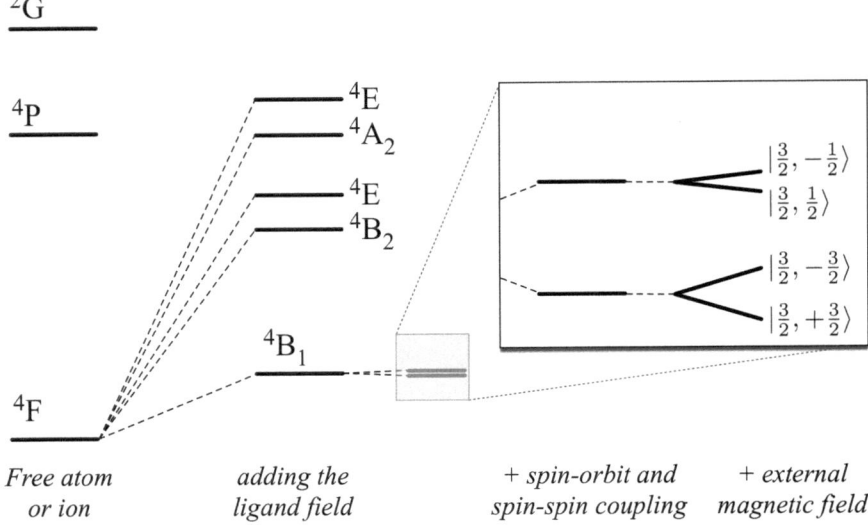

^{2}G

^{4}P

^{4}E
^{4}A$_2$

^{4}E
^{4}B$_2$

$\left|\frac{3}{2}, -\frac{1}{2}\right\rangle$
$\left|\frac{3}{2}, \frac{1}{2}\right\rangle$

$\left|\frac{3}{2}, -\frac{3}{2}\right\rangle$
$\left|\frac{3}{2}, +\frac{3}{2}\right\rangle$

^{4}B$_1$

^{4}F

| *Free atom or ion* | *adding the ligand field* | *+ spin-orbit and spin-spin coupling* | *+ external magnetic field* |

Fig. 2.1 Removal of the degeneracy of the energy levels of the d^7 manifold (*first column*) in a distorted tetrahedral ligand-field (*second column*), under the influence of spin-orbit coupling (first column of the inset) and in an external magnetic field (inset, second column). Only the lower states are shown in the figure. The labelling in the first two columns is $^{2S+1}\Gamma$, where Γ is the irreducible representation of the many-electron wave function. The labelling in the inset is $|J, M_J\rangle$

energy levels in these systems starts in general by addressing the zero field situation. In the absence of an external field and assuming a quenched orbital angular moment, the effect of spin-orbit coupling on the levels of the ground state can be qualitatively analysed with second-order perturbation theory. The perturbation operator takes the following form

$$\hat{V} = \zeta \hat{L} \cdot \hat{S} \qquad (2.11)$$

with ζ a tabulated atomic spin-orbit parameter, determined either by calculation or extracted from experimental data. For those cases that the orbital angular moment of the ground state is zero, it is convenient to derive a model Hamiltonian to describe the sub-levels of the ground state that only depends on the spin variables. Therefore, we write the unperturbed vectors as the product of the $|L, M_L\rangle$ spatial and $|S, M_S\rangle$ spin parts. The spatial part of the ground state is represented with $|0\rangle$, and $|\kappa\rangle$ denotes the spatial part of the excited states. The spin-only Hamiltonian that describes the zero-field splitting (no external magnetic field) of the levels is derived as the sum of first and second-order corrections. In first-order perturbation theory the energy correction equals

$$\langle 0|V|0\rangle = \langle S, M_S|\zeta \hat{S}|S, M_S\rangle \langle 0|\hat{L}|0\rangle \qquad (2.12)$$

Independent of the value of S or M_S, this product is strictly zero since we assumed that the ground state has no orbital angular moment. This is often referred to in the literature as the absence of a first-order angular moment. At second-order perturbation theory, the correction becomes slightly more involved

$$\sum_{\kappa \neq 0} \frac{\langle 0|\hat{V}|\kappa\rangle\langle\kappa|\hat{V}|0\rangle}{E_0 - E_\kappa} = \langle S, M_S|\zeta\hat{S}\sum_{\kappa \neq 0} \frac{\langle 0|\hat{L}|\kappa\rangle\langle\kappa|\hat{L}|0\rangle}{E_0 - E_\kappa}\zeta\hat{S}|S, M_S\rangle$$

$$= \langle S, M_S|\hat{S}\overline{\overline{D}}\hat{S}|S, M_S\rangle \qquad (2.13)$$

where the operator now only contains spin operators and the spin-anisotropy tensor $\overline{\overline{D}}$ contains all the information about the spatial anisotropy of the system

$$\overline{\overline{D}} = \zeta^2 \sum_{\kappa \neq 0} \frac{\langle 0|\hat{L}|\kappa\rangle\langle\kappa|\hat{L}|0\rangle}{E_0 - E_\kappa} \qquad (2.14)$$

In the most general case, the D-tensor has nine non-zero elements but one can always find an orientation in space such that the tensor becomes diagonal (as will be shown below in a numerical example) and only three parameters remain. Furthermore, the tensor may be written in a traceless form (sum of the diagonal elements equal to zero) and then the spin Hamiltonian takes the following form with only two parameters

$$\hat{H}_{ZFS} = D\left(\hat{S}_z^2 - \frac{1}{3}\hat{S}^2\right) + E(\hat{S}_x^2 - \hat{S}_y^2) \qquad (2.15)$$

where D and E are defined as the axial and rhombic anisotropy parameter, respectively.

$$D = D_{zz} - \frac{1}{2}(D_{xx} + D_{yy}) \qquad E = \frac{1}{2}(D_{xx} - D_{yy}) \qquad (2.16)$$

with $|D| \geqslant 3E \geqslant 0$.

> **2.5** Write out the $\hat{S}\overline{\overline{D}}\hat{S}$ operator. The only non-zero elements of the D-tensor are D_{xx}, D_{yy}, and D_{zz} on the diagonal. Determine the trace of the tensor and construct a traceless tensor $\overline{\overline{\Delta}}$. Write out the product $\hat{S}\overline{\overline{\Delta}}\hat{S}$ and check that the outcome coincides with Eq. 2.15.

The next step is to calculate the matrix elements of this zero-field splitting Hamiltonian with the $|S, M_S\rangle$ spin-functions of the ground state. The diagonalization of the resulting matrix gives the energies of the sub-levels under the influence of the spin-orbit coupling with the sub-levels of higher lying electronic states. The matrix elements for the different d^n configurations are well-documented in textbooks and articles and can be found in Appendix C. To illustrate the procedure we will derive the $\langle 1, 1|\hat{H}_{ZFS}|1, 1\rangle$ matrix element. To determine the effect of $\hat{H}_{ZFS}$ on $|1, 1\rangle$ we need to know how $\hat{S}_{x,y,z}^2$ act on this function. Whereas $|S, M_S\rangle$ are eigenfunctions of

$\hat{S}_z^2$, the effect of $\hat{S}_{x,y}^2$ is most easily determined via the ladder operators $S^{\pm}$. From the definitions $\hat{S}_x = \frac{1}{2}(\hat{S}^+ + \hat{S}^-)$ and $\hat{S}_y = -\frac{1}{2i}(\hat{S}^+ - \hat{S}^-)$ one arrives at

$$\hat{S}_x^2 = \frac{1}{4}\left(\hat{S}^+\hat{S}^+ + \hat{S}^-\hat{S}^- + \hat{S}^+\hat{S}^- + \hat{S}^-\hat{S}^+\right)$$
$$\hat{S}_y^2 = -\frac{1}{4}\left(\hat{S}^+\hat{S}^+ + \hat{S}^-\hat{S}^- - \hat{S}^+\hat{S}^- - \hat{S}^-\hat{S}^+\right)$$
(2.17)

Using Eq. 1.23, it is easily seen that the first and the last term give zero when applied on $|1, 1\rangle$ and that the other terms result in

$$\hat{S}^-\hat{S}^-|1, 1\rangle = 2|1, -1\rangle \qquad \hat{S}^+\hat{S}^-|1, 1\rangle = 2|1, 1\rangle \qquad (2.18)$$

which defines the action of $\hat{S}^2$ and $\hat{S}_{x,y,z}^2$ on $|1, 1\rangle$ as follows

$$\hat{S}^2|1, 1\rangle = S(S + 1)|1, 1\rangle = 2|1, 1\rangle \qquad \hat{S}_z^2|1, 1\rangle = |1, 1\rangle$$
$$\hat{S}_x^2|1, 1\rangle = \frac{1}{2}|1, -1\rangle + \frac{1}{2}|1, 1\rangle \qquad \hat{S}_y^2|1, 1\rangle = -\frac{1}{2}|1, -1\rangle + \frac{1}{2}|1, 1\rangle \quad (2.19)$$

Using the definition of Eq. 2.15 the matrix element becomes

$$\langle 1, 1|\hat{H}_{ZFS}|1, 1\rangle = D\left(1 - \frac{1}{3}\cdot 2\right) + \frac{1}{2}E - \frac{1}{2}E = \frac{1}{3}D \qquad (2.20)$$

The other matrix elements can be calculated following the same procedure and the final matrix representation of $\hat{H}_{ZFS}$ is

	$|1, 1\rangle$	$|1, 0\rangle$	$|1, -1\rangle$
$\langle 1, 1|$	$\frac{1}{3}D$	0	E
$\langle 1, 0|$	0	$-\frac{2}{3}D$	0
$\langle 1, -1|$	E	0	$\frac{1}{3}D$

(2.21)

The eigenvalues are $E_1 = -\frac{2}{3}D$; $E_{2,3} = \frac{1}{3}D \pm E$ and the corresponding eigenvectors $\Phi_1 = |1, 0\rangle$; $\Phi_{2,3} = (|1, 1\rangle \pm |1, -1\rangle)/\sqrt{2}$, which shows that the spin-orbit coupling between the $|S, M_S\rangle$ levels of the ground state with those of the excited states removes the degeneracy in the ground state spin manifold in the absence of an external field. D and E are related to the energy differences by

$$D = \frac{1}{2}(E_2 + E_3) - E_1 \qquad E = \frac{1}{2}(E_2 - E_3) \qquad (2.22)$$

In systems with an even number of electrons and both D and E different from zero, the zero-field splitting completely removes the degeneracy of the ground state manifold. On the contrary, the levels in systems with an odd number of electrons remain doubly degenerate at zero-field, often referred to as Kramers doublets. For systems with integer spin moment, the wave function of the lowest level is dominated by the $M_S = 0$ determinant when D is positive. This means that the projection of the spin moment on the magnetic z-axis is (practically) zero, while the projection on the x–y plane is maximal; the system has *easy plane* magnetism. When D is negative, the largest contributions to the lowest level arise from the determinants with $M_S = \pm M_{Smax}$, and hence, maximal projection of the spin moment on the z axis. This is known as *easy-axis* magnetism. The same applies for half-integer spin moment systems.

2.6 Demonstrate that the degeneracy of the $M_S = \pm 1/2$ sub-levels of the $S = 1/2$ manifold cannot be removed without an external magnetic field. Hint: Calculate the $\langle 1/2, \pm 1/2 | \hat{H}_{ZFS} | 1/2, \pm 1/2 \rangle$ matrix elements.

2.3.2 Splitting in an External Magnetic Field

The last column in Fig. 2.1 shows how an external magnetic field $\mathbf{H}$ affects the energies of the M_S sublevels of the electronic manifolds of a paramagnetic material. This effect is described by the Zeeman Hamiltonian

$$\hat{H}_{ZE} = \mu_B \mathbf{H} \cdot (\hat{L} + g_e \hat{S}) \tag{2.23}$$

When spin-orbit coupling is neglected and the ground state has no orbital moment, the expression reduces to its isotropic spin-only form

$$\hat{H}_{ZE} = \mu_B g_e \mathbf{H} \cdot \hat{S} \tag{2.24}$$

Defining the field direction as the z-axis, the Hamiltonian reduces to $\mu_B g_e H \hat{S}_z$ and the energies of the M_S sublevels vary linearly with the field strength as

$$E_n = M_S \mu_B g_e H \tag{2.25}$$

Typical examples of such paramagnetic systems are organic radicals where spin-orbit coupling plays a minor role, but Eq. 2.25 can also be used to describe the evolution of the energy of the M_S sublevels of the ground state in $3d$ transition metal complexes when the zero-field splitting is absent ($S = 1/2$) or significantly larger than the effect of the external field and the spin-orbit interaction with excited states is small. This splitting of the energy levels with the external field is not only at the very origin of electron paramagnetic resonance [4] techniques, but also manifests itself in the magnetic susceptibility of paramagnetic materials.

When a material is placed in a magnetic field, the sample becomes magnetized and the magnetization M is related to the field strength H by

$$\frac{\partial M}{\partial H} = \chi \tag{2.26}$$

The magnetic susceptibility χ is material dependent. It is a tensor, although the sample can be oriented with respect to the external field such that it becomes diagonal. In most cases, χ can be written as the sum of a diamagnetic (χ^D) and a paramagnetic contribution (χ^P). The latter contribution is temperature dependent and normally dominates in systems with unpaired electrons, i.e. in paramagnetic materials. The diamagnetic contribution does not depend on the temperature and can be estimated rather accurately from tabulated data for atoms and groups of atoms present in the material or by empirical formula [5]. Therefore, it is commonly assumed that the magnetic susceptibility data have been corrected for this contribution and one only has to analyze the paramagnetic part. For weak magnetic fields (and not too low temperatures), χ is independent of H and the magnetization can be related to the field as

$$M = \chi H \tag{2.27}$$

The link with the variation of the microscopic energy levels is given by statistical mechanics through the Boltzmann distribution

$$M = -\frac{\partial E}{\partial H} = N_A \frac{\sum\limits_n \frac{-\partial E_n}{\partial H} e^{-E_n/kT}}{\sum\limits_n e^{-E_n/kT}} \tag{2.28}$$

where T is the temperature, N_A Avogadro's number and k represents Boltzmann's constant. This expression can be significantly simplified by two assumptions originally proposed by van Vleck. In the first place, it is assumed that the energy of a given sublevel can be approximated by a Taylor series in the magnetic field strength

$$E_n = E_n^{(0)} + E_n^{(1)} H + E_n^{(2)} H^2 + \cdots \qquad -\frac{\partial E_n}{\partial H} = -E_n^{(1)} - 2E_n^{(2)} H + \cdots \tag{2.29}$$

The substitution of this expansion in the exponent of Eq. 2.28 leads to the second simplification when the series are limited to the first two terms

$$e^{-E_n/kT} = e^{(-E_n^{(0)} - E_n^{(1)} H)/kT} = e^{E_n^{(0)}/kT} e^{E_n^{(1)} H/kT} \approx e^{-E_n^{(0)}/kT} \left(1 - \frac{E_n^{(1)} H}{kT}\right) \tag{2.30}$$

Applying these two simplifications transforms Eq. 2.28 to

$$M = \frac{N_A \sum_n (E_n^{(1)} - 2E_n^{(2)} H)(1 - E_n^{(1)} H/kT) e^{-E_n^{(0)}/kT}}{\sum_n (1 - E_n^{(1)} H/kT) e^{-E_n^{(0)}/kT}} \tag{2.31}$$

If we limit ourselves to materials without spontaneous macroscopic magnetization, that is $M = 0$ at zero field, it is easily shown by substituting $H = 0$ that $\sum_n E_n^{(1)} e^{-E_n^{(0)}/kT} = 0$ and we arrive at

$$M = \frac{N_A H \sum_n (E_n^{(1)2}/kT - 2E_n^{(2)}) e^{-E_n^{(0)}/kT}}{\sum_n e^{-E_n^{(0)}/kT}} \tag{2.32}$$

Realizing that in the present case of negligible spin-orbit coupling the energies vary linearly with the field, the $E^{(2)}$-term can be neglected and the van Vleck equation for the magnetic susceptibility emerges from Eq. 2.27

$$\chi = \frac{N_A \sum_n E_n^{(1)2} e^{-E_n^{(0)}/kT}}{kT \sum_n e^{-E_n^{(0)}/kT}} \tag{2.33}$$

Under the assumption that the excited states are sufficiently far away from the ground state that their effect can be neglected, $E^{(0)}$ can be taken as reference point and put to zero. Then, Eqs. 2.25 and 2.29 can be used to obtain an analytical expression of $E_n^{(1)}$. The substitution of $E^{(0)} = 0$ and $E^{(1)} = \mu_B g_e M_S$ leads to

$$\chi = \frac{N_A (\mu_B g_e)^2}{kT} \frac{\sum\limits_{M_S=-S}^{S} M_S^2}{2S+1} \tag{2.34}$$

The summation over M_S^2 can be simplified using

$$\sum_{M_S=-S}^{S} M_S^2 = \frac{S(S+1)(2S+1)}{3} \tag{2.35}$$

and the final expression emerges

$$\chi = \frac{N_A (\mu_B g_e)^2}{3kT} S(S+1) \tag{2.36}$$

This expression shows that the magnetic susceptibility is inversely proportional to the temperature and is known as the Curie law:

$$\chi = \frac{C}{T} \tag{2.37}$$

where C is a constant that only depends on the spin quantum number of the ground state.

2.7 (a) Show that the denominator in Eq. 2.33 is equal to $2S + 1$ when $E^{(0)} = 0$ and verify the expression for the summation over M_S of M_S^2 for singlet, triplet and quintet states. (b) Many researchers in the field of molecular magnetism use the cgsemu (centimeter-gram-second electromagnetic units) system instead of the standard units defined by the international systems of units SI. In this alternative unit system, the value of $N_A \mu_B^2 / 3k$ is equal to 0.12505 (nearly 1/8). Calculate C for the S-values that can be found for the TM ions and the lanthanides.

Virtually always deviations to the Curie law are observed at low enough temperatures, because the magnetic centers in any real system are never truly isolated but interact with their environment. Moreover, spin-orbit coupling can also introduce extra interactions not covered by the Curie law. The interactions with other magnetic centers will be addressed in more detail in Chap. 3, but we describe here a mean-field approach to include their effect. In this rather crude approximation, each magnetic center experiences an internal field due to the average interaction with the other centers in addition to the uniform external field. This internal field depends on the average magnetization (M) of the material and is known as the Weiss field

$$\mathbf{H} = \mathbf{H}_{ext} + H_{int} = \mathbf{H}_{ext} + \lambda M \tag{2.38}$$

Combining this expression with the Eqs. 2.27 and 2.37 and assuming that H is aligned along z, one obtains

$$\chi = \frac{M}{H} = \frac{M}{H_{ext} + \lambda M} = \frac{C}{T} \tag{2.39}$$

Then, with $H_{ext} = M\big((T - C\lambda)/C\big)$, the measured susceptibility χ_{ext} can be written as

$$\chi_{ext} = \frac{M}{H_{ext}} = \frac{C}{T - \lambda C} \tag{2.40}$$

known as the Curie-Weiss law. λC is the Weiss constant and often written as Θ. Positive values of Θ are indicative of ferromagnetic interactions and a material with dominating antiferromagnetic interactions will show a negative Θ.

A more specific expression of Θ can be derived by extending the spin Hamiltonian of Eq. 2.24 with the Weiss field.

$$\hat{H} = \mu_B \mathbf{H} \cdot (\hat{L} + g_e \hat{S}_z) - nJ\langle S_z \rangle \hat{S}_z \qquad (2.41)$$

where n is the number of magnetic centers interacting with the center under consideration, J parametrizes the strength of the interactions and $\langle S_z \rangle$ is the average S_z value given by the Boltzmann distribution

$$\langle S_z \rangle = \frac{\displaystyle\sum_{M_S=-S}^{S} M_S e^{-E(S,M_S)}}{\displaystyle\sum_{M_S=-S}^{S} e^{-E(S,M_S)}} \qquad (2.42)$$

Taking the external field along the z-axis, the eigenvalues of this mean-field Hamiltonian are

$$E_n = M_S \mu_B g_e H - nJ\langle S_z \rangle M_S \qquad (2.43)$$

After expanding the exponents in Eq. 2.42 in a Taylor series and only maintaining the first two terms, the energy eigenvalues are inserted to arrive at

$$\langle S_z \rangle = \frac{\displaystyle\sum_{M_S=-S}^{S} M_S\left(1 - M_S(\mu_B g_e H - nJ\langle S_z \rangle)/kT\right)}{\displaystyle\sum_{M_S=-S}^{S} \left(1 - M_S(\mu_B g_e H - nJ\langle S_z \rangle)/kT\right)}$$
$$= \frac{-(S(S+1)(2S+1)/3)(\mu_B g_e H - nJ\langle S_z \rangle)/kT}{2S+1} \qquad (2.44)$$

using the simplification of the sum over M_S^2 used before (Eq. 2.35). This equation requires some rewriting but finally the average S_z value reduces to

$$\langle S_z \rangle = -\frac{S(S+1)\mu_b g_e H}{3kT - nJS(S+1)} \qquad (2.45)$$

which can be used to express the magnetization and the magnetic susceptibility

$$M = N_A \mu_B g_e \langle S_z \rangle \qquad (2.46)$$

$$\chi = \frac{M}{H} = N_A \mu_B^2 g_e^2 \frac{S(S+1)}{3kT - nJS(S+1)} = \frac{C}{T - \Theta} \qquad (2.47)$$

to obtain the Curie–Weiss law with an explicit expression for $\Theta = nJS(S+1)/3k$.

2.8 Derive Eq. 2.45 from Eq. 2.44.

Anisotropy of the g-tensor: Before combining the effect of the zero-field splitting and the external magnetic field, we have to establish how spin-orbit coupling affects the Zeeman effect. This is most easily done for a system with $S = \frac{1}{2}$ and a quenched orbital moment. As an example, we will consider one unpaired electron. The g-factor in the Zeeman Hamiltonian of Eq. 2.24 is now replaced by a tensor

$$\hat{H}_{ZE} = \mu_B \overline{\overline{g}} \mathbf{H} \cdot \hat{S} \qquad (2.48)$$

$\overline{\overline{g}}$ can be transformed to a diagonal form when the coordinate axis frames of the field and the g-tensor coincide. The axis frame that diagonalizes the g-tensor is not necessarily the same as the frame that diagonalizes the D-tensor introduced in the previous section, although this is often assumed to be the case. Furthermore, it should be noted that, strictly speaking S is not a good quantum number anymore when spin-orbit coupling is considered. Therefore, the spin operator in Eq. 2.48 is often replaced by an effective spin operator $\tilde{S}$ with the same formal properties.

To evaluate the effect of spin-orbit coupling we will start writing down the first-order corrected wave functions of the $M_S = \pm\frac{1}{2}$ sublevels, then calculate the matrix elements of the Zeeman Hamiltonian (Eq. 2.23) and compare these to the matrix elements of the (effective) spin-only Zeeman Hamiltonian given in Eq. 2.48 to find analytical expressions for the diagonal elements of $\overline{\overline{g}}$. With $\zeta \hat{L} \cdot \hat{S}$ as perturbation operator, the wave functions that describe the lowest two levels become

$$\psi^{(1)} = \psi^{(0)} + \zeta \sum_{i \neq 0} \frac{\langle \psi_i^{(0)} | \hat{L} \cdot \hat{S} | \psi_0^{(0)} \rangle}{E_0 - E_i} \psi_i^{(0)} \qquad (2.49)$$

where $\psi_i^{(0)}$ represent the different M_S components of excited states. The spin part of the wave function is not written explicitly and can either be α or β. Replacing $\hat{L} \cdot \hat{S}$ by $\hat{L}_z \hat{S}_z + \frac{1}{2}(\hat{L}^+ \hat{S}^- + \hat{L}^- \hat{S}^+)$ the expressions for the two wave functions can easily be derived

$$\psi^{(1)} = \psi_0^{(0)} + \frac{1}{2}\zeta \sum_{i \neq 0} \frac{\langle \psi_i^{(0)} | \hat{L}_z | \psi_0^{(0)} \rangle}{E_0 - E_i} \psi_i^{(0)} + \frac{1}{2}\zeta \sum_{i \neq 0} \frac{\langle \psi_i^{(0)} | \hat{L}^+ | \psi_0^{(0)} \rangle}{E_0 - E_i} \overline{\psi}_i^{(0)}$$

$$\overline{\psi}^{(1)} = \overline{\psi}_0^{(0)} - \frac{1}{2}\zeta \sum_{i \neq 0} \frac{\langle \psi_i^{(0)} | \hat{L}_z | \psi_0^{(0)} \rangle}{E_0 - E_i} \overline{\psi}_i^{(0)} + \frac{1}{2}\zeta \sum_{i \neq 0} \frac{\langle \psi_i^{(0)} | \hat{L}^- | \psi_0^{(0)} \rangle}{E_0 - E_i} \psi_i^{(0)} \quad (2.50)$$

and show that the first-order corrected wave functions are no longer spin eigenfunctions, but rather a mixture of α and β contributions.

2.9 Show that the $\hat{L} \cdot \hat{S}$ matrix elements of $\psi_0^{(0)}$ and an excited state $\psi_i^{(0)}$ are equal to $\frac{1}{2}\langle \psi_i^{(0)}|\hat{L}_z|\psi_0^{(0)}\rangle$ when the spin part of $\psi_0^{(0)}$ and $\psi_i^{(0)}$ is identical and equal to $\frac{1}{2}\langle \psi_i^{(0)}|\hat{L}^+|\psi_0^{(0)}\rangle$ when $\psi_0^{(0)}$ has α- and $\psi_i^{(0)}$ has β-spin.

For a magnetic field along the z-axis (the quantization axis of the system), **H** can be replaced by a scalar and the two expressions of the Zeeman Hamiltonian (Eqs. 2.23 and 2.48) reduce to $\mu_B H(\hat{L}_z + g_e\hat{S}_z)$ and $\mu_B H g_{zz}\hat{S}_z$. The corresponding matrix elements are

$$\mu_B H g_{zz}\langle \psi^{(1)}|\hat{S}_z|\psi^{(1)}\rangle = \frac{1}{2}\mu_B H g_{zz} = \mu_B B\langle \psi^{(1)}|\hat{L}_z + g_e\hat{S}_z|\psi^{(1)}\rangle \tag{2.51}$$

$$\mu_B H g_{zz}\langle \overline{\psi}^{(1)}|\hat{S}_z|\overline{\psi}^{(1)}\rangle = -\frac{1}{2}\mu_B H g_{zz} = \mu_B H\langle \overline{\psi}^{(1)}|\hat{L}_z + g_e\hat{S}_z|\overline{\psi}^{(1)}\rangle \tag{2.52}$$

The off-diagonal elements are zero in both Hamiltonians due to the spin-orthogonality. Now an expression for g_{zz} emerges from either equation as

$$g_{zz} = 2\left[g_e\langle \psi^{(1)}|\hat{S}_z|\psi^{(1)}\rangle + \langle \psi^{(1)}|\hat{L}_z|\psi^{(1)}\rangle\right] = g_e + 2\langle \psi^{(1)}|\hat{L}_z|\psi^{(1)}\rangle \tag{2.53}$$

The last term can be specified by substitution of the definitions given in Eq. 2.50

$$\langle \psi^{(1)}|\hat{L}_z|\psi^{(1)}\rangle = \langle \psi^{(0)}|\hat{L}_z|\psi^{(0)}\rangle + \frac{1}{2}\zeta\sum_{i\neq 0}\frac{\langle \psi_i^{(0)}|\hat{L}_z|\psi_0^{(0)}\rangle\langle \psi_0^{(0)}|\hat{L}_z|\psi_i^{(0)}\rangle}{E_0 - E_i}$$

$$+ \frac{1}{2}\zeta\sum_{i\neq 0}\frac{\langle \psi_i^{(0)}|\hat{L}_x + i\hat{L}_y|\psi_0^{(0)}\rangle\langle \psi_0^{(0)}|\hat{L}_z|\overline{\psi}_i^{(0)}\rangle}{E_0 - E_i} + \frac{1}{4}\zeta^2 \ldots$$

$$+ \langle \psi^{(0)}|\hat{L}_z|\psi^{(0)}\rangle + \frac{1}{2}\zeta\sum_{i\neq 0}\frac{\langle \psi_i^{(0)}|\hat{L}_z|\psi_0^{(0)}\rangle\langle \psi_0^{(0)}|\hat{L}_z|\psi_i^{(0)}\rangle}{E_0 - E_i}$$

$$+ \frac{1}{2}\zeta\sum_{i\neq 0}\frac{\langle \psi_i^{(0)}|\hat{L}_x - i\hat{L}_y|\psi_0^{(0)}\rangle\langle \psi_0^{(0)}|\hat{L}_z|\overline{\psi}_i^{(0)}\rangle}{E_0 - E_i} + \frac{1}{4}\zeta^2 \ldots \tag{2.54}$$

Here, $\hat{L}^\pm$ is replaced by the expression in terms of $\hat{L}_{x,y}$ and the minus sign in front of $i\hat{L}_y$ in the seventh term on the right arises from the fact that the complex conjugated function $\Psi_1^{\dagger(1)}$ was written. Moreover, the third and seventh terms are zero because of the spin-orthogonality. The terms that are quadratic in ζ are neglected. This somewhat awkward expression can be further simplified by taking into account that the zeroth-order wave function of the ground state has no orbital moment, $\langle \psi^{(0)}|\hat{L}_z|\psi^{(0)}\rangle = 0$.

Hence, we can write

$$g_{zz} = g_e + 2\zeta \sum_{i \neq 0} \frac{\langle \psi_i^{(0)} | \hat{L}_z | \psi_0^{(0)} \rangle \langle \psi_0^{(0)} | \hat{L}_z | \psi_i^{(0)} \rangle}{E_0 - E_i} \tag{2.55}$$

Next, we consider the case for a field along x, from which g_{xx} can be determined. The Hamiltonians can be written as $\mu_B H (\hat{L}_x + g_e \hat{S}_x)$ and $\mu_B H g_{xx} \hat{S}_x$ in this case and the matrix elements with $\psi^{(1)}$ and $\overline{\psi}^{(1)}$ are

$$\mu_B H g_{xx} \langle \psi^{(1)} | \hat{S}_x | \psi^{(1)} \rangle = \frac{1}{2} \mu_B H g_{xx} \langle \psi^{(1)} | \hat{S}^+ + \hat{S}^- | \psi^{(1)} \rangle = 0 \tag{2.56}$$

$$\mu_B H g_{xx} \langle \psi^{(1)} | \hat{S}_x | \overline{\psi}^{(1)} \rangle = \frac{1}{2} \mu_B H g_{xx} \langle \psi^{(1)} | \hat{S}^+ + \hat{S}^- | \overline{\psi}^{(1)} \rangle = \frac{1}{2} \mu_B H g_{xx}$$

$$= \mu_B H \langle \psi^{(1)} | \hat{L}_x + g_e \hat{S}_x | \overline{\psi}^{(1)} \rangle \tag{2.57}$$

and analogous for the other two matrix elements. Working out the expression for the off-diagonal matrix element gives

$$g_{xx} = 2 \langle \psi^{(1)} | \hat{L}_x | \overline{\psi}^{(1)} \rangle + 2 g_e \langle \psi^{(1)} | \hat{S}_x | \overline{\psi}^{(1)} \rangle = g_e + 2 \langle \psi^{(1)} | \hat{L}_x | \overline{\psi}^{(1)} \rangle \tag{2.58}$$

and one obtains the explicit equation for g_{xx} by substituting the definition of the first-order wave functions given in Eq. 2.50. Taking into account only the non-zero terms that are at most linear in ζ, we arrive at

$$\begin{aligned}
g_{xx} = g_e + 2 \Bigg(& \frac{1}{2} \zeta \sum_{i \neq 0} \frac{\langle \psi_i^{(0)} | \hat{L}_x - i \hat{L}_y | \psi_0^{(0)} \rangle \langle \psi_0^{(0)} | \hat{L}_x | \psi_i^{(0)} \rangle}{E_0 - E_i} \\
& + \frac{1}{2} \zeta \sum_{i \neq 0} \frac{\langle \psi_i^{(0)} | \hat{L}_x + i \hat{L}_y | \psi_0^{(0)} \rangle \langle \psi_0^{(0)} | \hat{L}_x | \psi_i^{(0)} \rangle}{E_0 - E_i} \Bigg) \\
= g_e + 2 \zeta & \sum_{i \neq 0} \frac{\langle \psi_i^{(0)} | \hat{L}_x | \psi_0^{(0)} \rangle \langle \psi_0^{(0)} | \hat{L}_x | \psi_i^{(0)} \rangle}{E_0 - E_i}
\end{aligned} \tag{2.59}$$

The expression for g_{yy} is obtained by replacing $\hat{L}_x$ with $\hat{L}_y$. Even the off-diagonal elements of the g-tensor, in case of a non-aligned sample, can be calculated with analogous equations combining the proper angular moment operators.

The procedure is best illustrated with a simple example. For this purpose, we fall back on the p^1 model system used before in the discussion of the orbital moment quenching. The external potential stabilizes the p_z orbital with respect to the degenerate p_x and p_y orbitals by an amount of ΔE as shown in Fig. 2.2. Using the explicit notation for spatial and spin part, the zeroth-order wave functions are

Fig. 2.2 stabilization of the p_z orbital by ΔE with respect to the degenerate p_x and p_y orbitals due to an external potential

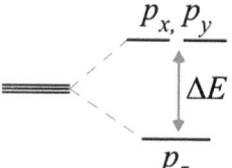

$$\psi_0^{(0)} = p_z\alpha = p_0\alpha \qquad \overline{\psi}_0^{(0)} = p_z\beta = p_0\beta \tag{2.60}$$

$$\psi_i^{(0)} = \{p_x\alpha, p_x\beta, p_y\alpha, p_y\beta\} = \{p_+\alpha, p_+\beta, p_-\alpha, p_-\beta\} \tag{2.61}$$

Now we apply Eq. 2.50 to obtain the expression of the first-order corrected wave functions for the lowest two levels

$$\psi_0^{(1)} = p_0\alpha + \frac{1}{2}\zeta\left[\frac{\langle p_+|\hat{L}_z|p_0\rangle}{\Delta E}p_+\alpha + \frac{\langle p_-|\hat{L}_z|p_0\rangle}{\Delta E}p_-\alpha\right]$$
$$+ \frac{1}{2}\zeta\left[\frac{\langle p_+|\hat{L}_+|p_0\rangle}{\Delta E}p_+\beta + \frac{\langle p_-|\hat{L}_+|p_0\rangle}{\Delta E}p_-\beta\right] = p_0\alpha + \frac{1}{2}\sqrt{2}\zeta p_+\beta \tag{2.62}$$

$$\overline{\psi}_0^{(1)} = p_0\beta - \frac{1}{2}\zeta\left[\frac{\langle p_+|\hat{L}_z|p_0\rangle}{\Delta E}p_+\beta + \frac{\langle p_-|\hat{L}_z|p_0\rangle}{\Delta E}p_-\beta\right]$$
$$+ \frac{1}{2}\zeta\left[\frac{\langle p_+|\hat{L}_-|p_0\rangle}{\Delta E}p_+\alpha + \frac{\langle p_-|\hat{L}_-|p_0\rangle}{\Delta E}p_-\alpha\right] = p_0\beta + \frac{1}{2}\sqrt{2}\zeta p_-\alpha \tag{2.63}$$

The values of g_{zz} and $g_{xx} = g_{yy}$ are determined from Eqs. 2.55 and 2.59 and lead to

$$g_{zz} = g_e + 2\zeta\left[\frac{\langle p_+|\hat{L}_z|p_0\rangle\langle p_0|\hat{L}_z|p_+\rangle}{\Delta E} + \frac{\langle p_-|\hat{L}_z|p_0\rangle\langle p_0|\hat{L}_z|p_-\rangle}{\Delta E}\right] = g_e \tag{2.64}$$

$$g_{xx} = g_e + 2\zeta\left[\frac{\langle p_+|\hat{L}_x|p_0\rangle\langle p_0|\hat{L}_x|p_+\rangle}{\Delta E} + \frac{\langle p_-|\hat{L}_x|p_0\rangle\langle p_0|\hat{L}_x|p_-\rangle}{\Delta E}\right]$$
$$= g_e + 2\zeta\left[\frac{\frac{1}{2}\sqrt{2}\cdot\frac{1}{2}\sqrt{2}}{\Delta E} + \frac{\frac{1}{2}\sqrt{2}\cdot\frac{1}{2}\sqrt{2}}{\Delta E}\right] = g_e + \frac{2\zeta}{\Delta E} \tag{2.65}$$

2.10 Demonstrate that the matrix element $\langle p_-|\hat{L}_x|p_0\rangle$ and all other matrix elements in Eq. 2.65 are equal to $\frac{1}{2}\sqrt{2}$. Hint: substitute $\hat{L}_x$ by $(\hat{L}^+ + \hat{L}^-)/2$.

2.3.3 Combining ZFS and the External Magnetic Field

The separate descriptions of the zero-field splitting and the effect of an external magnetic field on the atomic sublevels can now be combined into a unified description using the following spin Hamiltonian

$$\hat{H} = \mu_B \overline{\overline{g}} \cdot \mathbf{H}\hat{S} + \hat{S} \cdot \overline{\overline{D}} \cdot \hat{S} \tag{2.66}$$

In the coordinate frame that diagonalizes $\overline{\overline{g}}$ and $\overline{\overline{D}}$, which is assumed to be the same for both, the spin Hamiltonian simplifies to

$$\hat{H} = \mu_B(g_x H_x \hat{S}_x + g_y H_y \hat{S}_y + g_z H_z \hat{S}_z) + D\left(\hat{S}_z^2 - \frac{1}{3}\hat{S}^2\right) + E(\hat{S}_x^2 - \hat{S}_y^2) \tag{2.67}$$

In the first place we write down the explicit matrix representation of this Hamiltonian when the external field is aligned along the z-axis and assuming that the complex only presents axial anisotropy, that is $E = 0$. This means that both H_x and H_y are zero.

	$|1, 1\rangle$	$|1, 0\rangle$	$|1, -1\rangle$	
$\langle 1, 1|$	$\frac{1}{3}D + \mu_B g_z H$	0	0	(2.68)
$\langle 1, 0|$	0	$-\frac{2}{3}D$	0	
$\langle 1, -1|$	0	0	$\frac{1}{3}D - \mu_B g_z H$	

After shifting the diagonal by $\frac{2}{3}D$ to let the zero of energy coincide with the energy of the $|1, 0\rangle$ state, the resulting energies are

$$E_1 = 0 ; \quad E_{23} = D \pm \mu_B g_z H_z \tag{2.69}$$

The energies of the $|1, \pm 1\rangle$ states evolve linearly with H as shown on the left in Fig. 2.3.

The situation is slightly more complicated when the magnetic field is applied perpendicular to the principal magnetic axis. We will work out the matrix element between $|1, 1\rangle$ and $|1, 0\rangle$ and then give the full Hamiltonian for the field along the x-axis. The part of the Hamiltonian that accounts for the zero-field splitting does

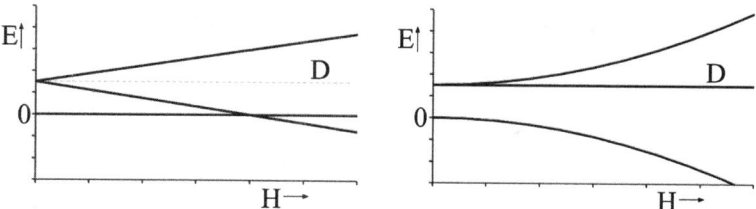

Fig. 2.3 Energies of the three components of a triplet state in an external field along the z-axis (*left*) and perpendicular to it (*right*)

not change, and hence, we can concentrate on the Zeeman interaction. The action of $\hat{S}_x$ on $|1, 0\rangle$ is easiest obtained by using the expression of $\hat{S}_x$ in terms of the ladder operators $\hat{S}^+$ and $\hat{S}^-$.

$$\langle 1, 1|\mu_B g_x H_x \frac{1}{2}(\hat{S}^+ + \hat{S}^-)|1, 0\rangle$$

$$= \mu_B g_x H_x \frac{1}{2}\left(\langle 1, 1|\sqrt{2}|1, 1\rangle + \langle 1, 1|\sqrt{2}|1 - 1\rangle\right) = \frac{\mu_B g_x H_x}{\sqrt{2}}$$
(2.70)

The other off-diagonal elements are the same except the interaction between $|1, 1\rangle$ and $|1 - 1\rangle$, which is zero. The full Hamiltonian takes this form

	$|1, 1\rangle$	$|1, 0\rangle$	$|1, -1\rangle$
$\langle 1, 1|$	$\frac{1}{3}D$	$\frac{1}{\sqrt{2}}\mu_B g_x H_x$	0
$\langle 1, 0|$	$\frac{1}{\sqrt{2}}\mu_B g_x H_x$	$-\frac{2}{3}D$	$\frac{1}{\sqrt{2}}\mu_B g_x H_x$
$\langle 1, -1|$	0	$\frac{1}{\sqrt{2}}\mu_B g_x H_x$	$\frac{1}{3}D$

(2.71)

After shifting the diagonal by $\frac{2}{3}D$, the energy eigenvalues can be determined as

$$E_1 = D; \quad E_{2,3} = \frac{1}{2}\left(D \pm \sqrt{D^2 + 4\mu_B^2 g_x^2 H_x^2}\right)$$
(2.72)

2.11 Confirm that the only effect of uniformly shifting the diagonal elements is the same shift of the energy eigenvalues.

The expressions for $E_{2,3}$ can be simplified by the Taylor expansion $\sqrt{p + q} = \sqrt{p} + \frac{1}{2}q/\sqrt{p} + \cdots$. Assuming that D^2 is (much) larger than $4\mu_B^2 g_x^2 H_x^2$, the expansion can be restricted to the first two terms only and the energies become

$$E_2 = D + \mu_B^2 g_x^2 H_x^2 / D \; ; \quad E_3 = -\mu_b^2 g_x^2 H_x^2 / D \qquad (2.73)$$

The evolution of the energies with increasing H_x is no longer linear and is depicted in the right part of Fig. 2.3. Applying the external field perpendicular to the z-axis implies of course not automatically that the field is oriented along the x-axis. It is therefore necessary to confront the above result to what is obtained when the field is applied along the y-axis. The Hamiltonian has the same general shape but the off-diagonal elements are slightly different now.

	$|1, 1\rangle$	$|1, 0\rangle$	$|1, -1\rangle$
$\langle 1, 1|$	$\frac{1}{3}D$	$-\frac{i}{\sqrt{2}}\mu_B g_y H_y$	0
$\langle 1, 0|$	$\frac{i}{\sqrt{2}}\mu_B g_y H_y$	$-\frac{2}{3}D$	$-\frac{i}{\sqrt{2}}\mu_B g_y H_y$
$\langle 1, -1|$	0	$\frac{i}{\sqrt{2}}\mu_B g_y H_y$	$\frac{1}{3}D$

$$(2.74)$$

However, this has no consequences for the eigenvalues of the matrix. Diagonalization of the (shifted) matrix gives exactly the same energies as derived from the Hamiltonian with the field along the x-axis as long as the system has no rhombic anisotropy; $g_x = g_y = g_\perp$ and $E = 0$. In the general case of axial and rhombic anisotropy, no analytical expressions for the energies can be derived and one commonly resorts to numerical approaches [6].

Problems

2.1 Extracting D and E for a Ni^{II} complex. The triplet ground state T_0 of a Ni^{II} complex has three M_S sublevels, which are degenerate in the absence of an external magnetic field and neglecting spin-orbit coupling. However, the interaction with the M_S sublevels of excited states (T_1, T_2, S_1, etc.) through the spin-orbit operator removes the degeneracy. Since the molecule is oriented in an arbitrary axes frame, the cartesian z-axis does not coincide with the magnetic z-axis and the wave functions of the three sublevels are complex functions, mixtures of the $M_S = 0, \pm 1$ components.

a. Construct the matrix representation of the $\hat{S} \cdot \overline{\overline{D}} \cdot \hat{S}$ spin Hamiltonian for an arbitrary axes frame, i.e., $\overline{\overline{D}}$ is not diagonal:

$$\hat{H} = \begin{pmatrix} \hat{S}_x & \hat{S}_y & \hat{S}_z \end{pmatrix} \begin{pmatrix} D_{xx} & D_{xy} & D_{xz} \\ D_{yx} & D_{yy} & D_{yz} \\ D_{zx} & D_{zy} & D_{zz} \end{pmatrix} \begin{pmatrix} \hat{S}_x \\ \hat{S}_y \\ \hat{S}_z \end{pmatrix}$$

Use $|S, M_S\rangle = \{|1, 1\rangle, |1, 0\rangle, |1, -1\rangle\}$ as basis.

b. The *ab initio* wave functions of the three states with the largest projection on the model space are

$$\Psi_1 = (0.686 - 0.024i)|1, -1\rangle + (-0.175 + 0.009i)|1, 0\rangle + (-0.685 + 0.046i)|1, 1\rangle$$
$$\Psi_2 = (-0.664 + 0.197i)|1, -1\rangle + (0.036 + 0.123i)|1, 0\rangle + (-0.667 + 0.187i)|1, 1\rangle$$
$$\Psi_3 = (0.110 - 01.08i)|1, -1\rangle + (0.957 - 0.147i)|1, 0\rangle + (-0.137 - 0.070i)|1, 1\rangle$$

Calculate the overlap matrix of $\widetilde{\Psi}_i$, the projections of Ψ on the model space.

c. The orthonormalized projections are given by

	$|1, -1\rangle$	$|1, 0\rangle$	$|1, 1\rangle$
$\widetilde{\Psi}'_1$	0.695504 − 0.023842i	−0.177003 + 0.008987i	−0.694343 + 0.046723i
$\widetilde{\Psi}'_2$	−0.672287 + 0.199091i	0.035909 + 0.124483i	−0.675035 + 0.189566i
$\widetilde{\Psi}'_3$	+0.110493 − 0.109038i	0.964328 − 0.147901i	−0.138092 − 0.070912i

$E_1 = 0.00$; $E_2 = 11.54 \text{ cm}^{-1}$; $E_3 = 37.55 \text{ cm}^{-1}$. Construct the effective Hamiltonian and check the consistency of the model Hamiltonian by comparing the numerical matrix elements of the effective Hamiltonian with the symbolic matrix elements of the model Hamiltonian.

d. Diagonalize the D-tensor, determine the axial (D) and rhombic (E) anisotropy parameters from Eq. 2.16 and compare the values with those obtained by extracting D and E from the energies differences (Eq. 2.22).

2.2 Extracting D and E for a Co^{II} complex. The ground state of a slightly distorted tetrahedral Co^{II} complex has quartet spin multiplicity. The fitting of the magnetic susceptibility shows that the complex has a rather large magnetic anisotropy, but it remains unclear whether the complex has an easy plane $(D > 0)$ or an easy axis $(D < 0)$ of magnetization.

1. Draw a level diagram showing the removal of the degeneracy of the M_S sublevels of the quartet state under the influence of (i) spin-orbit coupling and (ii) spin-orbit coupling and a small external magnetic field along the z-axis.
2. Can the anisotropy parameters D and E be determined from the energy differences at zero field? And the sign of D?
3. Construct the matrix representation of $\hat{S} \cdot \overline{\overline{D}} \cdot \hat{S}$ in an arbitrary frame in the $\{|\frac{3}{2}, \frac{3}{2}\rangle, |\frac{3}{2}, \frac{1}{2}\rangle, |\frac{3}{2}, -\frac{1}{2}\rangle, |\frac{3}{2}, -\frac{3}{2}\rangle\}$ basis.
4. Use the following data to construct the numerical effective Hamiltonian and extract D and E. Decide if this complex has easy-axis or easy-plane magnetism.

	$E(\text{cm}^{-1})$	$|\frac{3}{2}, -\frac{3}{2}\rangle$	$|\frac{3}{2}, -\frac{1}{2}\rangle$	$|\frac{3}{2}, \frac{1}{2}\rangle$	$|\frac{3}{2}, \frac{3}{2}\rangle$
$\widetilde{\Psi}'_1$	0.00	0.007808 + 0.008516i	0.058561 + −0.068873i	−0.207583 + −0.246313i	−0.709453 + 0.620169i
$\widetilde{\Psi}'_2$	0.00	0.942302 + 0.000087i	0.005791 + 0.322067i	−0.089417 + −0.013321i	0.000274 + −0.011550i
$\widetilde{\Psi}'_3$	32.47	0.090185 + −0.002021i	−0.000628 + 0.012098i	0.934946 + 0.117457i	−0.034599 + 0.320311i
$\widetilde{\Psi}'_4$	32.47	0.255568 + −0.196167i	−0.586647 + −0.737404i	0.010006 + −0.006830i	−0.045185 + −0.078075i

2.3 Anisotropic g values. EPR measurement on a Ti^{III} complex reveals a relatively large axial magnetic anisotropy by the application of a small magnetic external field.

1. What is the electronic configuration of the Ti^{III} ion? Assuming that the ligands have a closed-shell configuration, can the complex display a splitting of the M_S levels of the ground state at zero field?
2. Use Eqs. 2.55 and 2.59 to calculate the deviations of g_x and g_z from the free-electron value g_e based on the following computational results.

	$E(\text{cm}^{-1})$	$|h\overline{h}\phi_a|$	$|h\overline{h}\phi_b|$	$|h\overline{h}\phi_c|$	$|h\overline{h}\phi_d|$	$|h\overline{h}\phi_e|$
$\widetilde{\Psi}'_1$	0	0.6441	−0.7504	0.0179	0.1444	0.0304
$\widetilde{\Psi}'_2$	1005	−0.7562	−0.6105	0.1406	0.1875	−0.0215
$\widetilde{\Psi}'_3$	6662	−0.0772	−0.1649	−0.1597	−0.6456	0.7243
$\widetilde{\Psi}'_4$	11060	0.0293	−0.1168	0.4849	−0.6864	−0.5284
$\widetilde{\Psi}'_5$	13358	0.0805	0.1528	0.8481	0.2368	0.4414

	ϕ_a	ϕ_b	ϕ_c	ϕ_d	ϕ_e
$3d_{z^2}$	−0.0593	−0.1176	−0.6905	−0.0964	−0.7046
$3d_{x^2-y^2}$	0.1174	−0.3047	0.6775	−0.295	−0.5841
$3d_{xy}$	0.3816	0.182	0.1216	0.8439	−0.2986
$3d_{yz}$	0.8647	0.2572	−0.1397	−0.4	0.0825
$3d_{xz}$	−0.2916	0.8897	0.1583	−0.1724	−0.2576

Although the normalized projections $\widetilde{\Psi}'_i$ are not strictly orthogonal, the deviation is small enough to be neglected. $\zeta_{Ti} = 123\,\text{cm}^{-1}$. Matrix elements of $\hat{l}$ can be found in Appendix A.

3. Are the calculated g-values in line with the observed anisotropy? What is the effect of increasing/decreasing the gap between the ground state and the first excited state on g_z and g_x.

2.4 Barrier for spin reversal: Given a system with a total spin moment of $S = 5$ and easy-axis anisotropy, calculate the energies of the different M_S components of the ground state wave function.

References

1. A. Ceulemans, *Group Theory Applied to Chemistry*. Theoretical Chemistry and Computational Modeling (Springer, Dordrecht, 2013)
2. K.G. Dyall, K. Fægri Jr., *Introduction to Relativistic Quantum Chemistry* (Oxford University Press, Oxford, 2007)
3. M. Reiher, A. Wolf, *Relativistic Quantum Chemistry, The Fundamental Theory of Molecular Science* (Wiley-VCH, Weinheim, 2009)
4. A. Abragam, B. Bleaney, *Electron Paramagnetic Resonance of Transition Ions* (Dover Publications, New York, 1986)
5. O. Kahn, *Molecular Magnetism* (VCH Publishers, New York, 1993)
6. R. Boča, *Theoretical Foundations of Molecular Magnetism* (Elsevier, Amsterdam, 1999)

Chapter 3
Two (or More) Magnetic Centers

Abstract The description of the magnetic interactions is now extended to more than one magnetic center. First it is shown that the two-electron/two-orbital system can be approached from different viewpoints using (de-)localized, (non-)orthogonal orbitals. After this quantum chemical description of the magnetic interaction we discuss the more phenomenological approach based on spin operators. Starting with the standard Heisenberg Hamiltonian for isotropic bilinear interactions, the chapter discusses how biquadratic, anisotropic and four-center interactions can be accounted for within this spin formalism. Furthermore, it is shown how the microscopic electronic interaction parameters can be used to describe macroscopic properties by diagonalization of model Hamiltonians, Monte Carlo simulations and some other techniques.

3.1 Localized Versus Delocalized Description of the Two-Electron/Two-Orbital Problem

The most simple magnetic systems have only two magnetic sites, each with spin $\frac{1}{2}$. Examples are doubly bridged binuclear Cu^{II} complexes. The energy splitting between the lowest singlet and triplet spin states in such complexes turns out to depend strongly on the geometry of the bridging $Cu–L_2–Cu$ units (R_1, R_2, α, β, etc.) depicted in Fig. 3.1 and this magneto-structural correlation can be well explained by using simple quantum theoretical models that can be developed *"on the back of an envelope"*. The basis for such models is provided in this section, they are further elaborated in Chap. 4. We consider a many-electron system, which, in addition to closed shells of electrons, has two *magnetic* electrons which are mainly localized on two magnetic sites A and B. We assume that the orbitals of the complex, i.e. its molecular orbitals (MOs), have been determined by a self-consistent field procedure.

Configuration Interaction using delocalized orbitals: We first consider the case where the two magnetic electrons are coupled to a spin triplet, $S = 1$. The simplest description of the $M_S = 1$ component of the lowest lying triplet state is one single Slater determinant

© Springer International Publishing Switzerland 2016

C. Graaf and R. Broer, *Magnetic Interactions in Molecules and Solids*,
Theoretical Chemistry and Computational Modelling,
DOI 10.1007/978-3-319-22951-5_3

Fig. 3.1 Schematic
representation of a complex
with a bridging CuL$_2$Cu unit
and four external ligands L$_e$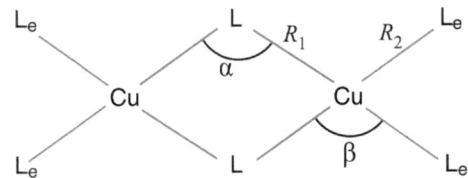

$$\Phi^{12}(S, M_S) = \Phi^{12}(1, 1) = |\ldots \phi_1 \phi_2| \qquad (3.1)$$

where the MOs ϕ_1 and ϕ_2 are bonding and antibonding combinations of *atomic* orbitals localized at/around the magnetic centers A and B. In other words: ϕ_1 and ϕ_2 are *molecular* orbitals that are delocalized over the two magnetic centers. In the case of two Cu^{II} centers they will mainly be built from bonding and antibonding combinations of Cu-3d orbitals. The dots in the determinant denote the other electrons of the system, in doubly occupied MOs. In the following these closed shell, *inactive*, electrons will be omitted from the notation for the Slater determinants:

$$\Phi^{12}(1, 1) = |\phi_1 \phi_2| \qquad (3.2a)$$

The orbitals ϕ_1 and ϕ_2 are occupied with one electron and are often referred to as the magnetic orbitals. The Slater determinant shown in Eq. 3.2a has the correct spin symmetry for the $M_S = 1$ component of an $S = 1$ manifold, the corresponding wave function of the $M_S = -1$ component is given by

$$\Phi^{12}(1, -1) = |\overline{\phi}_1 \overline{\phi}_2| \qquad (3.2b)$$

and for the $M_S = 0$ component we need two Slater determinants:

$$\Phi^{12}(1, 0) = \left(|\phi_1 \overline{\phi}_2| - |\phi_2 \overline{\phi}_1| \right)/\sqrt{2} \qquad (3.2c)$$

These three $S = 1$ wave functions belong to the electronic configuration $\ldots \phi_1^1 \phi_2^1$, which in addition gives rise to a spin-singlet wave function,

$$\Phi^{12}(0, 0) = \left(|\phi_1 \overline{\phi}_2| + |\phi_2 \overline{\phi}_1| \right)/\sqrt{2} \qquad (3.3)$$

In addition to the $\ldots \phi_1^1 \phi_2^1$ electronic configuration, two more configurations can be defined using the two MOs ϕ_1 and ϕ_2. Both are of closed-shell character and can be written as $\ldots \phi_1^2$ and $\ldots \phi_2^2$. They give rise to two more Slater determinants

$$\Phi^{11}(0, 0) = |\phi_1 \overline{\phi}_1| \qquad (3.4)$$

and

$$\Phi^{22}(0, 0) = |\phi_2 \overline{\phi}_2| \qquad (3.5)$$

Fig. 3.2 Schematic representation of the four $M_S = 0$ CSFs that can be generated with two electrons in two orbitals

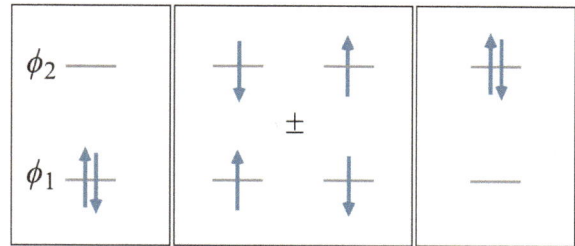

As we have seen in Eq. 3.2, the configuration $\ldots \phi_1^1 \phi_2^1$ leads also to three $S = 1$ wave functions, with $M_S = 1, 0$ and -1. It is customary to call the simplest wave functions built from one electronic configuration that obey the spin (and spatial) symmetry requirements a Configuration State Function, CSF. Hence the approximate wave functions of Eqs. 3.2a–3.5 constitute the six CSFs that can be formed by distributing two electrons over the two MOs ϕ_1 and ϕ_2. Summarizing: the *two electrons in two orbitals* model, i.e. distributing two electrons over two orbitals in all possible ways yields three electronic configurations, which give rise to six CSFs. Three of them form the $M_S = 1, 0, -1$ components of an $S = 1$ state, the other three represent each a separate $S = 0$ CSF. Figure 3.2 represents the four $M_S = 0$ CSFs. From left to right, we have $\Psi^{11}(0, 0)$, $\Psi^{12}(0, 0)$ (*plus* combination), $\Psi^{12}(1, 0)$ (*minus* combination), and $\Psi^{22}(0, 0)$.

The relative energies of the four states can of course be computed by performing a complete active space configuration interaction (CASCI) calculation with the active orbitals being ϕ_1 and ϕ_2, but here we are going to analyze the relative energies by considering the physics of the system.

If the splitting between the bonding ϕ_1 and the antibonding ϕ_2 is large, the ground state is expected to be a spin singlet state, which is rather well described by $\Phi^{11}(0, 0)$. This is the situation that we would encounter, for example, if the ions A and B would be two Li atoms without any environment forming a Li_2 molecule, or, even simpler, two H atoms forming H_2. In these cases the stabilization of ϕ_1 is so large that the two electrons pair to occupy jointly this strongly bonding orbital. However, in magnetic systems the interaction between the magnetic ions A and B is quite weak. Moreover, in most complexes they are separated by *bridging* ligands. Then, the splitting between ϕ_1 and ϕ_2 is small and the three configurations $\ldots \phi_1^2$, $\ldots \phi_1^1 \phi_2^1$ and $\ldots \phi_2^2$ are close in energy. The fact that we have three distinct low lying singlet CSFs suggests that we can use variation theory to generate improved descriptions of these three states, by forming linear combinations of $\Phi^{11}(0, 0)$, $\Phi^{12}(0, 0)$ and $\Phi^{22}(0, 0)$. On the contrary, in order to improve the description of the $S = 1$ state we would have to go beyond the *two electrons in two orbitals* model, but doing this only for the triplet and not for the singlet states would destroy the balance between them, preventing us from determining whether the ground state is magnetic ($S \neq 0$) or not ($S = 0$).

In many cases the $A-B$ system is centrosymmetric, i.e. there is at least one symmetry operation that transforms A into B and *vice versa*. Then ϕ_1 and ϕ_2 belong to different irreducible symmetry representations, and consequently the wave functions corresponding to $\ldots \phi_1^1 \phi_2^1$ belong also to a different representation compared to those of $\ldots \phi_1^2$ and $\ldots \phi_2^2$. As a result, in those cases there is no Hamiltonian matrix element of $\Phi^{11}(0,0)$ and $\Phi^{22}(0,0)$ with $\Phi^{12}(0,0)$:

$$\langle \Phi^{11}(0,0)|\hat{H}|\Phi^{12}(0,0)\rangle = \langle \Phi^{22}(0,0)|\hat{H}|\Phi^{12}(0,0)\rangle = 0 \tag{3.6}$$

Then, improved singlet variational wave functions can be formed by making linear combinations of only $\Phi^{11}(0,0)$ and $\Phi^{22}(0,0)$, whereas the singlet wave function corresponding to $\ldots \phi_1^1 \phi_2^1$ cannot be improved within this *two electrons in two orbitals* scheme. In other magnetic systems there is no strict but only approximate symmetry, giving rise to non-zero but still quite small matrix elements $\langle \Phi^{11}(0,0)|\hat{H}|\Phi^{12}(0,0)\rangle$ and $\langle \Phi^{11}(0,0)|\hat{H}|\Phi^{12}(0,0)\rangle$. In the following we will assume symmetry, hence assume that these matrix elements are zero and turn our attention to the Hamiltonian matrix element $\langle \Phi^{11}(0,0)|\hat{H}|\Phi^{22}(0,0)\rangle$ that leads to an improved singlet wave function

$$\Psi(0,0) = c_1 \Phi^{11}(0,0) + c_2 \Phi^{22}(0,0) \tag{3.7}$$

where c_1 and c_2 are chosen to minimize the energy of $\Psi(0,0)$. Using the Slater–Condon rules we find for the Hamiltonian matrix element between the two closed shell determinants

$$\langle \Phi^{11}(0,0)|\hat{H}|\Phi^{22}(0,0)\rangle = \langle \phi_1(1)\phi_1(2)|\frac{1}{r_{12}}|\phi_2(1)\phi_2(2)\rangle \tag{3.8}$$

By rearranging the (real) functions in this integral, it becomes clear that the matrix element is equal to the exchange integral K_{12}

$$\langle \phi_1(1)\phi_1(2)|\frac{1}{r_{12}}|\phi_2(1)\phi_2(2)\rangle = \langle \phi_1(1)\phi_2(1)|\frac{1}{r_{12}}|\phi_2(2)\phi_1(2)\rangle = K_{12} \tag{3.9}$$

3.1 Demonstrate that $\langle \Phi^{11}(0,0)|\hat{H}|\Phi^{22}(0,0)\rangle = K_{12}$.

The exchange integral K_{12} does not vanish, not even in case of weakly interacting A and B centers. This becomes clear if we introduce the *localized orthogonal* orbitals ψ_a and ψ_b that can be constructed from the delocalized molecular orbitals ϕ_1 and ϕ_2:

$$\psi_a = \frac{1}{\sqrt{2}}(\phi_1 + \phi_2) \qquad \psi_b = \frac{1}{\sqrt{2}}(\phi_1 - \phi_2) \tag{3.10a}$$

If A and B are strongly coupled, ψ_a will be mainly localized on A, but with important *orthogonalization tails* on B and *vice versa*. Only if A and B are weakly coupled, ψ_a and ψ_b are nearly completely localized on A and B, respectively.

3.2 (a) Demonstrate that ψ_a and ψ_b are normalized and orthogonal (b) Consider the example where the delocalized orbitals ϕ_1 and ϕ_2 are bonding and antibonding combinations of atom centered basis functions χ_a and χ_b:

$$\phi_1 = \frac{\chi_a + \chi_b}{\sqrt{(2(1 + S)}} \qquad \phi_2 = \frac{\chi_a - \chi_b}{\sqrt{(2(1 - S)}}$$

with $S = \langle \chi_a | \chi_b \rangle$. Compute the coefficient of the orthogonalization tail of ψ_a on atom B, for $S = 0.2$ and for $S = 0.003$.

Using the inverse relations of Eq. 3.10a

$$\phi_1 = \frac{1}{\sqrt{2}}(\psi_a + \psi_b) \qquad \phi_2 = \frac{1}{\sqrt{2}}(\psi_a - \psi_b) \tag{3.10b}$$

the exchange integral K_{12} can be rewritten as a sum of Coulomb integrals

$$K_{12} = \frac{1}{4}\langle(\psi_a + \psi_b)(\psi_a - \psi_b)|\hat{H}|(\psi_a - \psi_b)(\psi_a + \psi_b)\rangle$$

$$= \frac{1}{4}\langle\psi_a\psi_a|\hat{H}|\psi_a\psi_a\rangle - 2\langle\psi_a\psi_b|\hat{H}|\psi_a\psi_b\rangle + \langle\psi_b\psi_b|\hat{H}|\psi_b\psi_b\rangle$$

$$= \frac{1}{4}\left(J_{aa} - 2J_{ab} + J_{bb}\right) \tag{3.11}$$

Only the term J_{ab} approaches zero for small coupling between A and B. The other terms are local Coulomb integrals which are both positive and both occur with a positive coefficient: these two terms do not cancel each other. Note, that since we have assumed centrosymmetry, $J_{aa} = J_{bb}$. Figure 3.3 shows the localized orthogonal orbitals ψ_a and ψ_b and the product of these as they appear in the Coulomb integrals of the expression of K_{12}. From this pictorial representation it is obvious that J_{ab}—with $\psi_a\psi_a \times \psi_b\psi_b$ in the numerator—is small for weak interaction between A and B, while J_{aa} and J_{bb} do not strongly depend on the distance between A and B.

Hence, there is significant interaction between the configurations $\ldots \phi_1^2$ and $\ldots \phi_2^2$. Note that the smaller J_{ab}, the larger is K_{12}, and therefore, in the case of weak coupling between A and B we need to use the two-configuration wave function of Eq. 3.7 instead of simply $\Phi^{11}(0, 0)$. Clearly, the best two-configuration wave function is obtained by varying c_1/c_2 until the energy expectation value is minimal. Only in case of strong coupling between A and B (remember the case of Li_2 near equilibrium distance) J_{ab} may become so large that K_{12} and therewith c_2 becomes negligible so that the closed shell determinant $\Phi^{11}(0, 0)$ is a reasonable *ansatz* for the lowest singlet wave function. For intermediate and small couplings the lowest $S = 0$ and $S = 1$ states are competing, i.e. close in energy. An approximate yet balanced treatment is obtained by describing the lowest $S = 0$ state using the expression of

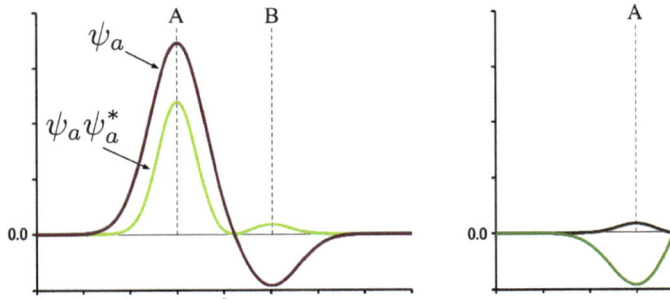

Fig. 3.3 Graphical representation of the localized orthogonal orbitals ψ_a (*left*) and ψ_b (*right*), and the respective products (charge densities) as they appear in the numerator of the Coulomb integrals

Eq. 3.7 and the $S = 1$ state using Eq. 3.2. Which state is the ground state, singlet or triplet, depends on the magnitudes of the two electron integrals J_{aa}, J_{ab} and K_{ab}.

Summarizing: A *Full CI* treatment of two electrons in two orbitals in a symmetric system leads to Eq. 3.7 for $S = 0$ and to Eq. 3.2 for $S = 1$. This approach forms a basis for the Hay–Thibeault–Hoffmann (HTH) model discussed in Chap. 4.

Valence Bond theory using localized orthogonal orbitals: For small couplings, i.e. nearly degenerate $S = 0$ and $S = 1$ states, Valence Bond (VB) theory provides a more intuitive starting point than the previous molecular orbital reasoning. For the $M_S = 0$ wave functions we make use of the local orthonormal orbitals ψ_a and ψ_b as defined in Eq. 3.10a and use them to construct two *neutral* determinants $|\psi_a \overline{\psi}_b|$ and $|\psi_b \overline{\psi}_a|$, where *neutral* does not mean that the centers A and B are uncharged, but maintain their oxidation state as in the ground configuration. Moreover, two *ionic* determinants are defined: $|\psi_a \overline{\psi}_a|$ and $|\psi_b \overline{\psi}_b|$. The simplest VB wave function for the lowest singlet state has the form

$$\Psi^{\text{cov}}(0, 0) = \frac{1}{\sqrt{2}} \left(|\psi_a \overline{\psi}_b| + |\psi_b \overline{\psi}_a| \right) \tag{3.12a}$$

which can also be written in terms of the symmetry-adapted, delocalized orbitals ϕ_1 and ϕ_2:

$$\Psi^{\text{cov}}(0, 0) = \frac{1}{\sqrt{2}} \left(|\phi_1 \overline{\phi}_1| - |\phi_2 \overline{\phi}_2| \right) \tag{3.12b}$$

The latter corresponds to a two-configuration CI wave function analogous to Eq. 3.7, be it with fixed coefficients $c_1 = -c_2 = 1/\sqrt{2}$. Note that it is tempting to characterize 3.12a as an open shell wave function, while the same wave function in 3.12b takes the form of a superposition of two closed shell determinants. The simplest VB representation for the $M_S = 0$ component of the lowest spin-triplet state is

$$\Psi^{\text{cov}}(1, 0) = \frac{1}{\sqrt{2}} \left(|\psi_a \overline{\psi}_b| - |\psi_b \overline{\psi}_a| \right) \tag{3.13a}$$

which in terms of ϕ_1 and ϕ_2 lead to $\Psi^{12}(1,0)$ of Eq. 3.2c. The other two $S = 1$ components are

$$\Psi^{\text{cov}}(1,1) = |\psi_a\psi_b| \tag{3.13b}$$

and

$$\Psi^{\text{cov}}(1,-1) = |\overline{\psi}_a\overline{\psi}_b| \tag{3.13c}$$

which written in terms of ϕ_1 and ϕ_2 yield the familiar Slater determinants $|\phi_1\phi_2|$ (Eq. 3.2a) and $|\overline{\phi}_1\overline{\phi}_2|$ (Eq. 3.2b).

3.3 Demonstrate the equivalence of the wave functions of Eqs. 3.12a and 3.12b, and those of Eqs. 3.2 and 3.13.

This simple Valence Bond *ansatz* with a common set of localized orthonormal orbitals for both states leads to a separation of the energy expectation values for singlet and triplet that reads

$$E_S^{\text{cov}} - E_T^{\text{cov}} = 2\langle\psi_a\psi_b|\frac{1}{r_{12}}|\psi_b\psi_a\rangle = 2K_{ab} \tag{3.14}$$

which is positive because the exchange integral

$$K_{ab} = \langle\psi_a(1)\psi_b(1)|\frac{1}{r_{12}}|\psi_a(2)\psi_b(2)\rangle \geqslant 0$$

3.4 Calculate the energy expectation values of the wave functions given in Eqs. 3.12a and 3.13a to demonstrate that $E_S^{\text{cov}} - E_T^{\text{cov}} = 2K_{ab}$.

$2K_{ab}$ is traditionally called the *direct* exchange. It favours high spin states. In this two electron case, treated with only covalent VB determinants, it favours $S = 1$ over $S = 0$. It is interesting to note that this positive energy difference between singlet and triplet spin states can be seen as a manifestation of Hund's rule for two electrons in two orbitals, be it in this case not for two degenerate orbitals on one site, but for two degenerate orbitals at two separate sites. In the single site case, i.e. for atoms, this Hund rule is almost always correct, however, in the two site case the sign of $E_S - E_T$ is often wrong. The simple covalent VB model using localized orthogonal orbitals is simply too crude.

The singlet VB wave function can be improved variationally by mixing in an ionic term

$$\Psi^{\text{ion}}(0,0) = \frac{1}{\sqrt{2}}\left(|\psi_a\overline{\psi}_a| + |\psi_b\overline{\psi}_b|\right) \tag{3.15}$$

leading to

$$\Psi(0,0) = C_{\text{cov}}\Psi^{\text{cov}}(0,0) + C_{\text{ion}}\Psi^{\text{ion}}(0,0) \tag{3.16}$$

which is the same as the CI wave function in Eq. 3.7, but now written in terms of the localized orthogonal orbitals ψ_a and ψ_b. Again, just as in the MO picture, there is no way to improve the $S = 1$ wave functions. A balanced VB treatment uses the wave functions of Eq. 3.16 for $S = 0$ and of Eq. 3.13 for $S = 1$, $M_S = 0$.

We can now draw the following conclusions:

- If we limit ourselves to using (apart from the doubly occupied orbitals) only two mutually orthogonal orbitals, either bonding and antibonding ϕ_a and ϕ_b, or localized ψ_a and ψ_b, then the best wave function for the lowest triplet state is the one given in Eq. 3.2 or, equivalently, 3.13a and the best wave function for the lowest singlet state is given in Eq. 3.7 or, equivalently, 3.16.
- It makes no difference whether we use MO theory and optimize the ratio c_1/c_2 in 3.7 or use VB theory and optimize the ratio $C_{\text{cov}}/C_{\text{ion}}$ in 3.16, both procedures lead to one and the same $S = 0$ wave function.
- For $\Psi(0,0)$ to become the ground state, the energy lowering due to mixing in of the ionic terms has to exceed the direct exchange $2K_{ab}$. This energy lowering is traditionally called the kinetic exchange. We will see in Chap. 4 that the kinetic exchange equals, to good approximation,[1] $4t_{ab}^2/(J_{aa} - J_{ab})$ with the *transfer* integral t_{ab} defined by $t_{ab} = \langle \psi_a \overline{\psi}_a | \hat{H} | \psi_a \overline{\psi}_b \rangle$.

Valence Bond theory using localized nonorthogonal orbitals: In the above we have used orthogonal localized orbitals ψ_a and ψ_b. What if we remove the orthogonality restriction? This nonorthogonal VB approach appeared for the first time in the work of Heitler and London [1]. It forms also the basis of the Kahn–Briat model discussed in the next chapter. We define normalized localized orbitals ϕ_a and ϕ_b with mutual overlap $S_{ab} = \langle \phi_a | \phi_b \rangle$. Then we write the covalent singlet wave function in terms of normalized Slater determinants that are now built from the nonorthogonal ϕ_a and ϕ_b:

$$\Psi(0,0) = \frac{1}{\sqrt{2 + 2S_{ab}^2}}\left(|\phi_a\overline{\phi}_b| + |\phi_b\overline{\phi}_a|\right) \tag{3.17}$$

Of course, we can formally express ϕ_a and ϕ_b in terms of our orthogonal localized orbitals ψ_a and ψ_b:

$$\phi_a = N(\psi_a + \nu\psi_b) \qquad \phi_b = N(\psi_b + \nu\psi_a)$$

$$\text{with} \quad N = \frac{1}{\sqrt{1 + \nu^2}} \quad \text{and} \quad S_{ab} = \frac{2\nu}{1 + \nu^2} \tag{3.18}$$

Once we have optimized ν or, equivalently, S_{ab}, to obtain the lowest energy possible for the singlet wave function in Eq. 3.18 we have retrieved once more our familiar

---

[1]This expression is reasonable for $J_{aa} - J_{ab} \gg t_{ab}$, see Chap. 4.

singlet wave function shown earlier in Eqs. 3.7 and 3.16. Instead of introducing an extra variational freedom by adding ionic contributions with optimized weight, the extra freedom is now obtained by allowing the localized orbitals to be mutually nonorthogonal and optimizing their overlap. Not surprisingly, we get no additional improvement if we now include ionic terms, so this VB approach with nonorthogonal orbitals gives no improvement as compared to the VB approach with orthogonal orbitals and ionic terms, but the nonorthogonal VB approach does allow us to write the singlet wave function $\Psi(0, 0)$ in terms of a covalent contribution alone.

3.5 Show that the $S = 1$, $M_S = 0$ functions

$$\frac{1}{\sqrt{2 + 2S^2}}\left(|\phi_a \overline{\phi}_b| + |\phi_b \overline{\phi}_a|\right) \text{ and } \frac{1}{\sqrt{2}}\left(|\psi_a \overline{\psi}_b| + |\psi_b \overline{\psi}_a|\right)$$

are identical. Hint: rewrite the first wave function in terms of orthogonal orbitals using Eq. 3.18.

It is not difficult to show that in order to make the singlet state the ground state, the magnetic orbitals have to overlap considerably. However, atomic magnetic orbitals such as first row transition metal $3d$ orbitals are quite compact and their mutual overlap, especially in compounds where they are separated by bridging ions, is negligible. This suggests that in such systems only the direct exchange, would play a role, leading to *parallel* or *ferromagnetic* spin coupling. But in reality we have many systems whose nearest neighbour paramagnetic ions have *antiferromagnetic* spin coupling.

Already in 1934 Kramers attempted to explain the magnetic interaction in antiferromagnetic ionic solids, by noting that it is possible to have a two-center spin coupling that is mediated by the bridging non-magnetic atoms. He called this bridge-mediated spin coupling *superexchange*. In 1959 Anderson described the physical basis responsible for the generation of this superexchange: It is simply that spin-paired electrons can gain energy by spreading into nonorthogonal overlapping orbitals, whereas unpaired electrons cannot. The model of Anderson is easiest illustrated by considering again a centrosymmetric system of two magnetic ions A and B, for example Cu^{2+} ions, with (apart from electrons in double occupied orbitals) one unpaired electron in a $3d$-orbital, which are separated by a closed shell anion such as Cl^- or O^{2-} with (formally) three doubly occupied valence p orbitals. We denote the relevant atomic $3d$ functions χ_a and χ_b. It turns out that for structural and symmetry reasons commonly only one of the three valence p orbitals plays a role, so we consider only one bridging atomic function χ_h. We now allow the localized magnetic orbitals to have optimum admixture of the bridging ligand function:

$$\phi_a = c_a \chi_a + c_h \chi_h \qquad \phi_b = c_b \chi_b + c_h' \chi_h \qquad (3.19)$$

as illustrated in Fig. 3.4.

Fig. 3.4 In the *left column*:
The localized magnetic
orbitals (χ_a and χ_b) and the
bridging atomic function
(χ_h, which has small
orthogonalization tails on the
magnetic centers). On the
right: Localized magnetic
orbitals with optimally
admixed ligand
delocalization (ϕ_a and ϕ_b)

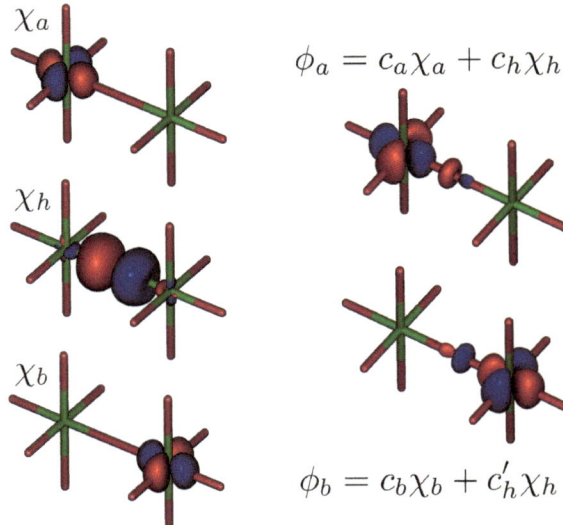

$$\phi_a = c_a \chi_a + c_h \chi_h$$

$$\phi_b = c_b \chi_b + c'_h \chi_h$$

This gives the magnetic orbitals non-vanishing amplitudes also at the intermediate
ligands and the exchange effects therefore need no longer be small. In quantum theoretical studies the name *Anderson model* is commonly associated with a CASSCF
approach, in which the active electrons are the magnetic electrons, the active orbitals
are predominantly the magnetic functions, but they are optimized in the SCF process.
This guarantees that they contain the optimum amount of intermediate ligand character, so that the superexchange is accounted for.

3.2 Model Spin Hamiltonians for Isotropic Interactions

Under the assumption of a common spatial part of the wave function, the lowest
energy levels of the two-electron/two-orbital problem discussed in the previous
section can be described with a model Hamiltonian that only contains spin operators. Starting from a general expression of the interaction of two spatially separated
spin moments S_1 and S_2 of arbitrary strength (not limiting ourselves to the $S = \frac{1}{2}$
case discussed before), the spin Hamiltonian can be written in terms of local spin
operators $\hat{S}_1$ and $\hat{S}_2$

$$\hat{H} = \begin{pmatrix} \hat{S}_{x,1} & \hat{S}_{y,1} & \hat{S}_{z,1} \end{pmatrix} \begin{pmatrix} A_{xx} & A_{xy} & A_{xz} \\ A_{yx} & A_{yy} & A_{yz} \\ A_{zx} & A_{zy} & A_{zz} \end{pmatrix} \begin{pmatrix} \hat{S}_{x,2} \\ \hat{S}_{y,2} \\ \hat{S}_{z,2} \end{pmatrix} \tag{3.20}$$

This expression is greatly simplified by orienting the system along the magnetic axis frame making all non-diagonal elements of the A-tensor equal to zero.

$$\hat{H} = A_{xx}\hat{S}_{x,1}\hat{S}_{x,2} + A_{yy}\hat{S}_{y,1}\hat{S}_{y,2} + A_{zz}\hat{S}_{z,1}\hat{S}_{z,2} \tag{3.21}$$

It is common practice to divide the interaction in a part that does not depend on the spatial orientation of spin—the isotropic part, parametrized by the scalar J—and another part that models the anisotropy of the interaction parametrizing it with a diagonal tensor D.

$$\hat{H} = -J\left(\hat{S}_{x,1}\hat{S}_{x,2} + \hat{S}_{y,1}\hat{S}_{y,2} + \hat{S}_{z,1}\hat{S}_{z,2}\right) \\ + D_{xx}\hat{S}_{x,1}\hat{S}_{x,2} + D_{yy}\hat{S}_{y,1}\hat{S}_{y,2} + D_{zz}\hat{S}_{z,1}\hat{S}_{z,2} \tag{3.22}$$

The minus sign in front of J is by convention, but be aware that other definitions are often used in the literature. Negative J-values indicate antiferromagnetic coupling and positive values are characteristic of ferromagnetic interactions in the definition that we use here.

3.2.1 Heisenberg Hamiltonian

For the moment, we leave aside the anisotropic part of the interaction and concentrate on the isotropic part. The equation can then be written in the following from

$$\hat{H} = -J\hat{S}_1 \cdot \hat{S}_2 \tag{3.23}$$

which is known as the *Heisenberg* or *Heisenberg-Dirac-van Vleck* Hamiltonian. For systems with two magnetic sites, the eigenvalues of the Hamiltonian are easily derived by rewriting the product of local operators using the relation

$$\hat{S}^2 = (\hat{S}_1 + \hat{S}_2)^2 = \hat{S}_1^2 + \hat{S}_2^2 + 2\hat{S}_1 \cdot \hat{S}_2 \tag{3.24}$$

from this follows

$$\hat{S}_1 \cdot \hat{S}_2 = \frac{1}{2}(\hat{S}^2 - \hat{S}_1^2 - \hat{S}_2^2) \tag{3.25}$$

which leads to an alternative formulation of the Heisenberg Hamiltonian

$$\hat{H} = -\frac{1}{2}J(\hat{S}^2 - \hat{S}_1^2 - \hat{S}_2^2) \tag{3.26}$$

for which the eigenvalues can be written down directly

$$E(S) = -\frac{1}{2}J\big(S(S+1) - S_1(S_1+1) - S_2(S_2+1)\big) \qquad (3.27)$$

3.6 Calculate the eigenvalues of the Heisenberg Hamiltonian of the spin eigenfunctions with maximum and minimum spin moment of a dimeric system with $S_1 = S_2$.

Since the reference point of energy can be chosen arbitrarily, the expression for the eigenvalues is usually simplified by adding a constant factor equal to $-\frac{1}{2}J\big(S_1(S_1 + 1) + S_2(S_2 + 1)\big)$, leading to

$$E(S) = -\frac{1}{2}JS(S+1) \qquad (3.28)$$

From this it is easily derived that the difference between two subsequent eigenvalues is given by

$$E(S-1) - E(S) = JS \qquad (3.29)$$

where S runs from $S_1 + S_2$ to $|S_1 - S_2|$. This regular Landé pattern gives the exact energy differences as long as the interaction with excited electronic configurations is negligible and is the basis for extracting magnetic coupling parameters from electronic structure calculations. The generalization of the Heisenberg Hamiltonian to multiple sites is straightforward

$$\hat{H} = \sum_{i>j} -J_{ij}\hat{S}_i\hat{S}_j \qquad (3.30)$$

In systems with multiple magnetic sites, the number of energy differences between the different spin functions is not always enough to determine all the J-values. An obvious example is the three-center/three-electron case as depicted in Fig. 3.5. The Hamiltonian has three different J-values while the quartet and the two doublet states only define two energy differences. The effective Hamiltonian theory described in Chap. 1 is a more general approach to extract J-values, because it not only uses the energies but also information contained in the wave function. To illustrate the procedure we first treat a simple biradical model with two $S = \frac{1}{2}$ spins and after that focus on the three center problem.

The following results for the singlet and triplet states were obtained from an *ab initio* calculation on a dimer with two $S = \frac{1}{2}$ centers:

Fig. 3.5 Three center
$S = 1/2$ system with a
quartet (Q) and two doublets
(D_1, D_2). The three J-values
cannot be determined from
the two energy differences

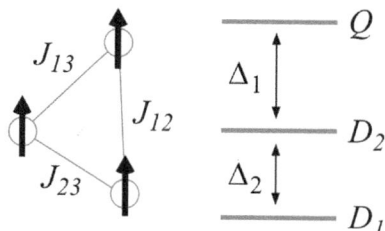

$$\Psi_S = 0.6776|\phi_1\bar{\phi}_2| - 0.6776|\bar{\phi}_1\phi_2| + 0.1287|\phi_1\bar{\phi}_1| + 0.1287|\phi_2\bar{\phi}_2| + \ldots$$
$$\Psi_T = 0.7029|\phi_1\bar{\phi}_2| + 0.7029|\bar{\phi}_1\phi_2| + \ldots$$
$$E_S = -29.441750 \; E_h$$
$$E_T = -29.438299 \; E_h$$

Here, ϕ_1 and ϕ_2 are basis functions localized on site 1 and 2, respectively. The model space of the Heisenberg Hamiltonian is spanned by the $M_S = 0$ determinants $|\phi_1\bar{\phi}_2|$ and $|\bar{\phi}_1\phi_2|$ and the matrix representation of the Hamiltonian can be obtained by calculating the matrix elements $\langle\phi_1\bar{\phi}_2|\hat{H}|\phi_1\bar{\phi}_2\rangle$ and $\langle\phi_1\bar{\phi}_2|\hat{H}|\bar{\phi}_1\phi_2\rangle$. For this purpose, we first substitute the $\hat{S}_x$ and $\hat{S}_y$ operators by the ladder operators $\hat{S}^\pm$

$$\hat{H} = -J\left(\frac{1}{2}\left\{\hat{S}_1^+\hat{S}_2^- + \hat{S}_1^-\hat{S}_2^+\right\} + \hat{S}_{z,1}\hat{S}_{z,2}\right) \tag{3.31}$$

The action of the different operators on the model space determinants gives:

$$\hat{S}_1^+\hat{S}_2^-|\phi_1\bar{\phi}_2| = 0 \qquad\qquad \hat{S}_1^+\hat{S}_2^-|\bar{\phi}_1\phi_2| = |\phi_1\bar{\phi}_2|$$
$$\hat{S}_1^-\hat{S}_2^+|\phi_1\bar{\phi}_2| = |\bar{\phi}_1\phi_2| \qquad\qquad \hat{S}_1^-\hat{S}_2^+|\bar{\phi}_1\phi_2| = 0 \tag{3.32}$$
$$\hat{S}_{z,1}\hat{S}_{z,2}|\phi_1\bar{\phi}_2| = -\frac{1}{4}|\phi_1\bar{\phi}_2| \qquad \hat{S}_{z,1}\hat{S}_{z,2}|\bar{\phi}_1\phi_2| = -\frac{1}{4}|\bar{\phi}_1\phi_2|$$

and from this the matrix elements can directly be written down:

$$\langle\phi_1\bar{\phi}_2|\hat{H}|\phi_1\bar{\phi}_2\rangle = -J\langle\phi_1\bar{\phi}_2|\left(0 + \frac{1}{2}|\bar{\phi}_1\phi_2\rangle - \frac{1}{4}|\phi_1\bar{\phi}_2\rangle\right) = \frac{1}{4}J$$
$$\langle\phi_1\bar{\phi}_2|\hat{H}|\bar{\phi}_1\phi_2\rangle = -J\langle\phi_1\bar{\phi}_2|\left(0 + \frac{1}{2}|\phi_1\bar{\phi}_2\rangle + 0\right) = -\frac{1}{2}J \tag{3.33}$$

In matrix form:

	$|\phi_1\overline{\phi}_2\rangle$	$|\overline{\phi}_1\phi_2\rangle$
$\langle\phi_1\overline{\phi}_2|$	$\frac{1}{4}J$	$-\frac{1}{2}J$
$\langle\overline{\phi}_1\phi_2|$	$-\frac{1}{2}J$	$\frac{1}{4}J$

$$\tag{3.34}$$

The diagonalization of this matrix gives $E_1 = \frac{3}{4}J$ and $E_2 = -\frac{1}{4}J$ and the corresponding eigenvectors are $\Psi_1 = \frac{1}{\sqrt{2}}\{|\phi_1\bar{\phi}_2| - |\bar{\phi}_1\phi_2|\}$ and $\Psi_2 = \frac{1}{\sqrt{2}}\{|\phi_1\bar{\phi}_2| + |\bar{\phi}_1\phi_2|\}$. Note that the eigenfunctions of the Heisenberg Hamiltonian are multideterminantal functions; linear combinations of the basis determinants $|\phi_1\bar{\phi}_2|$ and $|\bar{\phi}_1\phi_2|$. In the next step, we build an effective Hamiltonian that connects the *ab initio* results with the model Hamiltonian. In the first place the wave functions are projected on the model space and orthonormalized:

$$
\begin{aligned}
\text{Projections:} \quad & \widetilde{\Psi}_T = 0.7029|\phi_1\bar{\phi}_2| + 0.7029|\bar{\phi}_1\phi_2| \\
& \widetilde{\Psi}_S = 0.6776|\phi_1\bar{\phi}_2| - 0.6776|\bar{\phi}_1\phi_2| \\
\text{Norms:} \quad & \langle\widetilde{\Psi}_T|\widetilde{\Psi}_T\rangle = 0.7029^2 + 0.7029^2 = 0.9881 \\
& \langle\widetilde{\Psi}_S|\widetilde{\Psi}_S\rangle = 0.6776^2 + 0.6776^2 = 0.9183 \\
\text{Normalized projections:} \quad & \widetilde{\Psi}_T^N = 0.707107|\phi_1\bar{\phi}_2| + 0.707107|\bar{\phi}_1\phi_2| \\
& \widetilde{\Psi}_S^N = 0.707107|\phi_1\bar{\phi}_2| - 0.707107|\bar{\phi}_1\phi_2| \\
& \langle\widetilde{\Psi}_T^N|\widetilde{\Psi}_S^N\rangle = 0
\end{aligned}
$$

In the next step, the effective Hamiltonian is constructed by substituting these normalized projections and the corresponding energies in the Bloch equation as discussed in Sect. 1.4.

$$
\hat{H}^{eff} = \sum_i |\widetilde{\Psi}_i^N\rangle E_i \langle\widetilde{\Psi}_i^N| \tag{3.35}
$$

The basis of the effective Hamiltonian is the same as for the Heisenberg Hamiltonian. The use of orthonormal projections ensures that the effective Hamiltonian is hermitian with $\langle\phi_1\bar{\phi}_2|\hat{H}^{eff}|\bar{\phi}_1\phi_2\rangle = \langle\bar{\phi}_1\phi_2|\hat{H}^{eff}|\phi_1\bar{\phi}_2\rangle$.

$$
\begin{aligned}
\langle\phi_1\bar{\phi}_2|\hat{H}^{eff}|\phi_1\bar{\phi}_2\rangle = & \langle\phi_1\bar{\phi}_2|\{0.707107|\phi_1\bar{\phi}_2\rangle + 0.707107|\bar{\phi}_1\phi_2\rangle\} \cdot -29.438299 \\
& \cdot \{0.707107\langle\phi_1\bar{\phi}_2| + 0.707107\langle\bar{\phi}_1\phi_2|\}|\phi_1\bar{\phi}_2\rangle \\
& + \langle\phi_1\bar{\phi}_2|\{0.707107|\phi_1\bar{\phi}_2\rangle - 0.707107|\bar{\phi}_1\phi_2\rangle\} \cdot -29.441751 \\
& \cdot \{0.707107\langle\phi_1\bar{\phi}_2| - 0.707107\langle\bar{\phi}_1\phi_2|\}|\phi_1\bar{\phi}_2\rangle \\
= & \, 0.707107^2\langle\phi_1\bar{\phi}_2|\phi_1\bar{\phi}_2\rangle \cdot -29.438299 + 0.707107^2\langle\phi_1\bar{\phi}_2|\phi_1\bar{\phi}_2\rangle \cdot \\
& -29.441751 = 0.5\{-29.438299 - 29.441751\} = -29.440025 \, E_h
\end{aligned} \tag{3.36}
$$

The other diagonal matrix element, $\langle\bar{\phi}_1\phi_2|\hat{H}^{eff}|\bar{\phi}_1\phi_2\rangle$, has the same numerical value. The off-diagonal matrix element is calculated by the same procedure:

$$
\begin{aligned}
\langle\phi_1\bar{\phi}_2|\hat{H}^{eff}|\bar{\phi}_1\phi_2\rangle = & \, 0.707107\langle\phi_1\bar{\phi}_2|\phi_1\bar{\phi}_2\rangle \cdot -29.438299 \cdot 0.707107\langle\bar{\phi}_1\phi_2|\bar{\phi}_1\phi_2\rangle \\
& +0.707107\langle\phi_1\bar{\phi}_2|\phi_1\bar{\phi}_2\rangle \cdot -29.438299 \cdot -0.707107\langle\bar{\phi}_1\phi_2|\bar{\phi}_1\phi_2\rangle \\
= & \, 0.5(-29.438299 + 29.441751) = 0.001726 \, E_h \tag{3.37}
\end{aligned}
$$

Finally, the numerical effective Hamiltonian becomes:

$$
\begin{array}{c|cc}
 & |\phi_1\bar{\phi}_2\rangle & |\bar{\phi}_1\phi_2\rangle \\
\hline
\langle\phi_1\bar{\phi}_2| & -29.440025 & 0.001726 \\
\langle\bar{\phi}_1\phi_2| & 0.001726 & -29.440025
\end{array}
\tag{3.38}
$$

The two Hamiltonians (Eqs. 3.34 and 3.38) can only be compared when they have the same zero of energy. Therefore the diagonal matrix elements of the Heisenberg Hamiltonian are shifted by $-\frac{1}{4}J$ and those of the effective Hamiltonian by -29.440025 E_h. The comparison shows that there is a one-to-one correspondence between both matrices and that the magnetic coupling parameter is equal to -2×0.001726 $E_h = -757.6\,cm^{-1}$. In this simple case, the eigenfunctions of the Heisenberg Hamiltonian are the same as those of $\hat{S}^2$; Ψ_1 and Ψ_2 are directly the singlet and triplet functions, respectively. Hence J is also given by the difference of the singlet and triplet energies: $J = E_S - E_T = -29.441751 - -29.43830 = -0.003452$ $E_h = -757.6\,cm^{-1}$.

However, the extraction strategy based on effective Hamiltonians is generally applicable and in cases where the energies of the different spin states do not provide enough information to determine all the magnetic coupling parameters one necessarily has to rely on the effective Hamiltonian procedure. The three different magnetic coupling strengths J_{12}, J_{13} and J_{23} of the three center/three electron case of Fig. 3.5 cannot be extracted from the two energy differences defined by the quartet and the two doublets states. Instead an effective Hamiltonian has to be constructed from the *ab initio* wave functions and compared to the matrix representation of the Heisenberg Hamiltonian

$$
\hat{H} = -J_{12}\hat{S}_1 \cdot \hat{S}_2 - J_{13}\hat{S}_1 \cdot \hat{S}_3 - J_{23}\hat{S}_2 \cdot \hat{S}_3
\tag{3.39}
$$

$$
\begin{array}{c|ccc}
 & |\bar{\phi}_1\phi_2\phi_3\rangle & |\phi_1\bar{\phi}_2\phi_3\rangle & |\phi_1\phi_2\bar{\phi}_3\rangle \\
\hline
\langle\bar{\phi}_1\phi_2\phi_3| & \frac{1}{4}(J_{12}+J_{13}-J_{23}) & -\frac{1}{2}J_{12} & -\frac{1}{2}J_{13} \\
\\
\langle\phi_1\bar{\phi}_2\phi_3| & -\frac{1}{2}J_{12} & \frac{1}{4}(J_{12}-J_{13}+J_{23}) & -\frac{1}{2}J_{23} \\
\\
\langle\phi_1\phi_2\bar{\phi}_3| & -\frac{1}{2}J_{13} & -\frac{1}{2}J_{23} & \frac{1}{4}(-J_{12}+J_{13}+J_{23})
\end{array}
\tag{3.40}
$$

The quartet spin function $Q = |\alpha\alpha\alpha\rangle$ is an eigenfunction of the Heisenberg Hamiltonian of Eq. 3.39 with eigenvalue $-\frac{1}{4}(J_{12}+J_{13}+J_{23})$. However, the doublet functions D_1 and D_2 defined in Eq. 1.41 are not:

$$
\hat{H}\{|\alpha\alpha\beta\rangle - |\beta\alpha\alpha\rangle\} = -\frac{1}{4}(J_{12}-J_{23})(|\alpha\alpha\beta\rangle + |\beta\alpha\alpha\rangle)
$$
$$
+ \frac{3}{4}J_{13}(|\alpha\alpha\beta\rangle - |\beta\alpha\alpha\rangle) + \frac{1}{2}(J_{12}-J_{23})|\alpha\beta\alpha\rangle
\tag{3.41}
$$

$$\hat{H}\{2|\alpha\beta\alpha\rangle - |\alpha\alpha\beta\rangle - |\beta\alpha\alpha\rangle\} = \left(J_{12} + J_{23} - \frac{1}{2}J_{13}\right)|\alpha\beta\alpha\rangle$$

$$+ \left(\frac{1}{4}J_{12} - \frac{5}{4}J_{23} + \frac{1}{4}J_{13}\right)|\alpha\alpha\beta\rangle + \left(-\frac{5}{4}J_{12} + \frac{1}{4}J_{23} + \frac{1}{4}J_{13}\right)|\beta\alpha\alpha\rangle$$

$$\text{(3.42)}$$

In the special case of $J_{12} = J_{23} = J_1$; $J_{13} = J_2$, these two expression reduce to

$$\hat{H}D_1 = \frac{3}{4}J_2 D_1 \tag{3.43a}$$

$$\hat{H}D_2 = \left(J_1 - \frac{1}{4}J_2\right)D_2 \tag{3.43b}$$

and the two J-values can be directly extracted from the energy differences of the quartet and doublet states using the relations

$$J_1 = \frac{2}{3}\big(E(D_2) - E(Q)\big) \tag{3.44a}$$

$$J_2 = J_1 - \big(E(D_2) - E(D_1)\big) \tag{3.44b}$$

3.2.2 Ising Hamiltonian

The elimination of the anisotropic part in Eq. 3.22 leads to the Heisenberg Hamiltonian for isotropic magnetic interactions. The spins are considered as co-linear vectors whose principal quantization axis has no spatially preferred orientation. An even simpler model Hamiltonian can be obtained by putting A_{xx} and A_{yy} to zero in Eq. 3.21. Then, the spin reduces to a classical vector whose orientation in space is not defined and the resulting model Hamiltonian describes the isotropic coupling of two (anti-)parallel spins. Replacing A_{zz} by $-J$, the following expression is obtained

$$\hat{H} = -J\hat{S}_{z,1}\hat{S}_{z,2} \tag{3.45}$$

which is known as the Ising Hamiltonian. The big advantage of this simpler Hamiltonian is the fact that the eigenfunctions correspond to monodeterminantal functions, and therefore, this Hamiltonian can be used to determine the magnetic coupling strength from density functional theory (DFT) calculations and in extended systems treated in the periodic approximation. This is further discussed in the next chapters in Sects. 4.3.4 and 6.3. For now, we will restrict ourselves to some formal properties of the Ising Hamiltonian and a comparison with the Heisenberg Hamiltonian. To determine the magnetic coupling of the two-electron/two-orbital problem, two functions are needed describing parallel and anti-parallel coupling, $|\phi_1\phi_2|$ and $|\phi_1\bar{\phi}_2|$. By acting with the Ising Hamiltonian on these two function we get

$$-J\hat{S}_{z,1}\hat{S}_{z,2}|\phi_1\phi_2| = -\frac{1}{4}J|\phi_1\phi_2| \tag{3.46}$$

$$-J\hat{S}_{z,1}\hat{S}_{z,2}|\phi_1\bar{\phi}_2| = \frac{1}{4}J|\phi_1\bar{\phi}_2| \tag{3.47}$$

Assuming that the spatial part is identical in both functions (only spin degrees of freedom are taken into account for the moment), the energy difference between the two determinants gives an estimate of the magnetic coupling through $E_{LS} - E_{HS} = \frac{1}{2}J$. Here LS refers to the determinant with antiparallel alignment of the spins and HS to the parallel alignment. This expression is easily generalized to a pair of arbitrary spins

$$E_{LS} - E_{HS} = 2S_1 S_2 J \tag{3.48}$$

In any real case, the orbital part plays an important role and the influence of this on the extraction of the J-value from calculations will be discussed in the next chapter. The dimeric system $S_1 = S_2 = \frac{1}{2}$ only has HS and LS states, but for systems with higher spins there are several intermediate states. Remembering that the eigenfunctions of the Ising Hamiltonian are not necessarily spin eigenfunctions, we use a label to characterize the eigenfunctions that consists of the M_S-value of the two magnetic centers. Then the HS determinant is $|\pm\frac{1}{2}, \pm\frac{1}{2}\rangle$ and the LS state is represented as $|\pm\frac{1}{2}, \mp\frac{1}{2}\rangle$. For a system with two $S = 1$ spins, nine different determinants can be defined. The HS and LS states are separated by $2J$ as follows from the above equation, but in between these two, there is a set of five determinants with $M_S = 0$ functions on one or both magnetic centers with an expectation value equal to zero.

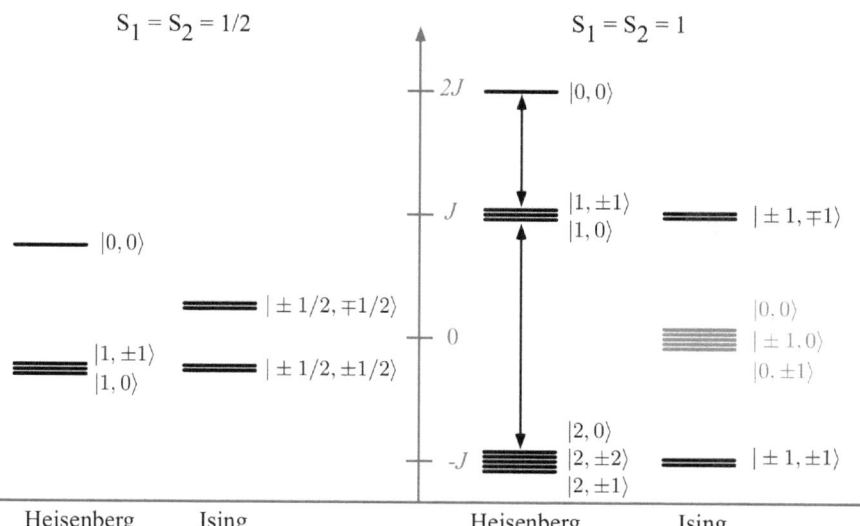

Fig. 3.6 Comparison of the Heisenberg and Ising eigenvalues for a dimeric system with $S_1 = S_2 = \frac{1}{2}$ (*left*) and 1 (*right*). The levels in *gray* are eigenfunctions of the Ising hamiltonian, but lie at much higher energy when the spatial part of the wave function is also considered

In the general case of $S_1 \neq S_2$, the gap between the lowest state and the group of degenerate first excited states of the Ising Hamiltonian is given by the value of the smallest spin. Figure 3.6 summarizes the energy levels of the Heisenberg and Ising Hamiltonians for the two systems.

3.2.3 Comparing the Heisenberg and Ising Hamiltonians

Table 3.1 compares some formal properties of the Ising and the Heisenberg Hamiltonian for symmetric dimeric systems with different spin moments. The spectral width W—defined as the difference between the lowest and highest eigenvalue—increases with the spin moment for both Hamiltonians. In absolute values the difference between Ising and Heisenberg grows larger, but it should be noted that the relative difference is significantly smaller for the system with $S = 5/2$ (16.7 %) than for the $S = 1/2$ case (50 %). For antiferromagnetic coupling (negative J), the first excited state of the Heisenberg Hamiltonian is always a threefold degenerate triplet state with a relative energy equal to J. In the Ising model, the gap depends linearly on the spin moment. For $S = 1/2$, the gap is smaller than in the Heisenberg model, but for the $S = 5/2$ system the separation of the ground state and the first excited state is much larger in the Ising model. In the case of ferromagnetic interaction, the spectrum of the Ising Hamiltonian is simply inverted, being symmetric around $E = 0$. This is not the case for the Heisenberg Hamiltonian. Except for the $S = 1/2$ system, the gap between ground and first excited state is larger, as is the degeneracy of the latter.

3.7 Complete the Table for atoms with six and seven unpaired electrons as can be found in the rare earth metal ions.

Before closing this section, a word of warning is needed concerning all the eigenstates of the Ising Hamiltonian between the ones with the highest and lowest energy.

Table 3.1 Spectral width (W), gap (Δ) and degeneracy of the first excited state of the Heisenberg and Ising Hamiltonian for a dimeric system with $S_1 = S_2 = 1/2 \ldots 5/2$ and $J = \pm 1$ K

Spin	Heisenberg						Ising		
	W	Δ		Degen.			W	Δ	Degen.
		AF	F	AF	F				
$1/2$	1	1	1	3	1		$1/2$	$1/2$	2
1	3	1	2	3	3		2	1	5
$3/2$	6	1	3	3	5		$9/4$	$3/2$	4
2	10	1	4	3	7		4	2	4
$5/2$	15	1	5	3	9		$25/2$	$5/2$	4

These eigenstates share the common feature that at least one of the local M_S values is not equal to $\pm M_S^{max}$. For example the Ising eigenstates in Fig. 3.6 with energy J have either on the left or the right (or both) centers an $\alpha\beta$ determinant. As soon as one adds the spatial part to the wave function, these determinants are raised in energy since they lack the stabilization by the exchange integral K present in the energy expression of the $|\alpha\alpha|$ and $|\beta\beta|$ determinants. Consequently, in any practical application focused on magnetic interactions one should only consider the eigenstates of the Ising Hamiltonian with $M_S = 0$ or M_S^{max}.

3.3 From Micro to Macro: The Bottom-Up Approach

In Sect. 2.3.2, we have shown how the temperature dependence of the magnetic susceptibility can be calculated based on the knowledge of the energy levels of the ion in a magnetic field. Substituting an analytical expression in the van Vleck equation, we derived the Curie law for paramagnetic systems without interaction between the magnetic centers. The same strategy can be followed for systems in which the interaction between the magnetic centers cannot be neglected, such as those discussed in this chapter so far. At difference with the derivation of Curie's law for isolated magnetic ions, we no longer can ignore the excited states and have to substitute $E^{(0)}$ by Eq. 3.28 in the van Vleck equation (Eq. 2.33). Using the same expression as before for $E^{(1)}$ we obtain

$$\chi = \frac{N_A \mu_B^2 g_e^2}{kT} \frac{\displaystyle\sum_{S=S_{min}}^{S_{max}} \sum_{M_S=-S}^{S} M_S^2 \exp(J S(S+1)/2kT)}{\displaystyle\sum_{S=S_{min}}^{S_{max}} \sum_{M_S=-S}^{S} \exp(J S(S+1)/2kT)} \tag{3.49}$$

which reduces to

$$\chi = \frac{N_A \mu_B^2 g_e^2}{3kT} \frac{\displaystyle\sum_{S=S_{min}}^{S_{max}} S(S+1)(2S+1) \exp(J S(S+1)/2kT)}{\displaystyle\sum_{S=S_{min}}^{S_{max}} (2S+1) \exp(J S(S+1)/2kT)} \tag{3.50}$$

by using Eq. 2.35. This so-called *Bleaney–Bowers* equation [2], which is normally used to fit experimental data to extract numerical values for J and g_e. The other way around is of course also possible; the equation can also be used to generate the $\chi(T)$ from an *ab initio* calculation of the microscopic parameters, J and sometimes g_e.

3.8 Confirm that the Bleaney–Bowers expression for a dimer with $S_1 = S_2 = \frac{1}{2}$ equals

$$\chi = \frac{2N_A\mu_B^2 g_e^2}{kT\left(3 + \exp(-J/kT)\right)}.$$

This is rather trivial as long as dimeric systems are concerned, because there is actually only one parameter in the analytical expression of χ and no new information is obtained by calculating $\chi(T)$ from the theoretical estimates of the J-value. The situation is different when polynuclear systems are considered. Most importantly, there are not many systems for which an exact expression of χ has been derived. In addition to the above stated expression for binuclear complexes, Boča derived expression for tri- and tetra-nuclear systems [3], which turn out to be rather lengthy. The situation is even more complicated for extended systems, which have (in principle) an infinite number of interacting magnetic centers.

In fact, the one-dimensional uniform Heisenberg chain is the only extended system for which an exact solution has been derived making use of the Bethe Ansatz [4]. Bonner and Fisher extended this $T = 0$ solution to finite temperatures by extrapolating the results obtained for small chains to chains of infinite length [5]. The Bonner-Fisher expression is still widely used to fit magnetic susceptibility data to determine the magnetic coupling strength in systems with a magnetic chain-like topology.

$$\chi(T) = \frac{N_A\mu_B^2 g_e^2}{kT} \frac{A + Bx + Cx^2}{1 + Dx + Ex^2 + Fx^3} \tag{3.51}$$

where the values of A–F are given in Appendix D and $x = |J|/2kT$. Similar strategies were used to derive expressions for $\chi(T)$ in magnetic chains in which the magnetic centers alternately interact through J_1 and J_2 [6]. Defining the Hamiltonian as

$$\hat{H} = -J \sum_{i=1} \left(\hat{S}_{2i} \cdot \hat{S}_{2i+1} + \alpha\hat{S}_{2i} \cdot \hat{S}_{2i-1}\right) \tag{3.52}$$

with the same quadratic/cubic equation as for the uniform Heisenberg chain for which A-F are also listed in the Appendix. Note that the Bonner-Fisher expression is only valid for $2kT/|J| > 0.5$ and hence the low-temperature data should not be included in the fitting procedure. Improvements upon the Bonner-Fisher expression for low temperatures have been published [7] and many more expressions for the magnetic susceptibility can be found in Ref. [8].

Magnetic susceptibility data in two-dimensional extended systems are often interpreted based on the work of Rushbrooke and Wood [9], who derived an expression for $\chi(T)$ valid for high temperatures. The discovery of the high T_c superconductors renewed the interest in the 2D Heisenberg lattices and the original work was extended to lower temperatures. A workable expression for a uniform lattice—characterized

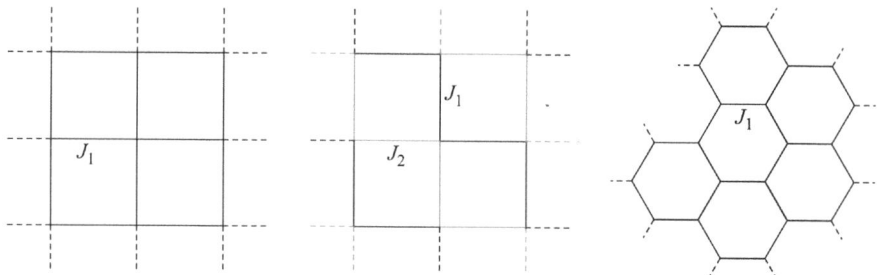

Fig. 3.7 Some examples of two-dimensional magnetic lattices for which analytical expressions have been derived to fit J from the temperature dependence of the magnetic susceptibility

by one single J, see Fig. 3.7 left—was given by Woodward and co-workers and reads

$$\chi(T) = \frac{N_A \mu_B^2 g_e^2}{kT} \sum_{n=1}^{5} \frac{a_n J/kT}{b_n J/kT} \tag{3.53}$$

More general expressions were derived by Curély for $S \neq 1/2$, for 2D lattices with different magnetic interaction paths (Fig. 3.7 middle) and to hexagonal (or honeycomb) lattices (Fig. 3.7 right) [10, 11].

When no analytical expression can be used to fit $\chi(T)$, the experimental data are interpreted by defining a magnetic model with the magnetic interactions that are considered *a priori* to be the most important ones. The corresponding Heisenberg Hamiltonian is then diagonalized and the resulting eigenvalues are substituted in the van Vleck equation. The J-values of the magnetic model are adjusted to give an optimal fit of the experimental data. However, one has to be aware that a multiparameter fit can have several solutions of equal quality and that this way of deriving experimental J-values can be subject to uncertainties. Actually, this is where computations can be helpful to discern the important interactions from less important ones and determine the sign and order of magnitude of the interactions. This would in principle lead to a well-founded magnetic model that will lead to reliable J-values from the fitting procedure.

A closely related procedure allows theoreticians to take the full journey from microscopic to macroscopic in a three-step strategy [12]. In the first stage, one calculates as exhaustive as possible the interactions among the different magnetic centers. This should not be restricted to nearest neighbors and preferably also include three- or four-body interactions, see Sect. 3.4. Secondly, a magnetic model is defined by writing down the Heisenberg Hamiltonian with the most important interactions. When dealing with an extended system, periodic boundary conditions can be applied. This is best illustrated taking the 1D Heisenberg chain as example. As illustrated in Fig. 3.8, the first center in the chain not only interacts with center 2 on the right, but also with the last center in the chain. In this way, there is no open end in the chain, exactly as in an infinite 1D chain. The topology of the model is actually a ring, but this turns out

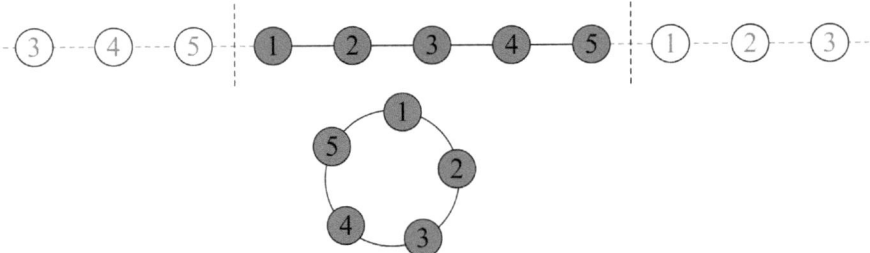

Fig. 3.8 Periodic boundaries in a one-dimensional chain. The central unit of five magnetic centers is repeated on the *left* and the *right* by introducing the interaction between center 1 and 5. The actual model is a closed ring of five centers

to be a very accurate representation of the 1D chain provided a large enough number of centers is considered. Finally, the Hamiltonian is diagonalized and the resulting eigenvalues are substituted in the van Vleck equation to obtain the magnetic susceptibility as function of the temperature from a rigorous *ab initio* treatment. The eigenvalues can of course also be used to derive any other macroscopic property such as the specific heat at constant magnetic field (C_B) by using the appropriate equation from standard statistical mechanics.

Inter- and intramolecular interactions: Generally speaking, transition metal based magnetic materials have large intramolecular interactions and weaker intermolecular interactions. Nevertheless, the control and understanding of the macroscopic properties depends critically on the knowledge of both types of interactions. Imagine a building block with two antiferromagnetically coupled spin moments as schematically depicted in the upper panel of Fig. 3.9. The interaction of the spin moments on the transition metals proceeds through the bridging ligand as will be profoundly analyzed in Chap. 5 and is also known as a through-bond interaction. Using transition metals with different spins ($S_1 \neq S_2$) causes that the unit has a net magnetic moment, despite the antiferromagnetic nature of the interaction. This is known as ferrimagnetism. The middle panel shows that a proper choice of the external ligands can link these building blocks into an *infinite* chain of antiferromagnetically coupled magnetic centers. Such entity is of course a very interesting object due to the net magnetic moment, however to take profit of this, one has to stick these chains together in a three-dimensional structure such that the chains are ferromagnetically coupled to each other as shown in the lower panel. This interchain coupling is typically much weaker as it does not involve magnetic centers that are connected by (covalent) bonds, and is usually referred to as through-space interaction. By carefully choosing the magnetic centers and the coordinating ligands, Kahn and co-workers were able to design and synthesize molecular-based magnets, initially with rather low critical temperatures for long-range order [13], but later many compounds have been reported with long-range order at much higher temperatures.

A different situation is encountered in most magnetic materials containing organic radicals. Typically, the building units are moieties with one unpaired electron, either

Fig. 3.9 *Upper* Antiferromagnetic coupling of two spins intermediated by a diamagnetic bridge (through-bond interactions). *Middle* After linking the units, a one-dimensional *ferrimagnetic* chain is obtained. *Lower* The 1D chains are linked together through weaker intramolecular (through-space) interactions, indicated by *dotted lines* (for simplicity only two dimensions are shown)

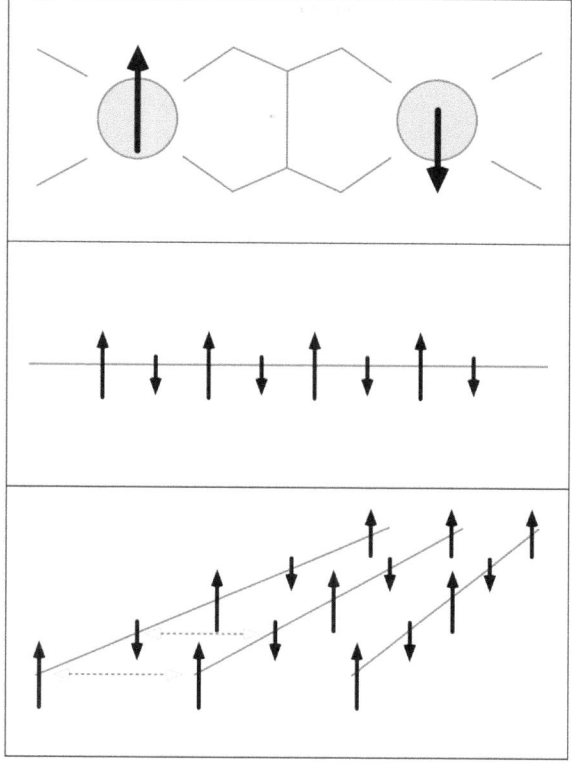

localized on a few atoms of the radical (N and O in nitroxides, central C in triarylmethyl, for example) or delocalized over a large part of the molecule (conjugated π-systems such as phenalenyl). The magnetic properties of these radical-based materials are determined by the through-space interactions between the units.

3.3.1 Monte Carlo Simulations, Renormalization Group Theory

There are powerful techniques to determine a few selected eigenvalues and eigenvectors of matrices of huge dimensions with millions or even billions of columns. This can be very efficiently exploited to calculate the electron correlation effects in the energy and wave function in electronic structure calculations where normally only the ground state and a few excited states are of interest. However, the accurate calculation of macroscopic properties such as the temperature dependence of the magnetic susceptibility cannot be done using only a few low lying eigenstates, but requires a much larger set. The eigenvalue spectrum of the Heisenberg Hamiltonian is very

Table 3.2 Dimension of the Heisenberg Hamiltonian for a system with N magnetic sites with $S = \frac{1}{2}$ and 1

S	$N = 2$	3	4	5	6	7	8	9	10	11	12
$\frac{1}{2}$	2	3	6	10	20	35	70	126	252	462	924
1	3	7	19	51	141	393	1107	3139	8953	25648	73764

dense and many levels are thermally occupied. Since selecting a balanced subset of states is nearly impossible, it is preferable to perform a full diagonalization of the Heisenberg Hamiltonian and include all states in the calculation of the macroscopic properties of the material under study.

However, the dimension g of the Heisenberg Hamiltonian grows rapidly with the number of magnetic sites N and the spin moment of these sites. For a model with all spin moments equal to $S = \frac{1}{2}$ the dimension is given by (Table 3.2)

$$g = \frac{(2NS)!}{\left((NS)!\right)^2} \qquad \text{if } N \text{ is even}$$

$$g = \frac{(2(NS + 1/2))!}{2\left((NS + 1/2)!\right)^2} \qquad \text{if } N \text{ is odd} \tag{3.54}$$

and for lattices with $S = 1$ spin moments the dimension is given by

$$g = 1 + \sum_{k=1}^{k < N/2} \binom{n}{2k}\binom{2k}{k} \tag{3.55}$$

for higher spin moments the increase is even steeper. Brute force diagonalization techniques can handle models with up to 16 $S = \frac{1}{2}$ magnetic sites. Using more powerful techniques such as those based on the Lanczos algorithm can push the limit up to 40 centers, which for most practical applications seems to be large enough. However, for larger models and for larger spin moment, it can be useful to consider more approximate techniques to obtain information on the macroscopic properties from the electronic structure parameters in a bottom-up approach. A good example is the family of polynuclear complexes intensively investigated for the possibility of single molecule magnet behaviour. Complexes with 19 Fe^{III} ions can hardly be expected to be treated via a full diagonalization of the Heisenberg Hamiltonian, but still has been studied in a bottom-up approach [14]. Among the many different approaches to have access to macroscopic properties starting at the microscopic description but without going through the full diagonalization of the Heisenberg Hamiltonian we will shortly mention two techniques, namely the renormalization group (RG) theory and classical Monte Carlo simulations.

Renormalization Group theory: The partition function Q is the central quantity of statistical mechanics and many thermodynamic functions can be derived from it. The partition function of the one-dimensional Ising chain is

$$Q = \sum_{M_S=\frac{1}{2},-\frac{1}{2}} \exp[J(M_S(1)M_S(2) + M_S(2)M_S(3) + M_S(3)M_S(4) + \ldots)/k_B T]$$

(3.56)

with $K = J/2k_B T$ and $M_S = \frac{1}{2}\sigma$ ($\sigma = \pm 1$), this can be rewritten to

$$Q = \sum_{\sigma_i=\pm 1} e^{K(\sigma_1\sigma_2+\sigma_2\sigma_3)} e^{K(\sigma_3\sigma_4+\sigma_4\sigma_5)} \ldots$$

(3.57)

After summing over $\sigma_2 = \pm 1$, we arrive at

$$Q = \sum_{\substack{\sigma_i=\pm 1 \\ i \neq 2}} [e^{K(\sigma_1+\sigma_3)} + e^{-K(\sigma_1+\sigma_3)}] e^{K(\sigma_3\sigma_4+\sigma_4\sigma_5)} \ldots$$

(3.58)

and when the summation is made over $\sigma_4, \sigma_6, \ldots$, the partition function becomes

$$Q = \sum_{\substack{\sigma_i=\pm 1 \\ i= \text{odd}}} \ldots [e^{K(\sigma_1+\sigma_3)} + e^{-K(\sigma_1+\sigma_3)}][e^{K(\sigma_3+\sigma_5)} + e^{-K(\sigma_3+\sigma_5)}] \ldots$$

(3.59)

If we can find a way to rewrite

$$[e^{K(\sigma_1+\sigma_3)} + e^{-K(\sigma_1+\sigma_3)}] \text{ as } f(K)e^{K'\sigma_1\sigma_3}$$

(3.60)

we can return to the original expression of the partition function but now with half the number of centers and replacing K, the interaction between magnetic centers by K', the effective interaction parameters between blocks containing two magnetic centers, as illustrated in Fig. 3.10. Substituting $\sigma_1 = \sigma_3 = \pm 1$ and $\sigma_1 = -\sigma_3 = \pm 1$, we obtain two equations from which $f(K)$ and K' can be determined

$$\left. \begin{array}{ll} \sigma_1 = \sigma_3 = \pm 1 & e^{2K} + e^{-2K} = fe^{K'} \\ \sigma_1 = -\sigma_3 = \pm 1 & 2 = fe^{K'} \end{array} \right\} \Rightarrow \begin{array}{l} K' = \frac{1}{2}\ln \cosh(2K) \\ f(K) = 2\cosh^{\frac{1}{2}}(2K) \end{array}$$

(3.61)

and

$$Q = \sum_{\substack{\sigma_i=\pm 1 \\ i= \text{odd}}} f(K)e^{K'\sigma_1\sigma_3} f(K)e^{K'\sigma_3\sigma_5} \ldots = f(K)^{N/2} \sum_{\substack{\sigma_i=\pm 1 \\ i= \text{odd}}} e^{K'\sigma_1\sigma_3} e^{K'\sigma_3\sigma_5} \ldots$$

(3.62)

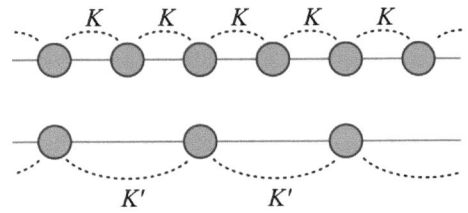

Fig. 3.10 Illustration of the renormalization procedure. In the *upper part* N particles are considered with an interaction K, while in the *lower part* the interaction K between sites is replaced by a larger-scale effective interaction K'

Hence, we have shown that the partition function of the whole system can be written in terms of properties that only depend on half the number of centers

$$Q(N, K) = f(K)^{N/2} Q(N/2, K') \tag{3.63}$$

and the recursive application of this formula connects the microscopic description with the thermodynamic large-scale properties. To elaborate the procedure a little more we use the relation of the free energy A and the partition function as given by statistical mechanics

$$\ln Q(N, K) = \frac{A}{-kT} = N\xi(K) \tag{3.64}$$

The free energy can be used to determine the specific heat and the temperature dependence of the specific heat can tell us something about the possible order-disorder phase transitions in magnetic systems. A is an extensive property and hence depends on the system size. It is here conveniently written as a product of the system size (N) and a system-size independent parameter ξ, which can be considered as the free energy per site.

$$\xi(K) = \frac{\ln Q}{N} = \frac{1}{2} \ln f(K) + \frac{1}{2}\xi(K') \tag{3.65}$$

where we have used $\ln x^a y = a \ln x + \ln y$ and $\ln Q(N/2, K') = (N/2)\xi(K')$, cf. Eq. 3.63. This brings us to the recursion relations to go from a description with N individual magnetic centers interacting through K to a description with ever increasing block size interacting through K'

$$K' = \frac{1}{2} \ln \cosh(2K)$$

$$\xi(K') = 2\xi(K) - \ln(2 \cosh^{\frac{1}{2}}(2K)) \tag{3.66}$$

The inverse relation can also be of use, especially in those cases where the property under study (here the free energy per site) is known in the thermodynamic limit, that is $K' \approx 0$

$$K = \frac{1}{2}\cosh^{-1}(e^{2K'})$$

$$\xi(k) = \frac{1}{2}\ln 2 + \frac{1}{2}K' + \frac{1}{2}\xi(K') \tag{3.67}$$

The one-dimensional Ising chain is not the most interesting magnetic system to study with renormalization theory, since it is known from the exact solution that there is no phase transition, the chain is disordered at any finite temperature. The two-dimensional Ising lattice does have an order/disorder phase transition, nicely reproduced with the renormalization procedure as discussed in Refs. [15, 16]. Such phase transition does not exist in a two-dimensional lattice described with the Heisenberg Hamiltonian. For this model, a non-zero interaction along the third dimension is needed to have an ordered (anti-)ferromagnetic system at finite temperature as stated by the Mermin-Wagner theorem.

Monte Carlo simulations: An alternative strategy to calculate thermodynamic properties is to explicitly follow the trajectory of a magnetic system by a computer simulation of the system. Along such trajectory, the system will adopt many conformations with different energy, magnetization and other microscopic observables. If the sampling of the conformational space is done correctly, a good estimate of the partition function can be made and with this all type of thermodynamic functions can be calculated.

There are basically two types of simulations to sample the conformational space. The first one is known as Molecular Dynamics and propagates a system in time by integrating the Newton's equations of motion. In its most rudimentary form the procedure can be described as follows. For a given set of atomic positions $r(t = t_0)$, one calculates the forces and from these the velocities $v(t_0)$, accelerations $a(t_0)$ and usually some higher derivatives. The atoms are then moved from $r(t_0)$ to $r(t_0 + \Delta t)$ by the formula $r(t_0 + \Delta t) = r(t_0) + v(t_0)\Delta t + (1/2)a(t_0)\Delta t^2 + \ldots$ and the time is updated from t_0 to $t_0 + \Delta t$. Then the cycle is repeated as long as one wants to follow the trajectory. The second method, the so-called Monte Carlo method, does not propagate the system in time but rather performs a *random* walk through the conformational space to calculate the partition function. Whereas numerical integration on a regular grid is much more efficient for low-dimensional functions, such approach is absolutely out of the question for extremely high dimensional functions, such as the partition function for any interesting N-particle system. In these cases a smart random walk is more effective and can be used to extract macroscopic properties as function of microscopic interactions.

To illustrate the procedure, we come back to the Ising model, but now focusing on the two-dimensional lattice with nearest neighbour interactions only. The sampling of the conformational space is usually done with the Metropolis algorithm, which starts by creating the initial spin conformation $\mathbb{S}_0$. This can be done in many ways, one of them is assigning a random spin direction $M_S = \pm\frac{1}{2}$ to each lattice point as shown in the left part of Fig. 3.11. After calculating the energy of this spin distribution, a trial step in conformational space is taken by inverting the spin at one of the lattice

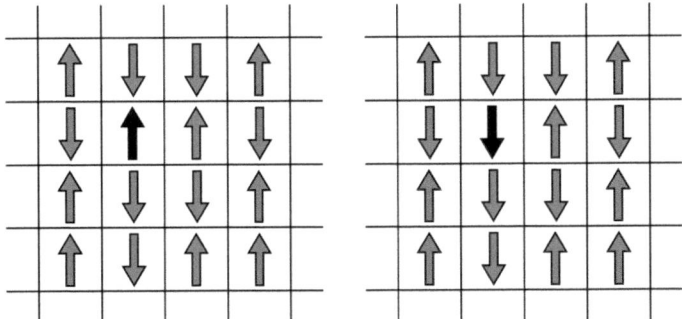

Fig. 3.11 *Left* Initial spin distribution $\mathbb{S}_0$ on a small part of the $N \times N$ spin lattice. *Right* Trial spin distribution $\mathbb{S}_t$ after inverting the M_S value of the *black* spin. Depending on the energy change, the new distribution can be accepted or rejected

sites to generate $\mathbb{S}_t$. The step is accepted when $\exp(-\Delta E/k_B T)$ is larger than a uniformly chosen random number between 0 and 1 and rejected otherwise (ΔE is the energy difference of $\mathbb{S}_t$ and $\mathbb{S}_0$). This means that trial spin distributions with lower energy are always accepted, while trial conformations with higher energies are accepted through an exponential weighting function. The closer the value of the exponential to 1 (that is, for small ΔE), the larger the chance for accepting the new conformation. Figure 3.12 shows how the ratio between accepted and rejected steps (γ) smoothly converges to an exponential function of the energy difference with an increasing number of steps in the conformational space.

In the trial distribution shown in the right panel of Fig. 3.11, the black spin was changed from $M_S = \frac{1}{2}$ to $-\frac{1}{2}$. The energy difference between $\mathbb{S}_t$ and $\mathbb{S}_0$ can easily be calculated by realizing that the only contributions to the energy difference arise

Fig. 3.12 Acceptance rate γ as function of the energy difference ΔE between the initial and trial spin conformation for three different simulation lengths N. The *black line* represents the weighting function $\exp(-0.5\Delta E)$

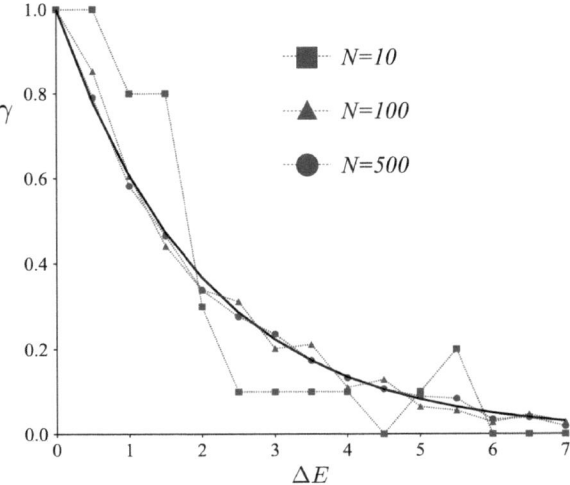

from the interaction involving the black spin. The differential part of the energy of $\mathbb{S}_0$ and $\mathbb{S}_t$ is

$$E'(\mathbb{S}_0) = -J \cdot \frac{1}{2} \left(\frac{1}{2} - \frac{1}{2} - \frac{1}{2} - \frac{1}{2} \right) = \frac{1}{2} J$$

$$E'(\mathbb{S}_t) = -J \cdot -\frac{1}{2} \left(\frac{1}{2} - \frac{1}{2} - \frac{1}{2} - \frac{1}{2} \right) = -\frac{1}{2} J \tag{3.68}$$

and from here the energy difference $\Delta E = -J$. When $J > 0$, that is for ferromagnetic interactions, the step is accepted because the energy of the system is lowered by the spin flip. Instead for antiferromagnetic interactions, $J < 0$, the energy difference is positive and the step will only be accepted when the $\exp(-\Delta E / k_B T)$ is larger than a random number between 0 and 1. Subsequently, the neighbouring spin is flipped and the accept/reject algorithm is repeated for all sites on the lattice. Then the total energy and magnetization (or other properties) are calculated and accumulated to determine the average properties after a certain amount of sweeps over the lattice.

In addition to the very basic application to the two-dimensional lattice with nearest neighbour interactions, this rather simple and intuitive approach to calculate thermodynamic properties can of course also be used to study magnetic systems with more complex magnetic structures. However, it fails badly when it comes to magnetic interactions between centers with spin moments different from $S = \frac{1}{2}$. In the basic form described above each lattice site can only adopt two states: *up* or *down*; α or β; positive or negative M_S. No distinction can be made between a lattice of magnetic sites with $S = \frac{1}{2}$ and any higher spin moment. For this purpose, the model Hamiltonian needs to be improved and a natural thing to do is to replace the Ising Hamiltonian with the Heisenberg Hamiltonian. An important drawback of using this more accurate model Hamiltonian is that the total energy of the lattice is no longer a simple sum of individual contributions as in the Ising case, and hence, the energy of a spin configuration cannot be calculated directly. Instead one can introduce two levels of accuracy in the Metropolis algorithm [17]. To decide on the acceptance of a spin flip the energy of a small cluster around the *active* lattice site is calculated with the Heisenberg Hamiltonian, while the rest of the lattice is considered as an Ising system. Keeping the cluster small enough, sweeping the lattice can be done rather efficiently in this half classic/half quantum treatment of the spin interactions. To study magnetic phenomena at low temperatures, one should definitely consider a full Quantum Monte Carlo approach [18].

3.4 Complex Interactions

The isotropic bilinear operator discussed so far is the most widely considered interaction in polynuclear magnetic systems since it accounts for an important part of the physics. However, it is not the whole story. In the very beginning of this chapter, we have set aside the spatial anisotropy in the interaction between two spin moments. Furthermore we have assumed that the interaction can be described with a simple vector product of linear operators and that more-than-two particle interactions are irrelevant. In this section, we will discuss refinements of the standard Hamiltonian and see how more complex interactions can be incorporated in the description of the magnetic couplings.

3.4.1 Biquadratic Exchange

The spin eigenfunctions for a binuclear complex with $S = 1$ magnetic centers are

$$
\begin{aligned}
Q &= \alpha\alpha\alpha\alpha \\
T &= \frac{1}{\sqrt{2}}(\alpha\alpha\beta\beta - \beta\beta\alpha\alpha) \\
S &= \frac{1}{2\sqrt{3}}(2(\alpha\alpha\beta\beta + \beta\beta\alpha\alpha) - \alpha\beta\alpha\beta - \alpha\beta\beta\alpha - \beta\alpha\alpha\beta - \beta\alpha\beta\alpha)
\end{aligned}
\tag{3.69}
$$

which are also eigenfunctions of the Heisenberg Hamiltonian, with eigenvalues of $-J$, J and $2J$, respectively.

$$
\hat{H}\Psi = -J\hat{S}_1 \cdot \hat{S}_2\Psi = -J\left[\frac{1}{2}(\hat{S}_1^+\hat{S}_2^- + \hat{S}_1^-\hat{S}_2^+) + \hat{S}_{z,1}\hat{S}_{z,2}\right]\Psi
\tag{3.70}
$$

with $\hat{S}_1 = \hat{s}(1) + \hat{s}(2)$ and $\hat{S}_2 = \hat{s}(3) + \hat{s}(4)$ (see Eq. 1.22), the eigenvalue of the quintet function arises from

$$
\begin{aligned}
\hat{H}Q &= -J\left[\frac{1}{2}\left\{(\hat{s}^+(1) + \hat{s}^+(2))(\hat{s}^-(3) + \hat{s}^-(4)) + (\hat{s}^-(1) + \hat{s}^-(2))(\hat{s}^+(3) + \hat{s}^+(4))\right\}\right. \\
&\quad \left. + (\hat{s}_z(1) + \hat{s}_z(2))(\hat{s}_z(3) + \hat{s}_z(4))\right]\alpha(1)\alpha(2)\alpha(3)\alpha(4) \\
&= -J\left[\frac{1}{2}\left\{(\hat{s}^+(1) + \hat{s}^+(2))\alpha\alpha\beta\beta + (\hat{s}^-(1) + \hat{s}^-(2))\cdot 0\right\}\right. \\
&\quad \left. + (\hat{s}_z(1) + \hat{s}_z(2))\left(\frac{1}{2} + \frac{1}{2}\right)\alpha\alpha\alpha\alpha\right] = -JQ
\end{aligned}
\tag{3.71}
$$

The calculation of the eigenvalues of the triplet and singlet functions is slightly more involved but follows exactly the same mechanics and can be derived as a useful exercise.

3.9 Calculate the outcome of $(\hat{s}^+(1) + \hat{s}^+(2))(\hat{s}^-(3) + \hat{s}^-(4))$, $(\hat{s}^-(1) + \hat{s}^-(2))(\hat{s}^+(3) + \hat{s}^+(4))$ and $(\hat{s}_z(1) + \hat{s}_z(2))(\hat{s}_z(3) + \hat{s}_z(4))$ acting on $\alpha\alpha\beta\beta$, $\beta\beta\alpha\alpha$, $\alpha\beta\alpha\beta$, $\alpha\beta\beta\alpha$, $\beta\alpha\beta\alpha$ and $\beta\alpha\alpha\beta$. Use the results to verify the Heisenberg Hamiltonian eigenvalues of the singlet and triplet spin functions.

As long as magnetic anisotropy can be neglected, the regular spacing between the energy levels, the Landé pattern of Eq. 3.29 gives a very accurate representation of the experimental situation. However, sometimes deviations have been observed, which are usually ascribed to biquadratic interactions and subsequently incorporated in the model by adding an extra term to the Heisenberg Hamiltonian

$$\hat{H} = -J\hat{S}_1 \cdot \hat{S}_2 + \lambda(\hat{S}_1 \cdot \hat{S}_2)^2 \tag{3.72}$$

Before calculating the eigenvalues of this new spin Hamiltonian, the second term has to be worked out a little more

$$
\begin{aligned}
(\hat{S}_1 \cdot \hat{S}_2)^2 &= \left[\frac{1}{2}(\hat{S}_1^+\hat{S}_2^- + \hat{S}_1^-\hat{S}_2^+) + \hat{S}_{z,1}\hat{S}_{z,2}\right]\left[\frac{1}{2}(\hat{S}_1^+\hat{S}_2^- + \hat{S}_1^-\hat{S}_2^+) + \hat{S}_{z,1}\hat{S}_{z,2}\right] \\
&= \frac{1}{4}\left[\hat{S}_1^+\hat{S}_2^-\hat{S}_1^+\hat{S}_2^- + \hat{S}_1^+\hat{S}_2^-\hat{S}_1^-\hat{S}_2^+ + \hat{S}_1^-\hat{S}_2^+\hat{S}_1^+\hat{S}_2^- + \hat{S}_1^-\hat{S}_2^+\hat{S}_1^-\hat{S}_2^+\right] \\
&\quad + \frac{1}{2}\left[\hat{S}_1^+\hat{S}_2^-\hat{S}_{z,1}\hat{S}_{z,2} + \hat{S}_1^-\hat{S}_2^+\hat{S}_{z,1}\hat{S}_{z,2} + \hat{S}_{z,1}\hat{S}_{z,2}\hat{S}_1^+\hat{S}_2^- + \hat{S}_{z,1}\hat{S}_{z,2}\hat{S}_1^-\hat{S}_2^+\right] \\
&\quad + \hat{S}_{z,1}\hat{S}_{z,2}\hat{S}_{z,1}\hat{S}_{z,2} \tag{3.73}
\end{aligned}
$$

The different $\hat{S}_1$ and $\hat{S}_2$ operators are again replaced by the sum of the one-electron operators $\hat{s}(1) + \hat{s}(2)$ and $\hat{s}(3) + \hat{s}(4)$ and the effect of the nine operators on the seven different determinants can be evaluated. Using the results summarized in Table 3.3, the effect of the biquadratic exchange operator on the spin functions listed in Eq. 3.69 is easily established:

$$\lambda(\hat{S}_1\hat{S}_2)^2\alpha\alpha\alpha\alpha = \lambda\alpha\alpha\alpha\alpha = \lambda Q \tag{3.74a}$$

$$
\begin{aligned}
\lambda(\hat{S}_1\hat{S}_2)^2\frac{(\alpha\alpha\beta\beta - \beta\beta\alpha\alpha)}{\sqrt{2}} &= \frac{\lambda}{\sqrt{2}}\left[\frac{1}{4}(4\alpha\alpha\beta\beta + 4\beta\beta\alpha\alpha) - \frac{1}{2}\kappa + \alpha\alpha\beta\beta\right. \\
&\quad \left. - \frac{1}{4}(4\alpha\alpha\beta\beta + 4\beta\beta\alpha\alpha) + \frac{1}{2}\kappa - \beta\beta\alpha\alpha\right] = \frac{\lambda}{\sqrt{2}}(\alpha\alpha\beta\beta - \beta\beta\alpha\alpha) = \lambda T
\end{aligned}
$$
$$\tag{3.74b}$$

Table 3.3 Effect of $(\hat{S}_1 \cdot \hat{S}_2)^2$ on the determinants that form the quintet, triplet and singlet spin functions of a binuclear system with $S = 1$

Operator	$\alpha\alpha\alpha\alpha$	$\alpha\alpha\beta\beta$	$\beta\beta\alpha\alpha$	$\alpha\beta\alpha\beta$	$\alpha\beta\beta\alpha$	$\beta\alpha\beta\alpha$	$\beta\alpha\alpha\beta$
$\hat{S}_1^+ \hat{S}_2^- \hat{S}_1^+ \hat{S}_2^-$	0	0	$4\alpha\alpha\beta\beta$	0	0	0	0
$\hat{S}_1^+ \hat{S}_2^- \hat{S}_1^- \hat{S}_2^+$	0	$4\alpha\alpha\beta\beta$	0	κ	κ	κ	κ
$\hat{S}_1^- \hat{S}_2^+ \hat{S}_1^+ \hat{S}_2^-$	0	0	$4\beta\beta\alpha\alpha$	κ	κ	κ	κ
$\hat{S}_1^- \hat{S}_2^+ \hat{S}_1^- \hat{S}_2^+$	0	$4\beta\beta\alpha\alpha$	0	0	0	0	0
$\hat{S}_1^+ \hat{S}_2^- \hat{S}_{z,1}\hat{S}_{z,2}$	0	0	$-\kappa$	0	0	0	0
$\hat{S}_1^- \hat{S}_2^+ \hat{S}_{z,1}\hat{S}_{z,2}$	0	$-\kappa$	0	0	0	0	0
$\hat{S}_{z,1}\hat{S}_{z,2}\hat{S}_1^+ \hat{S}_2^-$	0	0	0	$-\alpha\alpha\beta\beta$	$-\alpha\alpha\beta\beta$	$-\alpha\alpha\beta\beta$	$-\alpha\alpha\beta\beta$
$\hat{S}_{z,1}\hat{S}_{z,2}\hat{S}_1^- \hat{S}_2^+$	0	0	0	$-\beta\beta\alpha\alpha$	$-\beta\beta\alpha\alpha$	$-\beta\beta\alpha\alpha$	$-\beta\beta\alpha\alpha$
$\hat{S}_{z,1}\hat{S}_{z,2}\hat{S}_{z,1}\hat{S}_{z,2}$	$\alpha\alpha\alpha\alpha$	$\alpha\alpha\beta\beta$	$\beta\beta\alpha\alpha$	0	0	0	0

$\kappa = \alpha\beta\alpha\beta + \alpha\beta\beta\alpha + \beta\alpha\beta\alpha + \beta\alpha\alpha\beta$

$$\lambda(\hat{S}_1\hat{S}_2)^2 \frac{1}{2\sqrt{3}}\left(2(\alpha\alpha\beta\beta + \beta\beta\alpha\alpha) - \alpha\beta\alpha\beta - \alpha\beta\beta\alpha - \beta\alpha\alpha\beta - \beta\alpha\beta\alpha\right)$$

$$= \frac{\lambda}{2\sqrt{3}}\left[\frac{1}{4}(8\alpha\alpha\beta\beta + 8\beta\beta\alpha\alpha) - \frac{1}{2}2\kappa + 2\alpha\alpha\beta\beta + \frac{1}{4}(8\alpha\alpha\beta\beta + 8\beta\beta\alpha\alpha)\right.$$

$$\left. - \frac{1}{2}2\kappa + 2\beta\beta\alpha\alpha - 4(\frac{1}{4}2\kappa - \frac{1}{2}\alpha\alpha\beta\beta - \frac{1}{2}\beta\beta\alpha\alpha)\right]$$

$$= \frac{\lambda}{2\sqrt{3}}\left[8(\alpha\alpha\beta\beta + \beta\beta\alpha\alpha) + 4(-\alpha\beta\alpha\beta - \alpha\beta\beta\alpha - \beta\alpha\alpha\beta - \beta\alpha\beta\alpha)\right] = 4\lambda S$$

$$(3.74c)$$

and the eigenvalues of the Heisenberg Hamiltonian extended with a term for the biquadratic exchange are

$$\left(-J\hat{S}_1 \cdot \hat{S}_2 + \lambda(\hat{S}_1 \cdot \hat{S}_2)^2\right)Q = (-J + \lambda)Q \qquad (3.75a)$$

$$\left(-J\hat{S}_1 \cdot \hat{S}_2 + \lambda(\hat{S}_1 \cdot \hat{S}_2)^2\right)T = (J + \lambda)T \qquad (3.75b)$$

$$\left(-J\hat{S}_1 \cdot \hat{S}_2 + \lambda(\hat{S}_1 \cdot \hat{S}_2)^2\right)S = (2J + 4\lambda)S \qquad (3.75c)$$

3.4.2 Four-Center Interactions

Magnetic interactions are not restricted to the exchange of the spin moments on two magnetic centers, but can be extended to the simultaneous interaction of three or four magnetic centers. These interactions are in general smaller than the two-body interactions but can not always be neglected. A clear example is given by the magnetic

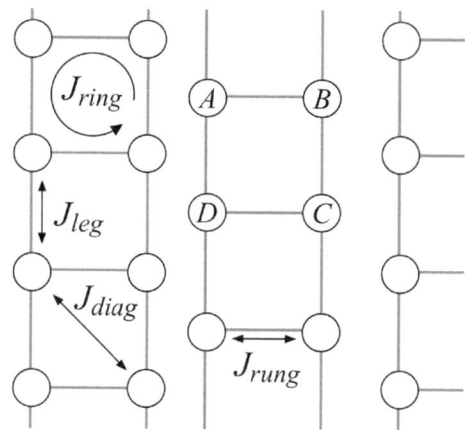

Fig. 3.13 Ladder-like structure formed by the Cu^{2+} ions in $SrCu_2O_3$. Oxygens on the *grey lines* between the copper ions are not shown. J_{leg}, J_{rung} and J_{diag} are the standard two-body interactions, J_{ring} is a four-body interaction that cyclically interchanges the four spins

interactions in the solid state compound $SrCu_2O_3$. This copper oxide has a layered structure, in which Cu_2O_3 layers are separated by Sr^{2+} ions. The Cu ions form a ladder-like structure as shown in Fig. 3.13 with oxygen ions between the magnetic centers. A straightforward fitting of the magnetic susceptibility with just the two-body interactions leads to the conclusion that the magnetic interaction along the legs is twice as large as the interactions along the rungs of the ladder. However, the local geometry does not support such a large difference; distances, angles, coordination are all very similar in both cases. Extending the model Hamiltonian used to fit experimental data with four-body interactions provides a more consistent picture: the interactions along leg and rung are similar and the four-body interaction is sizeable.

To get a hand on the four-body interactions, the four-center cluster ABCD shown in Fig. 3.13 is studied. The four magnetic centers, $A \ldots D$, have one unpaired electron, and therefore, a magnetic moment of $S = 1/2$. The Hamiltonian of this system is a sum of the standard two-body interactions plus $\hat{P}_{1234}$, a four-body operator that cyclically permutes the four spin functions.

$$\hat{H} = \sum_{i<j} -J_{ij}\hat{S}_i \cdot \hat{S}_j + J_r \hat{P}_{1234} \qquad (3.76)$$

To stay within a spin Hamiltonian formalism, the permutation operator has to be replaced by spin operators, which can be done in the following way [19]:

$$\hat{P}_{1234} = \kappa \left((\hat{S}_A \cdot \hat{S}_B)(\hat{S}_C \cdot \hat{S}_D) + (\hat{S}_A \cdot \hat{S}_D)(\hat{S}_B \cdot \hat{S}_C) - (\hat{S}_A \cdot \hat{S}_C)(\hat{S}_B \cdot \hat{S}_D) \right)$$
$$(3.77)$$

To check that this sum indeed cyclically permutes the spin functions, we compare the outcome of acting with $\hat{P}_{1234}$ and acting with the sum of bilinear operators on the wave function $\Psi = \alpha\beta\alpha\beta - \beta\alpha\beta\alpha$.

$$\hat{P}_{1234}(\alpha\beta\alpha\beta - \beta\alpha\beta\alpha) = \beta\alpha\beta\alpha - \alpha\beta\alpha\beta \tag{3.78}$$

Note that the wave function with only one of the terms is not an eigenfunction of the permutation operator $\hat{P}_{1234}$. To determine the result of the sum of four-spin operators, we will develop step-by-step the action of $(\hat{S}_A \cdot \hat{S}_D)(\hat{S}_B \cdot \hat{S}_C)$. The other two terms can be done by the reader as an exercise. In the first place, we need to establish the result of acting with $\hat{S}_i \cdot \hat{S}_j$ on the different two-electron determinants. By writing $\hat{S}$ as $\hat{S}_x + \hat{S}_y + \hat{S}_z$ and using Eq. 1.20a, the following relations are easily derived:

$$\hat{S}_1 \cdot \hat{S}_2 \alpha\alpha = \frac{1}{4}\alpha\alpha \qquad\qquad \hat{S}_1 \cdot \hat{S}_2 \alpha\beta = \frac{1}{2}\beta\alpha - \frac{1}{4}\alpha\beta$$

$$\hat{S}_1 \cdot \hat{S}_2 \beta\beta = \frac{1}{4}\beta\beta \qquad\qquad \hat{S}_1 \cdot \hat{S}_2 \beta\alpha = \frac{1}{2}\alpha\beta - \frac{1}{4}\beta\alpha \tag{3.79}$$

Next, we use these results to determine how $\hat{S}_B \cdot \hat{S}_C$ and $\hat{S}_A \cdot \hat{S}_D$ act on $\alpha\beta\alpha\beta$

$$\hat{S}_B \cdot \hat{S}_C \alpha(1)\beta(2)\alpha(3)\beta(4) = \left[\hat{S}_B \cdot \hat{S}_C \beta(2)\alpha(3)\right]\alpha(1)\beta(4)$$

$$= \left(\frac{1}{2}\alpha(2)\beta(3) - \frac{1}{4}\beta(2)\alpha(3)\right)\alpha(1)\beta(4) = \frac{1}{2}\alpha\alpha\beta\beta - \frac{1}{4}\alpha\beta\alpha\beta \tag{3.80}$$

with $\hat{S}_A \cdot \hat{S}_D \alpha\beta\alpha\beta = \frac{1}{2}\beta\beta\alpha\alpha - \frac{1}{4}\alpha\beta\alpha\beta$ and $\hat{S}_A \cdot \hat{S}_D \alpha\alpha\beta\beta = \frac{1}{2}\beta\alpha\beta\alpha - \frac{1}{4}\alpha\alpha\beta\beta$ the product $(\hat{S}_A \cdot \hat{S}_D)(\hat{S}_B \cdot \hat{S}_C)$ acting on $\alpha\beta\alpha\beta$ gives

$$(\hat{S}_A \cdot \hat{S}_D)(\hat{S}_B \cdot \hat{S}_C)\alpha\beta\alpha\beta = (\hat{S}_A \cdot \hat{S}_D)\left(\frac{1}{2}\alpha\alpha\beta\beta - \frac{1}{4}\alpha\beta\alpha\beta\right)$$

$$= \frac{1}{4}\beta\alpha\beta\alpha - \frac{1}{8}\alpha\alpha\beta\beta - \frac{1}{8}\beta\beta\alpha\alpha + \frac{1}{16}\alpha\beta\alpha\beta \tag{3.81}$$

Repeating this for the other two products of bilinear operators and summing the results of acting on $\beta\alpha\beta\alpha$ as well, we obtain

$$\frac{1}{4}\beta\alpha\beta\alpha - \frac{1}{8}\alpha\beta\beta\alpha - \frac{1}{8}\beta\alpha\alpha\beta + \frac{1}{16}\alpha\beta\alpha\beta - \frac{1}{4}\alpha\beta\alpha\beta + \frac{1}{8}\beta\alpha\alpha\beta$$

$$+ \frac{1}{8}\alpha\beta\beta\alpha - \frac{1}{16}\beta\alpha\beta\alpha + \frac{1}{4}\beta\alpha\beta\alpha - \frac{1}{8}\alpha\alpha\beta\beta - \frac{1}{8}\beta\beta\alpha\alpha + \frac{1}{16}\alpha\beta\alpha\beta$$

$$- \frac{1}{4}\alpha\beta\alpha\beta + \frac{1}{8}\beta\beta\alpha\alpha + \frac{1}{8}\alpha\alpha\beta\beta - \frac{1}{16}\beta\alpha\beta\alpha - \frac{1}{16}\alpha\beta\alpha\beta + \frac{1}{16}\beta\alpha\beta\alpha$$

$$= \frac{7}{16}(\beta\alpha\beta\alpha - \alpha\beta\alpha\beta) \tag{3.82}$$

This shows that, except for a constant that can be absorbed in the interaction constant J_r, the action of the cyclic permutation operator is identical to the linear combina-

Fig. 3.14 The three
different possibilities of the
cyclic permutations of the
spins on a *square* of *four*
magnetic centers

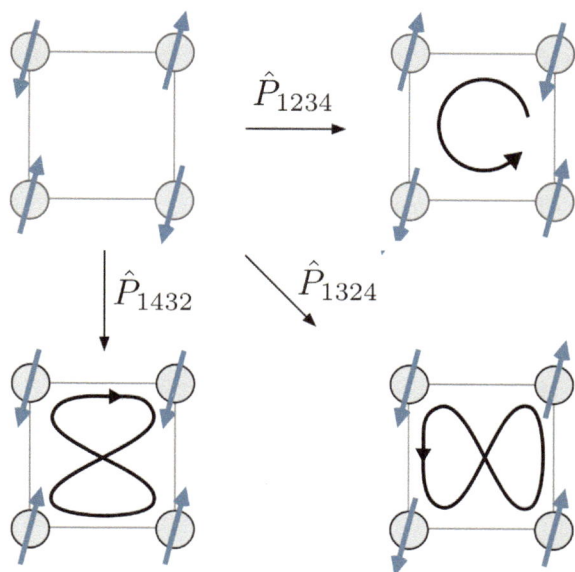

tion of products of bilinear operators. Therefore, we define the Hamiltonian for the
rectangle *ABCD* in Fig. 3.13 as

$$\hat{H} = -J_1(\hat{S}_A \cdot \hat{S}_B + \hat{S}_C \cdot \hat{S}_D) - J_2(\hat{S}_A \cdot \hat{S}_D + \hat{S}_B \cdot \hat{S}_C) - J_3(\hat{S}_A \cdot \hat{S}_C + \hat{S}_B \cdot \hat{S}_D)$$
$$+ J_r\left[(\hat{S}_A \cdot \hat{S}_B)(\hat{S}_C \cdot \hat{S}_D) + (\hat{S}_A \cdot \hat{S}_D)(\hat{S}_B \cdot \hat{S}_C) - (\hat{S}_A \cdot \hat{S}_C)(\hat{S}_B \cdot \hat{S}_D)\right]$$

$$(3.83)$$

where the subscripts 1, 2, 3, and *r* stand for *leg, rung, diag* and *ring*, respectively.
Before looking at the eigenvalues of this Hamiltonian, it should be mentioned that
$\hat{P}_{1234}$ is not the only way to cyclically permute the four spin functions. Alternatively,
one can apply the $\hat{P}_{1324}$ and $\hat{P}_{1423}$ operators to shift them around the rectangle as
illustrated in Fig. 3.14. These possibilities are carefully worked out in Ref. [20],
where the corresponding interaction parameters were shown to be so small that they
will be neglected here for simplicity.

The four unpaired electrons on the rectangle occupy the magnetic orbitals *a, b, c*
and *d*, respectively. They can be coupled to a quintet, three different triplets and two
singlets. A common basis for these six states is given by the six $M_S = 0$ determinants
$|a\bar{b}c\bar{d}|$, $|\bar{a}bc\bar{d}|$, $|a\bar{b}\bar{c}d|$, $|\bar{a}b\bar{c}d|$, $|ab\bar{c}\bar{d}|$ and $|\bar{a}\bar{b}cd|$. In the following, we will omit
the spatial part and return to a spin-only notation, $|\alpha\beta\alpha\beta|$, $|\beta\alpha\beta\alpha|$, etc.

3.10 Couple the spins of the four centers in a sequential fashion in all possible ways to check the existence of one quintet, three different triplets and two singlets for a system with four $S = 1/2$ magnetic moments.

The matrix representation of $\hat{H}$ is

	$\lvert\alpha\beta\alpha\beta\rangle$	$\lvert\beta\alpha\beta\alpha\rangle$	$\lvert\alpha\alpha\beta\beta\rangle$	$\lvert\beta\beta\alpha\alpha\rangle$	$\lvert\alpha\beta\beta\alpha\rangle$	$\lvert\beta\alpha\alpha\beta\rangle$
$\langle\alpha\beta\alpha\beta\rvert$	H_{11}					
$\langle\beta\alpha\beta\alpha\rvert$	$-\frac{1}{2}J_r$	H_{22}				
$\langle\alpha\alpha\beta\beta\rvert$	$-\frac{1}{2}J_1+\frac{1}{8}J_r$	$-\frac{1}{2}J_1+\frac{1}{8}J_r$	H_{33}			
$\langle\beta\beta\alpha\alpha\rvert$	$-\frac{1}{2}J_1+\frac{1}{8}J_r$	$-\frac{1}{2}J_1+\frac{1}{8}J_r$	0	H_{44}		
$\langle\alpha\beta\beta\alpha\rvert$	$-\frac{1}{2}J_2+\frac{1}{8}J_r$	$-\frac{1}{2}J_2+\frac{1}{8}J_r$	$-\frac{1}{2}J_3+\frac{1}{8}J_r$	$-\frac{1}{2}J_3+\frac{1}{8}J_r$	H_{55}	
$\langle\beta\alpha\alpha\beta\rvert$	$-\frac{1}{2}J_2+\frac{1}{8}J_r$	$-\frac{1}{2}J_2+\frac{1}{8}J_r$	$-\frac{1}{2}J_3+\frac{1}{8}J_r$	$-\frac{1}{2}J_3+\frac{1}{8}J_r$	0	H_{66}

with

$$
H_{11} = H_{22} = \frac{1}{2}(J_1 + J_2 - J_3) + \frac{1}{16}J_r
$$
$$
H_{33} = H_{44} = \frac{1}{2}(J_1 - J_2 + J_3) + \frac{1}{16}J_r
$$
$$
H_{55} = H_{66} = \frac{1}{2}(-J_1 + J_2 + J_3) + \frac{1}{16}J_r \tag{3.84}
$$

The diagonalization of this matrix should in principle give the necessary relations to extract the bilinear exchange parameters and the strength of the four-center interaction. There are five energy differences and only four parameters to be determined. However, the resulting equations are rather awkward and it is easier to extract the parameters by constructing a numerical effective Hamiltonian with the extra advantage that the assumption of very small contribution from the other type of permutations can be checked. For a square complex with $J_1 = J_2 = J$, the equations for the energies of the spin states are significantly more simple, giving

$$
E(Q) = 0 \tag{3.85}
$$
$$
E(T2) = E(T3) = J + J_3 \tag{3.86}
$$
$$
E(S2) = J + 2J_3 - \frac{1}{4}J_r \tag{3.87}
$$
$$
E(T1) = 2J - \frac{1}{2}J_r \tag{3.88}
$$
$$
E(S1) = 3J + \frac{3}{4}J_r \tag{3.89}
$$

3.11 Extract the magnetic coupling parameters for a four-center Cu^{2+} complex with a square geometry. (i) Under the assumption of equal coupling along the edges of the square, zero coupling along the diagonal and no four-center interactions; (ii) with a non-negligible ring exchange ($J_1 = J_2$; $J_r \neq 0$ and $J_3 = 0$); (iii) considering the three different interactions. The following total energies for the spin states were calculated: $E(Q) = -3953.38577312$ E_h; $E(T2) = E(T3) = -3953.39054141$ E_h; $E(S2) = -3953.39100763$ E_h; $E(T1) = -3953.39533075$ E_h; $E(S1) = -3953.39867933$ E_h. Are the estimates of J the same in the first case when extracted from different ΔE's?

3.4.3 Anisotropic Exchange

In Sect. 3.2 we have introduced the general expression (Eq. 3.20) to describe the interaction between two spin moments on different magnetic centers. So far, only the isotropic interactions have been considered in this chapter; the total spin moment (and the single-ion spin) in itself has no preferred orientation in space, only the relative orientation—parallel or antiparallel—of the local spins has been looked at. This is of course only part of the story. Due to relativistic effects, in many systems the spin moment is anisotropic as seen in the previous chapter for mononuclear complexes. The magnetic anisotropy is in some compounds accompanied by ferroelectricity. These so-called *multiferroic* compounds, often perovskite transition metal oxides, have potential applications as switches, sensors or memory devices. Coming back to Eq. 3.20, we will separate isotropic and anisotropic interactions before orienting the molecule in such a way that the magnetic frame coincides with the cartesian axes frame. Then, the Hamiltonian becomes

$$\hat{H} = -J\hat{S}_1 \cdot \hat{S}_2 + \hat{S}_1\overline{\overline{A}}\hat{S}_2 \tag{3.90}$$

As long as we are concerned with binuclear $S = 1/2$ complexes, no single-ion anisotropy has to be added and this Hamiltonian describes the lowest energy levels in the absence of an external magnetic field.

Symmetric anisotropy: The basis of this Hamiltonian can no longer be restricted to determinants with the same M_S value as was done for the isotropic interactions. The inclusion of magnetic anisotropy in the model causes the removal of the degeneracy of the different M_S levels and eventually mixing of the wave functions with different spin moment. Here, we have to consider four CSFs; the three components of the triplet plus the singlet. To facilitate the determination of the matrix elements of the model

Hamiltonian, it is common practice to consider the basis of *uncoupled* determinants and then transform to the basis of spin-adapted CSFs.

3.12 Perform the matrix multiplication of the anisotropic term in the model Hamiltonian.

The uncoupled basis is formed by the determinants $|\alpha\alpha|$, $|\alpha\beta|$, $|\beta\alpha|$ and $|\beta\beta|$. The result of acting with the isotropic part of Hamiltonian on these determinants can directly be written down with the help of Eqs. 3.79, but the anisotropic part requires a little more work. Based on the relations given in Eqs. 1.16a and 1.20a, the following is easily derived for the products of one-electron operators

- $\hat{S}_1 \overset{=}{A} \hat{S}_2 |\alpha\alpha|$

$$A_{xx}\hat{S}_x\hat{S}_x\alpha\alpha = \frac{1}{4}A_{xx}\beta\beta \qquad A_{xy}\hat{S}_x\hat{S}_y\alpha\alpha = -\frac{1}{4i}A_{xy}\beta\beta \qquad A_{xz}\hat{S}_x\hat{S}_z\alpha\alpha = \frac{1}{4}A_{xz}\beta\alpha$$

$$A_{yx}\hat{S}_y\hat{S}_x\alpha\alpha = -\frac{1}{4i}A_{yx}\beta\beta \qquad A_{yy}\hat{S}_y\hat{S}_y\alpha\alpha = -\frac{1}{4}A_{yy}\beta\beta \qquad A_{yz}\hat{S}_y\hat{S}_z\alpha\alpha = -\frac{1}{4i}A_{yz}\beta\alpha$$

$$A_{zx}\hat{S}_z\hat{S}_x\alpha\alpha = \frac{1}{4}A_{zx}\alpha\beta \qquad A_{zy}\hat{S}_z\hat{S}_y\alpha\alpha = -\frac{1}{4i}A_{zy}\alpha\beta \qquad A_{zz}\hat{S}_z\hat{S}_z\alpha\alpha = \frac{1}{4}A_{zz}\alpha\alpha$$

$$\text{(3.91a)}$$

- $\hat{S}_1 \overset{=}{A} \hat{S}_2 |\beta\beta|$

$$A_{xx}\hat{S}_x\hat{S}_x\beta\beta = \frac{1}{4}A_{xx}\alpha\alpha \qquad A_{xy}\hat{S}_x\hat{S}_y\beta\beta = \frac{1}{4i}A_{xy}\alpha\alpha \qquad A_{xz}\hat{S}_x\hat{S}_z\beta\beta = -\frac{1}{4}A_{xz}\alpha\beta$$

$$A_{yx}\hat{S}_y\hat{S}_x\beta\beta = \frac{1}{4i}A_{yx}\alpha\alpha \qquad A_{yy}\hat{S}_y\hat{S}_y\beta\beta = -\frac{1}{4}A_{yy}\alpha\alpha \qquad A_{yz}\hat{S}_y\hat{S}_z\beta\beta = -\frac{1}{4i}A_{yz}\alpha\beta$$

$$A_{zx}\hat{S}_z\hat{S}_x\beta\beta = -\frac{1}{4}A_{zx}\beta\alpha \qquad A_{zy}\hat{S}_z\hat{S}_y\beta\beta = -\frac{1}{4i}A_{zy}\beta\alpha \qquad A_{zz}\hat{S}_z\hat{S}_z\beta\beta = \frac{1}{4}A_{zz}\beta\beta$$

$$\text{(3.91b)}$$

- $\hat{S}_1 \overset{=}{A} \hat{S}_2 |\alpha\beta|$

$$A_{xx}\hat{S}_x\hat{S}_x\alpha\beta = \frac{1}{4}A_{xx}\beta\alpha \qquad A_{xy}\hat{S}_x\hat{S}_y\alpha\beta = \frac{1}{4i}A_{xy}\beta\alpha \qquad A_{xz}\hat{S}_x\hat{S}_z\alpha\beta = -\frac{1}{4}A_{xz}\beta\beta$$

$$A_{yx}\hat{S}_y\hat{S}_x\alpha\beta = -\frac{1}{4i}A_{yx}\beta\alpha \qquad A_{yy}\hat{S}_y\hat{S}_y\alpha\beta = \frac{1}{4}A_{yy}\beta\alpha \qquad A_{yz}\hat{S}_y\hat{S}_z\alpha\beta = \frac{1}{4i}A_{yz}\beta\beta$$

$$A_{zx}\hat{S}_z\hat{S}_x\alpha\beta = \frac{1}{4}A_{zx}\alpha\alpha \qquad A_{zy}\hat{S}_z\hat{S}_y\alpha\beta = \frac{1}{4i}A_{zy}\alpha\alpha \qquad A_{zz}\hat{S}_z\hat{S}_z\alpha\beta = -\frac{1}{4}A_{zz}\alpha\beta$$

$$\text{(3.91c)}$$

- $\hat{S}_1\overline{\overline{A}}\hat{S}_2|\beta\alpha|$

$$A_{xx}\hat{S}_x\hat{S}_x\beta\alpha = \frac{1}{4}A_{xx}\alpha\beta \qquad A_{xy}\hat{S}_x\hat{S}_y\beta\alpha = -\frac{1}{4i}A_{xy}\alpha\beta \qquad A_{xz}\hat{S}_x\hat{S}_z\beta\alpha = \frac{1}{4}A_{xz}\alpha\alpha$$

$$A_{yx}\hat{S}_y\hat{S}_x\beta\alpha = \frac{1}{4i}A_{yx}\alpha\beta \qquad A_{yy}\hat{S}_y\hat{S}_y\beta\alpha = \frac{1}{4}A_{yy}\alpha\beta \qquad A_{yz}\hat{S}_y\hat{S}_z\beta\alpha = \frac{1}{4i}A_{yz}\alpha\alpha$$

$$A_{zx}\hat{S}_z\hat{S}_x\beta\alpha = -\frac{1}{4}A_{zx}\beta\beta \qquad A_{zy}\hat{S}_z\hat{S}_y\beta\alpha = \frac{1}{4i}A_{zy}\beta\beta \qquad A_{zz}\hat{S}_z\hat{S}_z\beta\alpha = -\frac{1}{4}A_{zz}\beta\alpha$$

$$\tag{3.91d}$$

Following common practice, we write the anisotropic interaction as the sum of symmetric

$$D_{ij} = D_{ji} = \frac{1}{2}(A_{ij} + A_{ji}) \tag{3.92a}$$

and antisymmetric contributions

$$d_{ij} = -d_{ji} = \frac{1}{2}(A_{ij} - A_{ji}) \tag{3.92b}$$

For the moment we neglect the antisymmetric interaction and write down the matrix representation of the Hamiltonian as sum of isotropic and symmetric anisotropic interactions.

	$|\alpha\alpha\rangle$	$|\alpha\beta\rangle$	$|\beta\alpha\rangle$	$|\beta\beta\rangle$
$\langle\alpha\alpha|$	$-\frac{1}{4}(J + D_{zz})$	$\frac{1}{4}(D_{xz} - iD_{yz})$	$\frac{1}{4}(D_{xz} - iD_{yz})$	$\frac{1}{4}(D_{xx} - D_{yy} - 2iD_{xy})$
$\langle\alpha\beta|$	$\frac{1}{4}(D_{xz} + iD_{yz})$	$\frac{1}{4}(J + D_{zz})$	$-\frac{1}{2}J + \frac{1}{4}(D_{xx} + D_{yy})$	$-\frac{1}{4}(D_{xz} - iD_{yz})$
$\langle\beta\alpha|$	$\frac{1}{4}(D_{xz} + iD_{yz})$	$-\frac{1}{2}J + \frac{1}{4}(D_{xx} + D_{yy})$	$\frac{1}{4}(J + D_{zz})$	$-\frac{1}{4}(D_{xz} - iD_{yz})$
$\langle\beta\beta|$	$\frac{1}{4}(D_{xx} - D_{yy} + 2iD_{xy})$	$-\frac{1}{4}(D_{xz} + iD_{yz})$	$-\frac{1}{4}(D_{xz} + iD_{yz})$	$-\frac{1}{4}(J + D_{zz})$

The next step is the transformation from the uncoupled basis to a basis in which the two spin moments are coupled, i.e. a basis of the singlet and the three components of the triplet.

	$|T^+\rangle$	$|T^0\rangle$	$|T^-\rangle$	$|S\rangle$
$\langle T^+|$	$-\frac{1}{4}(J - D_{zz})$	$\frac{1}{2\sqrt{2}}(D_{xz} - iD_{yz})$	$\frac{1}{4}(D_{xx} - D_{yy} - 2iD_{yz})$	0
$\langle T^0|$	$\frac{1}{2\sqrt{2}}(D_{xz} + iD_{yz})$	$-\frac{1}{4}(J + 2D_{zz})$	$-\frac{1}{2\sqrt{2}}(D_{xz} - iD_{yz})$	0
$\langle T^-|$	$\frac{1}{4}(D_{xx} - D_{yy} + 2iD_{yz})$	$-\frac{1}{2\sqrt{2}}(D_{xz} + iD_{yz})$	$-\frac{1}{4}(J - D_{zz})$	0
$\langle S|$	0	0	0	$\frac{3}{4}J$

where the diagonal elements are simplified by the notion that $\overline{\overline{D}}$ can be written as a traceless tensor, that is $D_{xx} + D_{yy} + D_{zz} = 0$. For example,

$$
\begin{aligned}
\langle \alpha\beta + \beta\alpha | \hat{H} | \alpha\beta + \beta\alpha \rangle &= -\frac{1}{4}(J + D_{zz}) + 2\left(\frac{1}{2}J + \frac{1}{4}(D_{xx} + D_{yy})\right) \\
&- \frac{1}{4}(J + D_{zz}) = \frac{1}{4}J - \frac{1}{4}D_{zz} + \frac{1}{4}D_{xx} + \frac{1}{4}D_{yy}
\end{aligned}
\tag{3.93}
$$

which is simplified to $\frac{1}{4}(J - 2D_{zz})$ by subtracting $\frac{1}{4}(D_{xx} + D_{yy} + D_{zz})$, which equals zero.

3.13 (a) Show that transformation of the matrix representation in the uncoupled basis into the coupled basis can be done by applying the unitary transformation $\tilde{U}\hat{H}U$, where $\tilde{U}$ is the transpose of $U = \left(1, 0, 0, 0\,;\, 0, 1/\sqrt{2}, 0, 1/\sqrt{2}\,;\, 0, 1/\sqrt{2}, 0, -1/\sqrt{2}\,;\, 0, 0, 1, 0\right)$ (b) Show that the Hamiltonian of Eq. 3.90 is hermitian. Assume a diagonal D-tensor and show that the triplet part of the matrix is related to the D-tensor of an $S = 1$ mononuclear complex (Eq. 2.21) by a factor of $\frac{1}{2}$. Hint: the trace of the two matrices can be adjusted to simplify the comparison.

The construction of a numerical effective Hamiltonian from accurate electronic structure calculations permits us to determine the complete D-tensor and therewith the orientation of the magnetic axes frame of the system with its easy axis or easy plane, depending on the relative energies of the different M_S components of the triplet. When the magnetic axes frame coincides with the cartesian axes frame, $\overline{\overline{D}}$ is diagonal and the energy levels of the triplet can be described with two parameters; the axial anisotropy D and the rhombic anisotropy E as defined in Eq. 2.16. Hence, the symmetric anisotropic interaction of the $S = 1/2$ spin moments, which by themselves are isotropic by definition, makes that the total spin moment of the system is no longer fully isotropic.

Anti-symmetric anisotropy: The second ingredient of the anisotropic interaction is the asymmetric part, also known as the Dzyaloshinskii–Moriya (DM) interaction. It is held responsible for the appearance of ferromagnetism in antiferromagnetically coupled Cu^{2+} systems. Whereas the isotropic and symmetric anisotropic interactions do not affect the collinearity of the two local magnetic axes frames, the anti-symmetric interaction makes that the principal axis of the local moments are no longer parallel. In a pictorial description of the effect, shown in Fig. 3.15, the cancellation of antiferromagnetically coupled spin moments is no longer complete and a (small) ferromagnetic moment appears.

A rigorous description of the anti-symmetric interaction is obtained by including the d_{ij} in the matrix elements among the four determinants that span the model space.

Fig. 3.15 Schematic representation of the net ferromagnetic interaction due to non-collinear antiferromagnetically coupled spin moments

As example we construct two matrix elements to illustrate the difference with the matrix elements when only the symmetric interaction is considered.

$$\langle\alpha\alpha|\hat{H}|\alpha\beta\rangle = \frac{1}{4}A_{zx} + \frac{1}{4i}A_{zy} = \frac{1}{4}(D_{xz} + d_{zx}) + \frac{1}{4i}(D_{yz} + d_{zy})$$

$$= \frac{1}{4}(D_{xz} - d_{xz}) - \frac{1}{4}i(D_{yz} - d_{yz}) \tag{3.94a}$$

$$\langle\alpha\alpha|\hat{H}|\beta\alpha\rangle = \frac{1}{4}A_{xz} + \frac{1}{4i}A_{yz} = \frac{1}{4}(D_{xz} + d_{xz}) - \frac{1}{4}i(D_{yz} + d_{yz}) \tag{3.94b}$$

3.14 Use Eq. 3.92 to express A_{ij} and A_{ji} in terms of D_{ij} and d_{ij}.

The complete matrix representation of the Hamiltonian with isotropic and (anti-) symmetric anisotropic interactions in the uncoupled basis is directly obtained from the operations listed in Eq. 3.91 and using the definitions of D_{ij} in d_{ij} in Eq. 3.92

	$\|\alpha\alpha\rangle$	$\|\alpha\beta\rangle$	$\|\beta\alpha\rangle$	$\|\beta\beta\rangle$
$\langle\alpha\alpha\|$	$-\frac{1}{4}(J - D_{zz})$	$\frac{1}{4}(D_{xz} - d_{xz}$ $-i(D_{yz} - d_{yz}))$	$\frac{1}{4}(D_{xz} + d_{xz}$ $-i(D_{yz} + d_{yz}))$	$\frac{1}{4}(D_{xx} - D_{yy}$ $-2iD_{xy})$
$\langle\alpha\beta\|$	$\frac{1}{4}(D_{xz} - d_{xz}$ $+i(D_{yz} - d_{yz}))$	$\frac{1}{4}(J - D_{zz})$	$-\frac{1}{2}J + \frac{1}{4}(D_{xx}$ $+D_{yy} + 2id_{xy})$	$-\frac{1}{4}(D_{xz} + d_{xz}$ $-i(D_{yz} + d_{yz}))$
$\langle\beta\alpha\|$	$\frac{1}{4}(D_{xz} + d_{xz}$ $+i(D_{yz} + d_{yz}))$	$-\frac{1}{2}J + \frac{1}{4}(D_{xx}$ $+D_{yy} - 2id_{xy})$	$\frac{1}{4}(J - D_{zz})$	$-\frac{1}{4}(D_{xz} - d_{xz}$ $-i(D_{yz} - d_{yz}))$
$\langle\beta\beta\|$	$\frac{1}{4}(D_{xx} - D_{yy}$ $+2iD_{xy})$	$-\frac{1}{4}(D_{xz} + d_{xz}$ $+i(D_{yz} + d_{yz}))$	$-\frac{1}{4}(D_{xz} - d_{xz}$ $+i(D_{yz} - d_{yz}))$	$-\frac{1}{4}(J - D_{zz})$

and transformed to the coupled basis, the following Hamiltonian is obtained.

	$\lvert T^+\rangle$	$\lvert T^0\rangle$	$\lvert T^-\rangle$	$\lvert S\rangle$
$\langle T^+\rvert$	$-\frac{1}{4}(J - D_{zz})$	$\frac{1}{2\sqrt{2}}(D_{xz} - iD_{yz})$	$\frac{1}{4}(D_{xx} - D_{yy}$ $+2iD_{yz})$	$-\frac{1}{2\sqrt{2}}(d_{xz} - id_{yz})$
$\langle T^0\rvert$	$\frac{1}{2\sqrt{2}}(D_{xz} + iD_{yz})$	$-\frac{1}{4}(J - D_{xx}$ $-D_{yy} + D_{zz})$	$-\frac{1}{2\sqrt{2}}(D_{xz} - iD_{yz})$	$-\frac{1}{2}id_{xy}$
$\langle T^-\rvert$	$\frac{1}{4}(D_{xx} - D_{yy}$ $+2iD_{yz})$	$-\frac{1}{2\sqrt{2}}(D_{xz} + iD_{yz})$	$-\frac{1}{4}(J - D_{zz})$	$-\frac{1}{2\sqrt{2}}(d_{xz} + id_{yz})$
$\langle S\rvert$	$-\frac{1}{2\sqrt{2}}(d_{xz} + id_{yz})$	$\frac{1}{2}id_{xy}$	$-\frac{1}{2\sqrt{2}}(d_{xz} - id_{yz})$	$\frac{3}{4}J - \frac{1}{4}(D_{xx}$ $+D_{yy} + D_{zz})$

The triplet block and the diagonal elements are exactly the same as in the Hamiltonian that only considers the symmetric part of the anisotropic interaction. The anti-symmetric interaction introduces non-zero matrix elements for the coupling between singlet and triplet and causes a mixing between both spin states. The total spin quantum number is (at least formally) no longer a good quantum number. The number of parameters is now larger than the number of energy differences, even when the system is oriented in the coordinate frame that diagonalizes $\overline{\overline{D}}$. Therefore, a complete determination of the six parameters—J, D, E, d_{xy}, d_{xz} and d_{yz}—necessarily goes through the construction of a numerical effective Hamiltonian.

To close this section, we rewrite the Hamiltonian in the form that is most often used in the literature. The A-tensor in Eq. 3.90 is separated in a symmetric and anti-symmetric part.

$$\hat{H} = -J\hat{S}_1 \cdot \hat{S}_2 + \hat{S}_1\overline{\overline{D}}\hat{S}_2 + \hat{S}_1\overline{\overline{d}}\hat{S}_2 \tag{3.95}$$

where $\overline{\overline{D}}$ is diagonal if the orientation is chosen conveniently, and $\overline{\overline{d}}$ always has the following structure

$$\overline{\overline{d}} = \begin{pmatrix} 0 & d_{12} & -d_{13} \\ -d_{12} & 0 & d_{23} \\ d_{13} & -d_{23} & 0 \end{pmatrix} \tag{3.96}$$

This suggest that a shorter notation can be used by writing $\overline{\overline{d}}$ as a pseudovector $\mathbf{d} = (d_x, d_y, d_z)$ with $d_x = d_{23}$; $d_y = -d_{13}$ and $d_z = d_{12}$. The Hamiltonian then reads

$$\hat{H} = -J\hat{S}_1 \cdot \hat{S}_2 + \hat{S}_1\overline{\overline{D}}\hat{S}_2 + \mathbf{d}\hat{S}_1 \times \hat{S}_2 \tag{3.97}$$

Now, it also becomes clear that the DM interaction can only be non-zero when the local principal magnetic axis are not parallel. The situation becomes slightly more complicated when magnetic centers are considered with more than one unpaired electron. Then the single-ion anisotropy discussed in Chap. 2 has to be included in the model

$$\hat{H} = -J\hat{S}_1 \cdot \hat{S}_2 + \hat{S}_1\overline{\overline{D}}_1\hat{S}_1 + \hat{S}_2\overline{\overline{D}}_2\hat{S}_2 + \hat{S}_1\overline{\overline{D}}_{12}\hat{S}_2 + \mathbf{d}\hat{S}_1 \times \hat{S}_2 \qquad (3.98)$$

and it has been shown that even biquadratic anisotropic interactions can play an important role in the description of the low-energy physics of the complex [21]. The corresponding operator is

$$\hat{\kappa} = (\hat{S}_1\hat{S}_1)\mathbf{D}_{aabb}(\hat{S}_2\hat{S}_2) \qquad (3.99)$$

where $\mathbf{D}_{aabb}$ is tensor of rank 4 with 81 (3^4) parameters. However by choosing the proper magnetic axes frame this number is strongly reduced and when the system has a certain degree of symmetry one can eventually characterize the tensor with not more than nine parameters. Again one can resort to the numerical effective Hamiltonian to determine these parameters.

Problems

3.1 Overlap: Demonstrate that c_1/c_2 in Eq. 3.7 is equal to $1 - S_{ab}/1 + S_{ab}$, where $S_{ab} = \langle\phi_a|\phi_b\rangle$ and ϕ_a and ϕ_b are the orbitals of Eq. 3.17.

3.2 From delocalized to localized: Transform the following determinants and CSFs from a delocalized to a localized orbital basis. Determine the percentage of ionic and neutral character of the wave function. Are the wave functions eigenfunctions of $\hat{S}^2$?

a. $\Phi_1 = |g_1\bar{g}_1|; \Phi_2 = |g_1g_2|; \Phi_3 = |g_1\bar{u}_1|$
b. $\Psi_1 = (|g_1\bar{g}_1| + |u_1\bar{u}_1|)/\sqrt{2}; \Psi_2 = (|g_1\bar{g}_1| - |u_1\bar{u}_1|)/\sqrt{2}$
c. $\Phi_4 = |g_1u_1|; \Phi_5 = |g_1u_1v_1|$
d. $\Psi_3 = (2|g_1u_1\bar{v}_1| - |g_1\bar{u}_1v_1| - |\bar{g}_1u_1v_1|)/\sqrt{6}$

with $g_i = \frac{1}{\sqrt{2}}(a_i + b_i); u_i = \frac{1}{\sqrt{2}}(a_i - b_i); v_i = c_i. a_i, b_i$ and c_i are orbitals localized on centers A, B and C, respectively.

3.3 Singlet and triplet eigenvalues: Calculate the eigenvalues of the Heisenberg Hamiltonian given in Eq. 3.31 of $\Phi(T) = |\alpha\alpha|$ and $\Phi(S) = (|\alpha\beta| - |\beta\alpha|)/\sqrt{2}$.

3.4 Extracting J-values for a three-center system: The following wave functions Ψ_k were obtained from an *ab initio* calculation on a system with three $S = 1/2$ magnetic centers. Each magnetic orbital ϕ_i is localized on center i and has the same spatial part in all five wave functions.

	Ψ_1	Ψ_2	Ψ_3	Ψ_4	Ψ_5
$|\phi_1\phi_2\bar{\phi}_3|$	−0.4426	−0.6583	0.5774	−0.1465	0.1135
$|\phi_1\bar{\phi}_2\phi_3|$	0.7706	−0.0661	0.5774	0.0367	−0.2476
$|\bar{\phi}_1\phi_2\phi_3|$	−0.3280	0.7243	0.5774	0.1098	0.1341
$|\phi_1\bar{\phi}_1\phi_2|$	0.0102	0.0234	0.0000	−0.0440	0.0017
$|\phi_1\bar{\phi}_1\phi_3|$	−0.0725	−0.0495	0.0000	0.1244	0.0341
$|\phi_1\phi_2\bar{\phi}_2|$	0.2243	−0.1120	0.0000	0.7653	−0.5685
$|\phi_2\bar{\phi}_2\phi_3|$	0.2017	0.1336	0.0000	−0.5805	−0.7636
$|\phi_1\phi_3\bar{\phi}_3|$	−0.0789	0.0407	0.0000	−0.1472	0.0147
$|\phi_2\phi_3\bar{\phi}_3|$	0.0076	0.0508	0.0000	0.0579	0.0127

The energies (in E_h) are $E_1 = -27.9611962$, $E_2 = -27.9601927$, $E_3 = -27.9596947$, $E_4 = -27.8326257$, $E_5 = -27.83169141$.

a. Determine the M_S quantum numbers of the determinants and identify Ψ_3 as a spin eigenfunction with $S = 3/2$.
b. Extract the J-values from the energies of the lowest three states under the assumption that $J_{12} = J_{23} \neq J_{13}$ (see Eq. 3.44).
c. Write down the determinants that span the model space of the Heisenberg Hamiltonian and determine the norm of the projections of Ψ_k on this model space.
d. Select the three roots with the largest norm and orthogonalize the projections $\tilde{\Psi}_k$
e. Construct the 3×3 effective Hamiltonian and extract the different J-values by comparing with the matrix elements of the Heisenberg Hamiltonian given in Eq. 3.39.

3.5 Heisenberg twice. (a) Use the eigenvalues of Q, T and S for $\hat{H} = -J\hat{S}_1 \cdot \hat{S}_2$ to compute the eigenvalues of Q, T and S for the operator $\hat{S}_1 \cdot \hat{S}_2$. (b) From this, compute the eigenvalues of Q, T and S for the biquadratic operator $(\hat{S}_1 \cdot \hat{S}_2)^2$ and check the validity of Eq. 3.75.

3.6 Biquadratic interactions: Do the following total energies follow the regular spacing predicted by the Heisenberg Hamiltonian? $E_Q = -139.48992180\,E_h$, $E_T = -139.49305142\,E_h$ and $E_S = -139.49443101\,E_h$. Calculate J and λ (in meV) from the energy differences.

References

1. W. Heitler, F. London, Z. Phys. **44**, 455 (1927)
2. B. Bleaney, K.D. Bowers, Proc. R. Soc. Lond. Ser. A **214**, 451 (1952)
3. R. Boča, *Theoretical Foundations of Molecular Magnetism* (Elsevier, Amsterdam, 1999)
4. H. Bethe, Z. Phys. **71**, 205 (1931)
5. J.C. Bonner, M.E. Fisher, Phys. Rev. **135**(3A), A640 (1964)
6. J.W. Hall, W.E. Marsh, R.R. Weller, W.E. Hatfield, Inorg. Chem. **20**, 1033 (1981)
7. S. Eggert, I. Affleck, M. Takahashi, Phys. Rev. Lett. **73**(2), 332 (1994)
8. R. Georges, J.J. Borrás-Almenar, E. Coronado, J. Curély, M. Drillon, in *Magnetism: Molecules to Materials*, ed. by J.S. Miller, M. Drillon (Wiley-VCH, Weinheim, 2001), pp. 1–47, chap. 1

9. G.S. Rushbrooke, P.J. Wood, Mol. Phys. **1**, 257 (1958)
10. J. Curély, F. Lloret, M. Julve, Phys. Rev. B **58**, 11465 (1998)
11. J. Curély, J. Rouch, Physica B **254**, 298 (1998)
12. M. Deumal, M.J. Bearpark, J.J. Novoa, M.A. Robb, J. Phys. Chem. A **106**, 1299 (2002)
13. O. Kahn, Y. Pei, M. Verdaguer, J.P. Renard, J. Sletten, J. Am. Chem. Soc. **110**, 782 (1988)
14. E. Ruiz, A. Rodríguez-Fortea, J. Cano, S. Alvarez, J. Phys. Chem. Solids **65**, 799 (2004)
15. H.J. Maris, L.P. Kadanoff, Am. J. Phys. **46**, 652 (1978)
16. D. Chandler, *Introduction to Modern Statistical Mechanics* (Oxford University Press, Oxford, 1987)
17. J. Cano, R. Costa, S. Alvarez, E. Ruiz, J. Chem. Theory Comput. **3**, 782 (2007)
18. W. Schattke, R. Díez Muiño, *Quantum Monte Carlo Programming* (Wiley-VCH, Weinheim, 2013)
19. M. Roger, J.H. Hetherington, J.M. Delrieu, Rev. Mod. Phys. **55**, 1 (1983)
20. C.J. Calzado, C. de Graaf, E. Bordas, R. Caballol, J.P. Malrieu, Phys. Rev. B **67**, 132409 (2003)
21. R. Maurice, N. Guihéry, R. Bastardis, C. de Graaf, J. Chem. Theory Comput. **6**, 55 (2010)

Chapter 4
From Orbital Models to Accurate Predictions

Abstract Basic understanding and qualitative prediction of the isotropic magnetic coupling between two magnetic centers can be obtained with two well-established valence-only models. This chapter discusses the Kahn–Briat and Hay–Thibeault–Hoffmann models, which have been (and still are) of fundamental importance for understanding the basics of magnetism in polynuclear transition metal complexes. After shortly presenting the basic model for magnetism in organic radicals, we review the most evident magnetostructural relations and then move to the accurate prediction of the magnetic coupling. An overview of the most widely used quantum chemical methods is given, including wave function based methods and approaches within the spin-unrestricted setting such as density functional theory. The last part of the chapter is dedicated to the calculation of the interactions beyond the isotropic magnetic coupling.

4.1 Qualitative Valence-Only Models

The simplest electronic structure models for magnetic interactions only consider the unpaired electrons and their orbitals. All other electrons are taken as inactive and not included in the description. This leads to very simple wave functions, especially in the case of two identical $S = \frac{1}{2}$ magnetic centers. Such valence-only models, where *valence* is not used in its usual chemical context, are numerically not competitive with large-scale all-electron calculations, but have provided chemists and other scientists working in the field with important insights to control the magnetic interactions in transition metal complexes and materials with organic radicals.

4.1.1 The Kahn–Briat Model

Based on valence bond reasoning with nonorthogonal atomic-like orbitals, Kahn and Briat derived an elegant model that is capable of explaining and predicting magnetic behavior of transition metal complexes based on the shape of the localized magnetic

© Springer International Publishing Switzerland 2016   105
C. Graaf and R. Broer, *Magnetic Interactions in Molecules and Solids*,
Theoretical Chemistry and Computational Modelling,
DOI 10.1007/978-3-319-22951-5_4

orbitals [1]. Let ϕ_a and ϕ_b be the optimal local orbitals for the unpaired electrons on site A and B. These orbitals are normalized but not orthogonal

$$\langle \phi_a | \phi_b \rangle = S \qquad \langle \phi_a | \phi_a \rangle = \langle \phi_b | \phi_b \rangle = 1 \tag{4.1}$$

Multiplying the spatial part of the wave function $|\phi_a \phi_b| = |ab|$ with the singlet and triplet ($M_S = 0$) spin functions, the following normalized wave functions are obtained

$$\Psi_S = \frac{|a\bar{b}| + |b\bar{a}|}{\sqrt{2 + 2S^2}} \qquad \Psi_T = \frac{|a\bar{b}| - |b\bar{a}|}{\sqrt{2 - 2S^2}} \tag{4.2}$$

4.1 Confirm that the norms of Ψ_S and Ψ_T are equal to 1.

As shown in the previous chapter, the energy difference between singlet and triplet is proportional to the magnetic coupling strength. The energy expectation values of Ψ_S and Ψ_T are

$$E_{S,T} = \frac{\langle a\bar{b} \pm b\bar{a} | \hat{H} | a\bar{b} \pm b\bar{a} \rangle}{\langle a\bar{b} \pm b\bar{a} | a\bar{b} \pm b\bar{a} \rangle} = \frac{\langle a\bar{b} \pm b\bar{a} | \hat{H} | a\bar{b} \pm b\bar{a} \rangle}{2 \pm 2S} \tag{4.3}$$

with

$$\hat{H} = \hat{h}_1(1) + \hat{h}_1(2) + \frac{1 - \hat{P}_{12}}{r_{12}} \tag{4.4}$$

where $\hat{P}_{12}$ is the permutation operator. To avoid lengthy equations, some parameters will be introduced to facilitate the derivation.

$$\varepsilon = \langle a | \hat{h}_1 | a \rangle = \langle b | \hat{h}_1 | b \rangle \tag{4.5a}$$

$$\beta = \langle a | \hat{h}_1 | b \rangle = \langle b | \hat{h}_1 | a \rangle \tag{4.5b}$$

$$J^C = \langle ab | \frac{1}{r_{12}} | ab \rangle \tag{4.5c}$$

$$K = \langle ab | \frac{1}{r_{12}} | ba \rangle \tag{4.5d}$$

This results in the following expressions for the energy of the singlet and triplet states.

$$E_S = \frac{4\varepsilon + 4\beta S + 2J^C + 2K}{2 + 2S^2} = \frac{2\varepsilon + 2\beta S + J^C + K}{1 + S^2} \qquad (4.6a)$$

$$E_T = \frac{4\varepsilon - 4\beta S + 2J^C - 2K}{2 - 2S^2} = \frac{2\varepsilon - 2\beta S + J^C - K}{1 - S^2} \qquad (4.6b)$$

The energy difference is

$$E_S - E_T = \frac{(2\varepsilon + 2\beta S + J^C + K)(1 - S^2)}{(1 + S^2)(1 - S^2)} - \frac{(2\varepsilon - 2\beta S + J^C - K)(1 + S^2)}{(1 - S^2)(1 + S^2)}$$

$$= \frac{4\beta S + 2K - 4\varepsilon S^2 - 2J^C S^2}{1 - S^4} \qquad (4.7)$$

In general the overlap between the orbitals a and b is rather small given the fact that the magnetic centers are separated in space. Hence, the S^4 term can safely be discarded, and often the terms that are quadratic in the overlap are also neglected.

$$E_S - E_T \approx 2K - 4\varepsilon S^2 + 4\beta S - 2J^C S^2 \qquad (4.8)$$

$$\approx 2K + 4\beta S \qquad (4.9)$$

The second equation is the basis of the Kahn–Briat model. Given that K is positive and S opposite in sign to β, the energy difference between singlet and triplet can be interpreted as the sum of two opposite contributions. The direct exchange interaction between the electrons on both magnetic sites is dominant in case of negligible or zero overlap, for example due to different symmetries of the orbitals a and b. This favors a ferromagnetic alignment of the spin moments, while a large overlap between the magnetic orbitals favors the singlet, and hence, enhances the antiferromagnetic character of the coupling.

The generalization to two magnetic centers with more than one unpaired electron can be made by the introduction of *exchange pathways*. The total magnetic coupling parameter J of the Heisenberg Hamiltonian is decomposed as a sum of all the possible pairwise interactions weighted by the product of the number of unpaired electrons

$$J = \frac{1}{n_a n_b} \sum_{i \in A} \sum_{j \in B} J_{ij} \qquad (4.10)$$

where each J_{ij} is evaluated with the equation derived for two unpaired electrons (Eq. 4.9) and n_a and n_b make reference to the number of the unpaired electrons on the magnetic centers A and B.

4.1.2 The Hay–Thibeault–Hoffmann Model

The second valence-only model starts from a molecular orbital viewpoint and was derived in the mid 1970s by Hay, Thibeault and Hoffmann (HTH) [2], approximately at the same time as the Kahn–Briat model. The magnetic orbitals are defined as linear combinations of orthogonal atomic-like orbitals

$$\phi_1 = \frac{1}{\sqrt{2}} (\psi_a + \psi_b) \qquad \phi_2 = \frac{1}{\sqrt{2}} (\psi_a - \psi_b) \tag{4.11}$$

Similar to ϕ_a and ϕ_b of the Kahn–Briat model, the atomic-like orbitals of the HTH model have the largest amplitudes on the magnetic centers, but in contrast ψ_a and ψ_b show delocalization tails on the ligands to ensure the orthogonality between them. Therefore, in general ψ_a and ψ_b are slightly more delocalized than the nonorthogonal ϕ_a and ϕ_b.

In the original derivation, three determinants were constructed with the molecular orbitals ϕ_1 and ϕ_2

$$T = |\phi_1 \phi_2| \qquad S_1 = |\phi_1 \overline{\phi_1}| \qquad S_2 = |\phi_2 \overline{\phi_2}| \tag{4.12}$$

with the following energy expectation values

$$
\begin{aligned}
E_T &= \langle \phi_1 | \hat{h}_1 | \phi_1 \rangle + \langle \phi_2 | \hat{h}_1 | \phi_2 \rangle + \langle \phi_1 \phi_2 | \frac{1}{r_{12}} | \phi_1 \phi_2 \rangle - \langle \phi_1 \phi_2 | \frac{1}{r_{12}} | \phi_2 \phi_1 \rangle \\
&= h_1 + h_2 + J_{12} - K_{12} \\
E_{S_1} &= 2 \langle \phi_1 | \hat{h}_1 | \phi_1 \rangle + \langle \phi_1 \phi_1 | \frac{1}{r_{12}} | \phi_1 \phi_1 \rangle = 2h_1 + J_{11} \\
E_{S_2} &= 2 \langle \phi_2 | \hat{h}_1 | \phi_2 \rangle + \langle \phi_2 \phi_2 | \frac{1}{r_{12}} | \phi_2 \phi_2 \rangle = 2h_2 + J_{22}
\end{aligned}
\tag{4.13}
$$

S_1 and S_2 have the same spin and spatial symmetry and to obtain the energy of the lowest singlet a 2×2 matrix has to be diagonalized with E_{S_1} and E_{S_2} on the diagonal and the interaction between the two determinants as off-diagonal element

$$\langle S_1 | \hat{H} | S_2 \rangle = \langle \phi_1 \phi_1 | \frac{1}{r_{12}} | \phi_2 \phi_2 \rangle = \langle \phi_1 \phi_2 | \frac{1}{r_{12}} | \phi_2 \phi_1 \rangle = K_{12} \tag{4.14}$$

The second-order equation that arises from the condition that the secular determinant is equal to zero can be solved straightforwardly and gives the energy of the singlet

$$E_S = h_1 + h_2 + \frac{1}{2} (J_{11} + J_{22}) - \frac{1}{2} \sqrt{(2h_1 - 2h_2 + J_{11} - J_{22})^2 + 4K_{12}^2} \tag{4.15}$$

and the energy difference between singlet and triplet becomes

$$E_S - E_T = \frac{1}{2}(J_{11} + J_{22}) - \frac{1}{2}\sqrt{(2h_1 - 2h_2 + J_{11} - J_{22})^2 + 4K_{12}^2} - J_{12} + K_{12} \quad (4.16)$$

The square root term in the difference can be simplified by assuming that $J_{11} - J_{22}$ is small and that $4K_{12}^2$ is significantly larger than $(h_1 - h_2)^2$. The term then reduces to

$$\sqrt{(2h_1 - 2h_2 + J_{11} - J_{22})^2 + 4K_{12}^2} \approx \sqrt{(2h_1 - 2h_2)^2 + 4K_{12}^2}$$

$$\approx 2K_{12} + \frac{(h_1 - h_2)^2}{K_{12}} \quad (4.17)$$

using the Taylor series $\sqrt{p + q} = \sqrt{p} + \frac{1}{2}q/\sqrt{p} + \ldots$ with $p \gg q$. The expression for the energy difference now reads

$$E_S - E_T = \frac{1}{2}(J_{11} + J_{22}) - \frac{(h_1 - h_2)^2}{2K_{12}} - J_{12} \quad (4.18)$$

which is further simplified by introducing the orbital energies of the magnetic orbitals, which for the triplet state are defined as

$$\varepsilon_1 = h_1 + J_{12} - K_{12} \qquad \varepsilon_2 = h_2 + J_{12} - K_{12} \quad (4.19)$$

and makes that $h_1 - h_2$ can be replaced by $\varepsilon_1 - \varepsilon_2$, which is a much easier quantity to work with. The expression shows that in the HTH model the magnetic coupling can be obtained from the outcomes of one single restricted Hartree–Fock (RHF) calculation for the triplet state. Furthermore, by expressing the integrals using the local orbitals ψ_a and ψ_b instead of the molecular orbitals ϕ_1 and ϕ_2, the expression can be written even more compact. Through a somewhat tedious but straightforward derivation it can be shown that

$$J_{11} = \frac{1}{2}(J_{aa} + J_{ab}) + K_{ab} + 2\langle aa|\frac{1}{r_{12}}|ab\rangle$$

$$J_{22} = \frac{1}{2}(J_{aa} + J_{ab}) + K_{ab} - 2\langle aa|\frac{1}{r_{12}}|ab\rangle \quad (4.20)$$

$$J_{12} = \frac{1}{2}(J_{aa} + J_{ab}) - K_{ab}$$

$$K_{12} = \frac{1}{2}(J_{aa} - J_{ab})$$

4.2 Use the definitions of ϕ_1 and ϕ_2 given in Eq. 4.11 to express the integral J_{11} in terms of local orbitals. Remember that $\int \phi_a(1)\phi_a(2)(1/r_{12})\phi_b(1)\phi_b(2)d\tau = \int \phi_a(1)\phi_b(1)(1/r_{12})\phi_a(2)\phi_b(2)d\tau = K_{ab}$.

This brings us to the final expression of the HTH model for the singlet-triplet splitting

$$E_S - E_T = 2K_{ab} - \frac{(\varepsilon_1 - \varepsilon_2)^2}{J_{aa} - J_{ab}} \tag{4.21}$$

where immediately the two opposite contributions to the magnetic coupling can be recognized. The direct exchange K_{ab} favors the triplet, and hence, the parallel alignment of the spin moments. On the other hand, a large splitting between the orbital energies of ϕ_1 and ϕ_2 favors the antiferromagnetic component of the coupling $J_{aa} > J_{ab}$.

The magnetic coupling in systems with m unpaired electrons per magnetic center can also be studied with the HTH model. The direct exchange is written as the sum of exchange integrals between orbitals on center A and center B

$$K = \sum_{i \in A} \sum_{j \in B} K_{ij} \tag{4.22}$$

To evaluate the antiferromagnetic part of the coupling, the magnetic orbitals are grouped in pairs of *bonding* and *antibonding* orbitals and the total contribution is defined as the sum of the individual couplings divided by m^2

$$J^{AF} = -\frac{1}{m^2} \sum_{i=1}^{m/2} \frac{(\varepsilon_i - \varepsilon_{2i})^2}{J_{a_i a_i} - J_{a_i b_i}} \tag{4.23}$$

where ε_i is the orbital energy of the binding combination of ψ_a and ψ_b, and ε_{2i} the orbital energy of the antibonding combination.

4.1.3 McConnell's Model

The valence-only models discussed so far have been developed in the field of transition metal compounds, either molecular based or in extended systems. The dominant magnetic interactions in these systems typically involve atoms that are bonded through bridging diamagnetic ligands, the so-called through-bond interactions. In magnetic materials based on organic radicals the mechanism is fundamentally different; there is no diamagnetic bridge between the magnetic centers and the description of the interaction given in Sect. 3.1 (and further analyzed in Chap. 5) does not directly apply. Traditionally the magnetism caused by through-space interactions in such organic materials is rationalized with the McConnell I model [3]. To describe the interaction between two radicals, the model takes an atomic viewpoint and starts with the Heisenberg Hamiltonian in the following form

$$\hat{H} = -\sum_{i<j} J_{ij}\hat{S}_i\hat{S}_j \tag{4.24}$$

where the sum runs over all the atoms in the two radicals. The J_{ij} parameters can be interpreted as the parameter for the coupling of an electron in atomic orbital ϕ_i on site i and another electron in ϕ_j on site j. In a valence bond setting with non-orthogonal orbitals, the interaction can be written as the sum of a positive two-electron exchange integral and a one-electron integral

$$J_{ij} = \langle\phi_i\phi_j|\frac{1}{r_{12}}|\phi_j\phi_i\rangle + \langle\phi_i|\phi_j\rangle\langle\phi_i|\hat{h}(1)|\phi_j\rangle \tag{4.25}$$

The one-electron integral is dominated by the electron-nucleus attraction in most cases, and hence, negative in sign. From this it is concluded that, unless the overlap between the orbitals ϕ_i and ϕ_j is very small, the J_{ij} parameter is negative, favoring singlet coupling of the electrons. This expression is not very easy to handle and in all practical applications to rationalize the magnetic properties of radicals a series of simplifications is introduced. In the first place the summation is restricted to pairs of electrons on different units

$$\hat{H} = -\sum_{i\in A}\sum_{j\in B} J_{ij}\hat{S}_i\hat{S}_j \tag{4.26}$$

assuming that the interactions within a unit do not depend on the coupling of the total spin moment of the two radicals. The second and most fundamental approximation of McConnell's model is made by replacing the spin operators by a product of the total spin operator for each unit and the atomic spin populations ρ_i

$$\hat{H} = -\hat{S}_A \cdot \hat{S}_B \sum_{i\in A}\sum_{j\in B} J_{ij}\rho_i\rho_j \tag{4.27}$$

The third simplification lies in the restriction of the sum over i and j to the shortest contacts only. Thus, second nearest neighbour interactions (and beyond) between the units, which in many cases oppose the nearest neighbour interactions, are neglected. These simplifications lead to a very simple model to rationalize or predict magnetic properties of molecular crystals based on organic radicals. When regions of opposite spin densities overlap, $\rho_i\rho_j < 0$ one can expect ferromagnetic interactions and when close contacts have spin populations with the same sign, $\rho_i\rho_j > 0$, antiferromagnetism prevails. To illustrate its application, we consider two stacked benzyl radicals with the CH_2 groups in *para* and *meta* as illustrated in Fig. 4.1. The spin populations

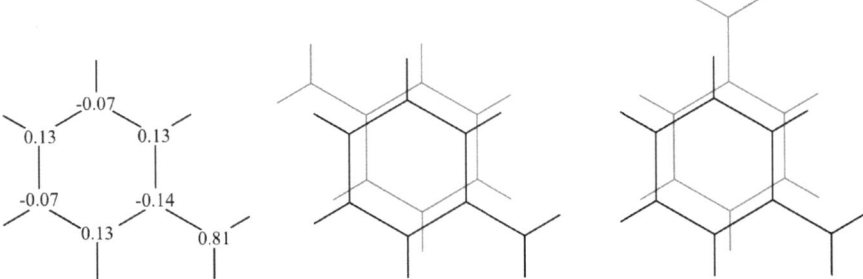

Fig. 4.1 *Left* Benzyl radical with the spin populations of the carbon atoms. Two benzyl radicals stacked with the CH_2 group in *para* (*middle*) and *meta* position (*right*)

shown in the left part of the figure have been calculated with a simple CASSCF calculation on the doublet spin state of the monomer, although in this case the alternation of the sign of the spin population can be anticipated. The closest contacts in the stacked dimers are formed by the aligned carbon atoms of the benzene ring. The $\rho_i\rho_j$ products for these atoms are all negative in the case of the *para* conformer (middle of Fig. 4.1) and therefore this dimer is expected to have a triplet ground state. The aligned carbon atoms of the *meta* conformer (right) have spin populations of the same sign, the $\rho_i\rho_j$ product is therefore positive, predicting an antiferromagnetic (singlet) ground state.

The conclusions on the nature of the ground state in the benzyl dimer extracted from the model of McConnell are in line with those of accurate ab initio calculations. Also in many organic magnetic materials, the model has proven its ability to correctly reproduce the dominant magnetic interactions. However, the careful step-by-step analysis of the model by Novoa and co-workers [4, 5] showed that the success of the model is at least partially due to a fortunate cancellation of errors. The analysis shows that there is no firm theoretical foundation for replacing the spin operators by atomic spin densities. Moreover, the model was shown to fail to predict the dominant magnetic interactions in several crystals with nitronyl nitroxide radicals and cannot reproduce the angle dependence of the magnetic interaction in the model system containing two H_2NO radicals. Hence, despite its numerous successes and versatility, the McConnell model should be applied with caution.

4.3 Use the reasoning of McConnell's model (Eq. 4.27) to predict the ground state of the benzyl dimer with the CH_2 groups in *ortho* and for the conformer where the two units are perfectly aligned.

4.2 Magnetostructural Correlations

From the very beginning of the study of the magnetic interactions in transition metal complexes a large part of the effort has been dedicated to derive relations between the geometrical structure of the complex and the nature and magnitude of the coupling of the localized spin moments. These magnetostructural correlations can be extremely useful to rationalize the variations in the magnetic behaviour of a family of similar complexes or to design new complexes with the desired properties. Magnetostructural relations can be extracted from experimental studies by comparing a large group of compounds and relate geometric parameters with the observed magnetic behaviour. This requires a large set of data, but it is often difficult to separate different (opposing) effects. On the other hand, theoretical studies can take a (model) complex and modify the geometry at will to establish the influence of a certain geometric parameter on the magnetic interaction. Combined with the qualitative valence-only models discussed in the previous sections one can boil down the complicated magnetic behaviour to very simple concepts and straightforward magnetostructural correlations. These concepts and correlations can yield design rules that can be utilized in the synthesis of materials with pre-defined magnetic properties.

M–L–M angle: One of the most famous magnetostructural correlations concerns the dependence of J on the M–L–M angle in transition metal complexes with a double bridge as depicted in Fig. 3.1. For angles close to 90° the coupling of the spin moments on the metal ions is ferromagnetic and for larger (and smaller) angles the coupling becomes antiferromagnetic. The curve shown in Fig. 4.2 is a typical example of this correlation and was obtained by calculating J from the singlet-triplet energy difference (see Eq. 3.29) using the wave functions discussed in Sect. 3.1, Eqs. 3.2a and 3.7. The change from ferromagnetic to antiferromagnetic interaction can be explained with the Hay–Thibeault–Hoffmann model. The largest contributions to

Fig. 4.2 Magnetic coupling strength of the copper dimer shown in the inset versus the angle α

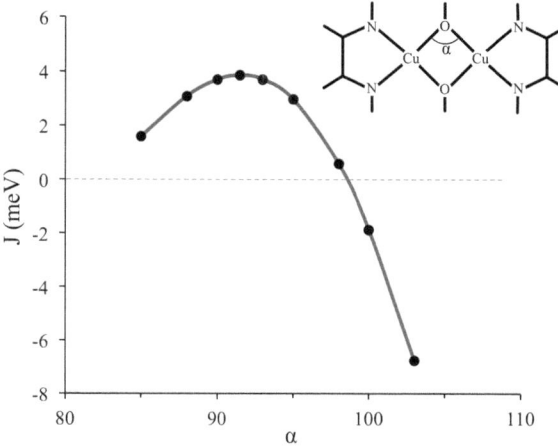

Fig. 4.3 Molecular orbital
diagram showing the
interaction of the plus and
minus combinations of the
$3d_{xy}$ orbitals on the metal
centers with the p_x and p_y
orbitals on the ligands

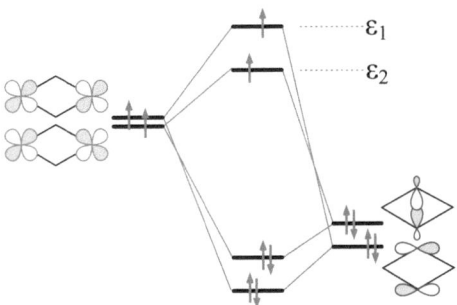

the magnetic orbitals arise from the plus and minus combinations of the Cu-$3d_{xy}$
orbitals, shown on the left in Fig. 4.3. The plus and minus combinations of the p_x and
p_y orbitals on the bridge in the right column of the MO diagram interact with the $3d_{xy}$
orbitals to form bonding and antibonding combinations as shown in the middle of the
figure. The bonding orbitals are doubly occupied and not relevant for the magnetic
properties, but the antibonding combinations correspond to the magnetic orbitals,
in which we readily recognize the large contribution from the $3d_{xy}$ orbitals with
non-negligible tails on the ligand. In the reasoning of the HTH model, the difference
in orbital energy ε of the two magnetic orbitals is directly related to the magnetic
coupling strength, cf. Eq. 4.21 and numerically proven by Ruiz and co-workers in
Ref. [6]. For (nearly) degenerate magnetic orbitals ($\varepsilon_1 \approx \varepsilon_2$) the antiferromagnetic
term is small and the direct exchange K_{ab} dominates. However, when the orbital
energies are sufficiently different, the antiferromagnetic term is the largest term and
J will become negative.

The upper part of Fig. 4.4 shows that the interaction of the p_x and p_y bridge
orbitals with the $3d_{xy}$ orbitals on the metal is approximately equal around $\alpha = 90°$.
Therefore, the near degeneracy of the plus and minus combination of the 3d orbitals
is maintained and one can expect a small ferromagnetic interaction of the spins. On
the contrary, for larger angles, the interaction along the x-direction becomes stronger
than for the y orbitals. This is reflected in a larger delocalization onto the ligand
in the *gerade* orbital than in the *ungerade* orbital,[1] see the lower part of Fig. 4.4.
The energies of the two magnetic orbitals are no longer similar and a considerable
antiferromagnetic contribution exists, which for large enough angles overcomes the
ferromagnetic contribution and turns the net coupling in an antiferromagnetic one.

Out-of plane angle: A second interesting magnetostructural relation that can easily
be explained with the HTH model is the increase in ferromagnetic coupling when the
side group of the bridging atoms is rotated out of the M–(L)$_2$–M plane. This relation
was described in detail in Ref. [6] and it was found that ferromagnetic coupling can
be obtained even in those molecules that have a rather large M–L–M angle. Figure 4.5

---

[1] *Gerade* and *ungerade* (odd and even in German) make reference to the effect of the sign of the
orbital under the action of the inversion operator. The *gerade* orbital does not change sign, while
the *ungerade* orbital is converted to its opposite.

Fig. 4.4 *Upper part gerade* and *ungerade* magnetic molecular orbitals for a M–L–M angle of 90°. The d_{xy} orbitals on the metal centers have an equal overlap with the p_x and p_y orbitals on the ligand. *Lower part* In a system with a larger M–L–M angle, the overlap is larger for p_x than for p_y

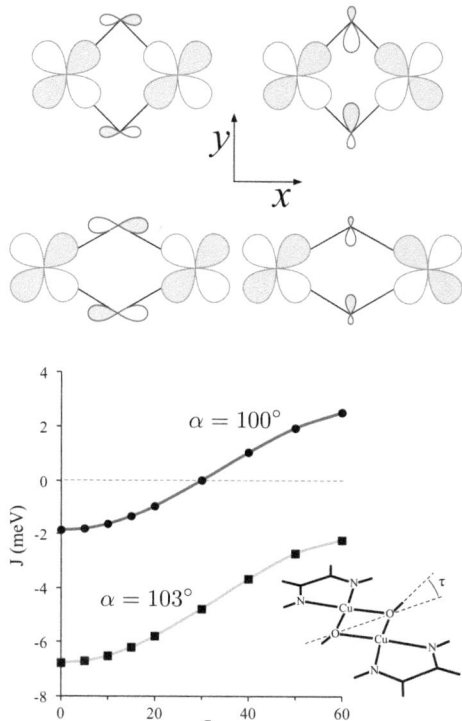

Fig. 4.5 Magnetic coupling strength $J = E_S - E_T$ as function of the out-of-plane angle τ for two different Cu–O–Cu angles

shows how the magnetic coupling varies when the hydrogen atom of the bridging OH groups is moved out of the plane formed by the Cu and O ions. In the case of the 103° Cu–O–Cu angle (squares), the magnetic coupling is diminished by approximately 4.5 meV but the ferromagnetic regime is not reached. Considering a slightly smaller M–L–M angle (circles), a similar change in the coupling is observed but now the behaviour is changed from antiferromagnetic to ferromagnetic near $\tau = 30°$.

The increased ferromagnetic character of the coupling upon the out-of-plane movement of the side group of the bridging ligand (in this simple case a hydrogen atom, but the same tendency is observed for bigger residues) is easily explained with the MO diagram represented in Fig. 4.3. In the case of a completely flat magnetic core, that is $\tau = 0°$, the ligand orbital in the xy-plane oriented along the y-axis (ϕ_1) is typically composed of sp hybrids, mixtures of s and p_y orbitals. When τ is different from zero, the xy-plane is no longer a symmetry plane of the complex and the p_z orbitals can also contribute to ϕ_1. This means that the hybridization is no longer purely sp, but has also some sp^2 character. The increased p-character of the hybrid increases the ligand orbital energy and reduces the gap with the $3d_{xy}$ orbitals in the left of Fig. 4.3. Consequently, the interaction becomes stronger and the antibonding combination, the magnetic orbital with energy ε_2, will be higher in energy. This reduces $(\varepsilon_1 - \varepsilon_2)^2$ and weakens the antiferromagnetic contribution to the coupling.

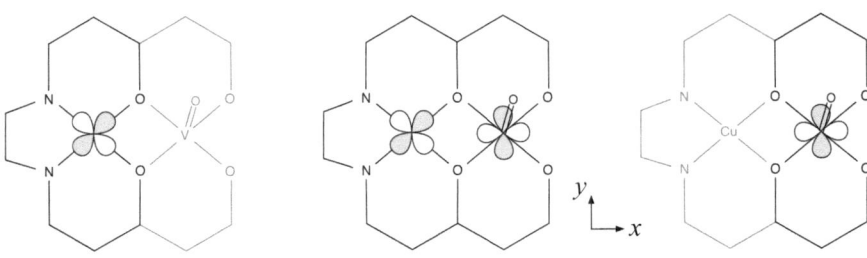

Fig. 4.6 Schematic representation of the Cu/V binuclear complex with a double alkoxo bridge. *Left* and *right* Magnetic orbitals for the Cu site and the V site, respectively. *Middle* superposition of the two magnetic orbitals

Exchange pathways: We will now further expand the relation between geometry and magnetic coupling strength by exploiting the concept of the *exchange pathway*, which was already briefly mentioned at the end of Sect. 4.1.1. For systems with more than one unpaired electron per magnetic center the Kahn–Briat model decomposes the total coupling in pairwise contributions as given in Eq. 4.10. These exchange pathways provide a very powerful tool to predict the nature of the magnetic coupling (ferro- or antiferromagnetic; weak, strong) in nearly all combinations of d^n magnetic ions. Many examples were discussed in the book by Kahn [7] and the concept has recently been reviewed by Launay and Verdaguer [8]. Here we will shortly discuss two examples to clarify the way of reasoning to rationalize or predict the nature of the coupling between two transition metals bridged by one or more diamagnetic ligands. For a full account on this subject we refer to the books of Kahn, and Launay and Verdaguer.

The first step in the procedure consist of an inspection of the coordination sphere of the magnetic centers to determine the shape and symmetry of the optimal local magnetic orbitals. This can either be done through calculation or by ligand field reasoning. Our first example is a binuclear complex of Cu^{2+} and V^{4+} with a double alkoxo-bridge. The copper ion has a d^9 electronic configuration. This means that all $3d$-orbitals are doubly occupied except the $3d_{xy}$ orbital, which is highest in energy because it directly points to the atoms of the first coordination sphere. The vanadium ion is covalently bound to the apex oxygen and the resulting vanadyl group has a formal oxidation state of VO(II) with one unpaired electron in the orbital of lowest energy, the largely non-bonding V-$3d_{x^2-y^2}$ orbital. Figure 4.6 shows the two magnetic orbitals of the two magnetic centers, the left panel corresponds to the magnetic orbital on Cu and the right panel to the VO site. The superposition of these two pictures in the middle defines the exchange pathway and can help us to decide upon the overlap between the two orbitals as they appear in the main equation of the Kahn–Briat model, see Eq. 4.9. Note, that this does not define a molecular orbital, it is merely a construction by superimposing the two magnetic orbitals. The product of the two functions is an odd function with respect to the xz-plane, and hence, integrating over the cartesian coordinates gives a zero overlap integral S of these two magnetic orbitals. When S is equal or close to zero, the first term in the Kahn–Briat equation determines

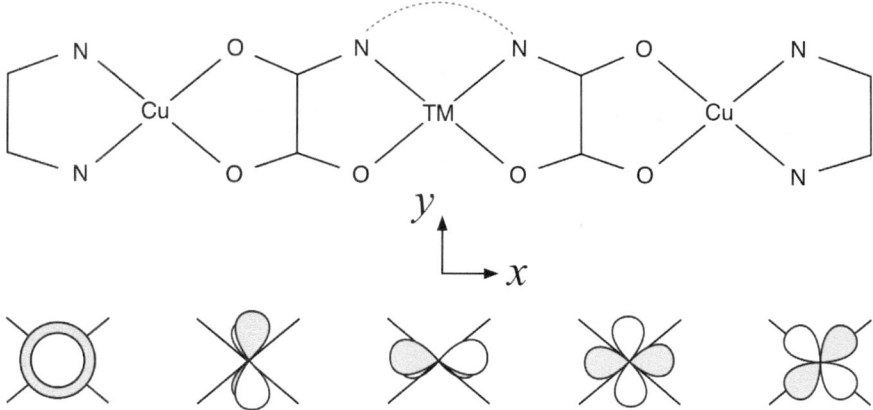

Fig. 4.7 Trinuclear model complex with C_{2v} symmetry. TM is one of the transition metals with an incomplete d^n electronic configuration with high spin coupling. The orbitals in the lower part are $3d_{z^2}$ (a_1), $3d_{yz}$ (b_2), $3d_{xz}$ (b_1), $3d_{x^2-y^2}$ (a_1) and $3d_{xy}$ (a_2) and are ordered from *left* to *right* by increasing orbital energy

the nature of the coupling. Therefore, the magnetic coupling in this Cu/V dimer is expected to be ferromagnetic, in line with the triplet ground state and singlet-triplet gap of approximately $100 \, \text{cm}^{-1}$ observed experimentally [9].

In complexes with more than one unpaired electron on at least one of the magnetic sites, the overall magnetic coupling is the sum of the couplings along all exchange paths weighted by the product of the number of unpaired electrons on each magnetic center (the number of paths). To illustrate the potential of the Kahn–Briat model for predicting the nature of the magnetic coupling, we will focus on the trinuclear Cu^{II} complex schematically depicted in the upper part of Fig. 4.7 and discuss the effect of replacing the copper ion in the middle by other transition metals. The complex has approximate C_{2v} symmetry and the five $3d$ orbitals belong to the a_1 (2x), a_2, b_1 and b_2 irreducible representations as shown in the lower part of the figure. The copper ions on the left and right sides of the complex with their $3d^9$ electronic configuration have only one unpaired electron, which resides in the $3d_{xy}$ orbital of a_2 symmetry. When the magnetic center in the middle is also occupied by a Cu^{2+} ion, the three magnetic orbitals are all of the same symmetry and hence there is a non-zero overlap leading to an antiferromagnetic coupling between the TM ions in the complex, in line with experiment [10].

Keeping track of the relative energy of the five $3d$ orbitals (see Fig. 4.7), we now consider the complexes that contain transition metals with other electronic configurations. Starting with the $3d^1$ configuration (for example, Ti^{3+}), the natural magnetic orbital in the middle is $3d_{z^2}$ with a_1 symmetry and the Cu orbitals on the outside are $3d_{xy}$ of a_2 symmetry. Hence, the exchange path includes orthogonal orbitals and ferromagnetic interactions are expected. Putting a transition metal with two d-electrons in the middle leads to an electronic configuration with the unpaired electrons in the

$3d_{z^2}$ and $3d_{yz}$ orbitals belonging to the a_1 and b_2 irreducible representations, respectively. The total coupling is a sum of four exchange paths, which appear in two pairs because of the left–right symmetry of the complex. Exchange path type 1 goes from the $3d_{z^2}$ orbital in the middle to the $3d_{xy}$ orbital on the Cu. Since they transform differently under the symmetry operations of the C_{2v} group, the overlap integral of the two orbitals is zero and this path contributes in a ferromagnetic way to the coupling. The same holds for the second pair of exchange paths involving the $3d_{yz}$ and $3d_{xy}$ orbitals. Hence, again a ferromagnetic coupling can be anticipated. The situation changes when the middle position is occupied by an ion with a d^5 electronic configuration. Five different exchange paths are now active; four of them involve orthogonal orbitals, but the fifth connects the $3d_{xy}$ natural magnetic orbitals of the centers. The latter gives an antiferromagnetic contribution and counterbalances the four weaker ferromagnetic exchange paths. When TM ions are placed in the center with more than 5 d-electrons, the ferromagnetic exchange paths disappear, the coupling gets gradually more antiferromagnetic until we arrive again at the strong antiferromagnetic coupling in the case of three ions with d^9 electronic configurations.

4.4 Consider the complex sketched in this box and predict the nature of the coupling when site A is occupied by Ni^{2+} and site B by Cr^{3+}. The out of plane TM-ligand distances are larger than the in-plane distances.

What is the number of exchange paths when Cr^{3+} is replaced by Mn^{2+}? What coupling can be expeced?

Counter-complementarity: Another relation between structure and magnetic coupling strength is covered by the concept of counter-complementarity. In systems with two magnetic centers connected by two different ligands the total magnetic coupling is in general not equal to the sum of the magnetic coupling via the two individual bridges but often significantly smaller. This anti-synergistic effect can most efficiently be explained for a system with two $S = 1/2$ spin moments based on two molecular orbital diagrams using the HTH model. Figure 4.8 shows the interaction of the atomic-like orbitals on the magnetic centers A and B with those of the bridge (L_1) that is expected to give the largest contribution to the coupling. The molecule is in the xy-plane and the A–B 'bond' is along the x-axis. The interaction of the $L_1 - p_x$ orbital with the *gerade* combination of d_{xy} orbitals is stronger than the interaction of the $L_1 - 2p_y$ with the *ungerade* d_{xy} orbitals. Therefore a gap is opened between the magnetic orbitals φ_1 and φ_2 with the antibonding combination of *gerade* d_{xy} and $L_1 - 2p_x$ at higher energy. Equation 4.21 shows that this gap (Δ_1) is directly proportional to the magnetic coupling through L_1.

Fig. 4.8 Molecular orbital diagram of the interaction of the atomic-like orbitals on the magnetic centers A and B and the orbitals on the lower bridge parallel and perpendicular to the A–B 'bond'

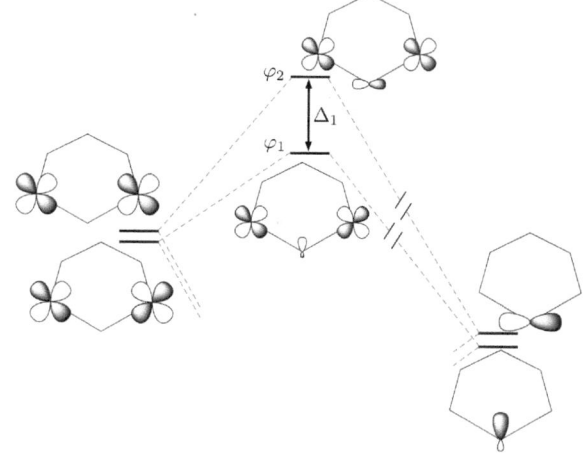

Fig. 4.9 Molecular orbital diagram of the interaction of the magnetic orbitals φ_1 and φ_2 (obtained by the interaction of the magnetic centers with the lower bridge) and those of the upper ligand. $\Delta_1 > \Delta_2$

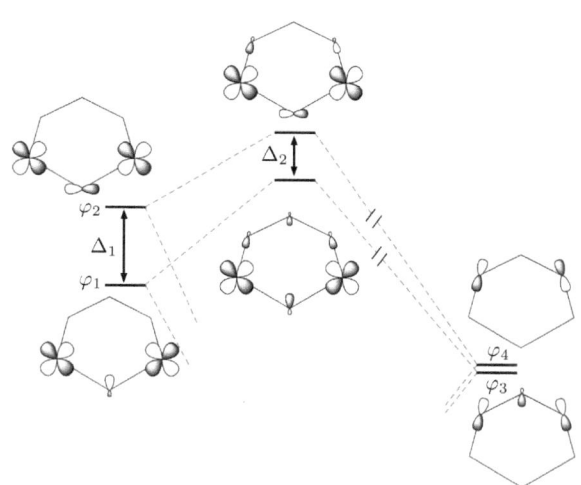

The next step takes into account the interaction with the second bridge (L_2) and is schematically depicted in Fig. 4.9. For symmetry reasons, φ_1 interacts with φ_3, and φ_2 with φ_4. If we take a perturbational point of view, the interaction strength is determined not only by the energy separation of the levels but also by the interaction matrix elements. The shape of the orbitals on the left and the right of the figure strongly suggest that the matrix elements can be assumed to be nearly the same, and hence, the final result of the interaction is solely determined by the differences in the orbital energies. φ_3 and φ_4 lie at lower energy than φ_1 and φ_2, but the separation between these two is much smaller than Δ_1. From this directly follows that $\varepsilon(\varphi_1) - \varepsilon(\varphi_3)$ is smaller than $\varepsilon(\varphi_2) - \varepsilon(\varphi_4)$. Consequently, the destabilization in the antibonding combination of φ_1 and φ_3 is larger than for φ_2 and φ_4 and leads to a smaller gap between the magnetic orbitals than after the considering only L_1: $\Delta_1 > \Delta_2$.

According to the HTH model this gives a reduction of the antiferromagnetic contribution to the magnetic coupling (see Eq. 4.21) and illustrates the anti-synergistic effect or counter-complementarity of the two ligands.

4.3 Accurate Computational Models

Although the qualitative models discussed so far are very useful for a basic understanding of the magnetic interactions between two spin moments, more quantitative predictions can only be obtained by going beyond the *valence*-only description considered so far. As shown in the previous chapter, the magnetic interaction parameter J of the Heisenberg Hamiltonian can in many cases be related to the energy difference of electronic states with different spin couplings. Hence, precise theoretical estimates of the magnetic coupling strengths are intimately related to the correct application of high-level computational schemes.

As shown in Sect. 3.1, the basic description of the magnetic coupling problem is intrinsically multideterminantal and in most cases one needs a multiconfigurational description for minimally accurate results. Before discussing the different computational schemes that can be used for quantitative estimates, we want to stress that a multideterminantal wave function is not necessarily a multiconfigurational wave function. This is best illustrated for the 2-electrons/2-orbitals case discussed before. The simplest representation of the triplet state is obtained with a single Slater determinant

$$\Phi_T = |\phi_a \phi_b| \tag{4.28}$$

where all the doubly occupied orbitals have been omitted. On the other hand, the most basic description of the open-shell singlet requires a wave function with two Slater determinants to fulfill the requirements of the spin symmetry.

$$\Phi_S = \frac{|\phi_a \overline{\phi}_b| - |\overline{\phi}_a \phi_b|}{\sqrt{2}} \tag{4.29}$$

This multideterminantal wave function only describes one electronic configuration since the occupation of the orbitals is the same in both determinants and is in general known as a configuration state function. In this simple monoconfigurational description, the energy of the triplet is lower than the singlet by twice the exchange integral K_{ab}. A more satisfactory description is obtained with a multiconfigurational singlet wave function by adding the Slater determinants with two electrons in the same orbital

$$\Phi'_S = c_1\{|\phi_a \overline{\phi}_b| - |\overline{\phi}_a \phi_b|\} + c_2\{|\phi_a \overline{\phi}_a| + |\phi_b \overline{\phi}_b|\} \tag{4.30}$$

where c_1 is much larger than c_2 for biradicalar systems, and their precise value has to be determined in a configuration interaction calculation. Wave functions of this type are often used as multideterminantal-multiconfigurational reference—mostly multireference (MR), for short—wave function.

MR-CISD is one of the most accurate ab initio computational schemes that can be used to describe the electronic structure of systems with a markedly multireference character. Although the ever increasing computing power constantly pushes the frontiers forward, the applicability of MR-CISD remains limited to small (model) systems. Moreover, the method suffers from the size-extensivety problem inherent to any truncated CI method. For these reasons, MR-CISD is hardly ever used in computational studies of molecules with unpaired electrons. There are, however, several alternative wave function based schemes that can provide very useful information about the magnetic interactions. In the following sections we will first discuss the ins-and-outs of a good reference wave function and introduce the difference dedicated CI (DDCI) method. Thereafter a short account will be given of two implementations of MR perturbation theory, and the chapter will be closed with a discussion of the consequences of lifting the restrictions of the spin symmetry as done in density functional theory (DFT).

4.3.1 The Reference Wave Function and Excited Determinants

An important factor in the accurate prediction of magnetic coupling (and other electronic structure) parameters is the proper choice of the reference wave function. There are many possible ways to construct the reference, but the complete active space (CAS) approach has emerged as one of the most versatile methods. The molecular orbitals are divided in three groups: the inactive, the active and the virtual orbitals. The orbitals in the first group are doubly occupied in all the Slater determinants of the multireference wave function, while the orbitals in the last group remain always unoccupied. The orbitals in the second class span the active space. The multiconfigurational wave function is generated by distributing N_{act} electrons—where N_{act} is the total number of electrons minus two times the number of inactive orbitals—over the M_{act} active orbitals. This is schematically outlined in Fig. 4.10. The doubly occupied or empty Hartree Fock orbitals shown on the left are divided in inactive, active and virtual orbitals. The multiconfigurational wave function is constructed by making a linear combination of Slater determinants Φ_1, Φ_2, etc. that differ by the occupation of the active orbitals. The CAS procedure generates a MR wave function in which all possible distributions of the active electrons over the active orbitals are considered. Although this approach often generates many determinants that are very high in energy and are not specially important in the final wave function, it has several important practical and conceptual advantages like the good convergence properties, size extensivity, orbital invariance, etc. [11]. Moreover, it has the advantage that selecting the active orbital space (although far from being trivial) is in most cases easier than making an unbiased selection of the most important configurations.

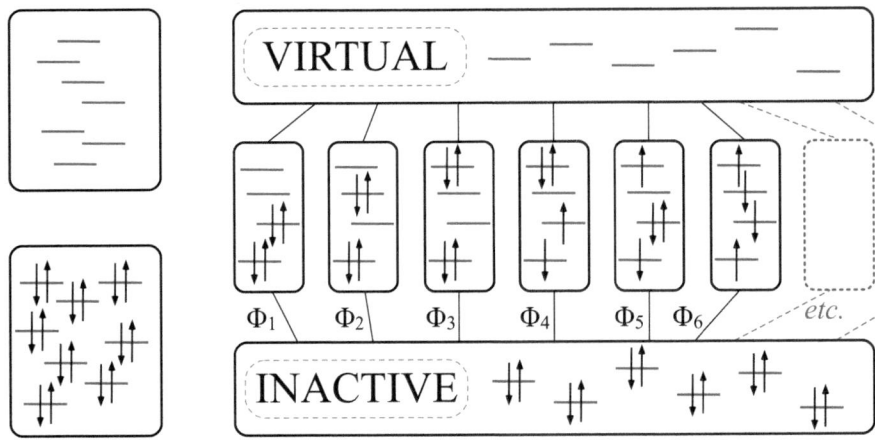

Fig. 4.10 Complete Active Space procedure to generate a multireference wave function. The occupied and virtual orbitals from a Hartree–Fock calculation (*left*) are divided in three groups (*right*): Inactive, active and virtual orbitals. A linear combination of Slater determinants is formed in which the inactive orbitals are always doubly occupied, the virtual orbitals are always empty and the active orbitals can be doubly occupied, singly occupied or unoccupied

4.5 How many determinants with $M_S = 0$ can be generated for the active space with 4 active orbitals and 4 electrons as shown in Fig. 4.10.

In virtually all calculations of magnetic interactions or related electronic structure parameters, the wave function expansion is restricted to singly and doubly excited determinants with respect to the reference. These determinants are often classified in eight different groups depending on how many holes/particles are created in the inactive/virtual orbitals. This can be very useful to decompose the wave function in smaller contributions and in this way facilitate the analysis of the results. Table 4.1 overviews the different classes and lists the labels used in some post Hartree–Fock methods that will be described in the remainder of this chapter. In the Table we have used $\hat{E}_{rs}$ to define the excitation operator $\hat{a}_s^\dagger \hat{a}_r$, eliminating an electron in orbital r and creating one in orbital s.

It is rather complicated to give generally applicable formulas for the number of determinants in each class, but rough estimates are rather easily calculated. Consider a system with k inactive orbitals, l virtual orbitals and n determinants in the reference wave function, the number of electrons is even and we restrict ourselves to the $M_S = 0$ subspace without any further spin or spatial symmetry. The approximate number of 2h-2p replacements is given by the product of the number of ways in which 2 holes can be created in the inactive orbitals (k^2) and the ways in which two particles can be placed in the virtual orbitals (l^2) multiplied with the number of determinants in the

Table 4.1 Classification of the singly and doubly excited determinants by the number of holes/particles created in the inactive (h, h')/virtual (p, p') orbitals

Excitation operator(s)	CASPT2	DDCI	NEVPT2
$\hat{E}_{ha}$; $\hat{E}_{ha}\hat{E}_{bc}$	Internal	$1h$	$\hat{V}_h^{+1}$
$\hat{E}_{ha}\hat{E}_{h'b}$		$2h$	$\hat{V}_{hh'}^{+2}$
$\hat{E}_{ap}$; $\hat{E}_{ap}\hat{E}_{bc}$	Semi-internal	$1p$	$\hat{V}_p^{-1}$
$\hat{E}_{hp}$; $\hat{E}_{hp}\hat{E}_{ab}$		$1h$-$1p$	$\hat{V}_{h,p}^{0}$
$\hat{E}_{hp}\hat{E}_{h'a}$		$2h$-$1p$	$\hat{V}_{hh',p}^{+1}$
$\hat{E}_{ap}\hat{E}_{bp'}$	External	$2p$	$\hat{V}_{pp'}^{-2}$
$\hat{E}_{hp}\hat{E}_{ap'}$		$1h$-$2p$	$\hat{V}_{h,pp'}^{-1}$
$\hat{E}_{hp}\hat{E}_{h'p'}$		$2h$-$2p$	$\hat{V}_{hh',pp'}^{0}$

a, b and c are active orbitals. The nomenclature used in some post Hartree–Fock methods is also listed

reference wave function, that is $n \times k^2 l^2$. A similar reasoning can be used to estimate the number of excitations with $2h$-$1p$ ($n \times k^2 l$), $1h$-$2p$ ($n \times k l^2$) and so forth.

4.6 Compute the number of $1h$-$1p$ determinants in the case of k inactive orbitals, l virtual orbitals and a (2,2) CAS space for $M_S = 0$.

4.3.2 Difference Dedicated Configuration Interaction

The majority of the excited determinants belong to the class of the $2h$-$2p$ excitations. This class easily constitutes 90 % of the determinants in medium-sized molecules using basis sets of reasonable quality, and hence, the contribution to the correlation energy is extremely large. However, including this class of excitations in the configuration interaction expansions has only a small effect on the vertical excitation energies (that is, the relative energies of the different electronic states at a fixed geometry). Hence, this differential effect can be neglected in the calculation of the relative energies of the spin states needed to extract J, the magnetic coupling parameter of the Heisenberg Hamiltonian, and various other electronic structure parameters. The elimination of the $2h$-$2p$ determinants leads to a drastic shortening of the configuration interaction expansion and widens the field of applicability of variational wave function based methods. The resulting variant of MRCI is generally known as the difference dedicated configuration interaction (DDCI) [12], which provides accurate vertical energy differences but cannot be used to compare total energies at different geometries.

4.7 Make a rough estimate of the total number of determinants in the MR-CISD wave function for a system with 74 electrons, 154 orbitals and a CAS(2,2)CI reference wave function. Calculate the percentage of $2h$-$2p$ excitations in the MR-CISD wave function (neglect the $1h$, $1p$, $1h$-$1p$, $2h$ and $2p$ excitations, they give rise to a very small number of determinants).

The justification for eliminating the $2h$-$2p$ determinants relies on second-order perturbation theory in its quasi-degenerate formulation as exposed in Chap. 1. Although it can be done for an arbitrary number of unpaired electrons, we will elaborate the 2-electrons/2-orbitals case for simplicity. The model space is spanned by the neutral and ionic determinants

$$\Phi_I = \{|h\bar{h}a\bar{b}|, |h\bar{h}b\bar{a}|, |h\bar{h}a\bar{a}|, |h\bar{h}b\bar{b}|\} \tag{4.31}$$

where h is one of the inactive orbitals, doubly occupied in all determinants of the model space. The lowest two eigenstates of the model space are the singlet and triplet spin functions whose energy difference is related to J. However, before diagonalizing we will first evaluate the effect of the $2h$-$2p$ external determinants on the matrix elements between the determinants of the model space with QDPT. First, we take a look at the off-diagonal elements and calculate the second-order contributions of $\Phi_R = |p\bar{p}a\bar{b}|$ and $\Phi_S = |p\bar{p}b\bar{a}|$ to the *dressed* matrix element of $\Phi_I = |h\bar{h}a\bar{b}|$ and $\Phi_J = |h\bar{h}b\bar{a}|$ according to the expression given in Eq. 1.86. The $2h$-$2p$ determinant Φ_R is obtained by making a double replacement in Φ_I, exciting the electrons in h to the unoccupied orbitals p and Φ_S arises from Φ_J in an analogous way. The contributions to the effective matrix element are

$$\Phi_R: \quad \frac{\langle h\bar{h}a\bar{b}|\hat{V}|p\bar{p}a\bar{b}\rangle \langle p\bar{p}a\bar{b}|\hat{V}|h\bar{h}b\bar{a}\rangle}{E_J - E_R} \tag{4.32a}$$

$$\Phi_S: \quad \frac{\langle h\bar{h}a\bar{b}|\hat{V}|p\bar{p}b\bar{a}\rangle \langle p\bar{p}b\bar{a}|\hat{V}|h\bar{h}b\bar{a}\rangle}{E_J - E_S} \tag{4.32b}$$

For the contribution of Φ_R, the second matrix element in the numerator is zero because the determinants on the left and the right of the operator have more than two different columns, and the same occurs for the first matrix element in the Φ_S contribution. This eliminates any second-order perturbation contribution from the $2h$-$2p$ determinants to the off-diagonal elements of the model space.

4.8 Write down the second-order contribution of Φ_Q to $\langle \Phi_I | \hat{H}^{eff} | \Phi_L \rangle$, where Φ_Q arises from a double excitation from orbital h to orbital p acting on the ionic determinant $\Phi_L = |h\bar{h}b\bar{b}|$. Argue that this contribution is equal to zero.

On the contrary, the diagonal elements do have a contribution from the $2h$-$2p$ excitations. Continuing with the external determinants Φ_R and Φ_S, it is easily shown that the former only contributes to $\langle \Phi_I | \hat{H}^{eff} | \Phi_I \rangle$ and the latter to $\langle \Phi_J | \hat{H}^{eff} | \Phi_J \rangle$

$$\Phi_R: \quad \frac{|\langle h\overline{h}a\overline{b} | \hat{V} | p\overline{p}a\overline{b} \rangle|^2}{E_I - E_R} \neq 0 \qquad \frac{|\langle h\overline{h}b\overline{a} | \hat{V} | p\overline{p}a\overline{b} \rangle|^2}{E_J - E_R} = 0 \tag{4.33a}$$

$$\Phi_S: \quad \frac{|\langle h\overline{h}a\overline{b} | \hat{V} | p\overline{p}b\overline{a} \rangle|^2}{E_I - E_S} = 0 \qquad \frac{|\langle h\overline{h}b\overline{a} | \hat{V} | p\overline{p}b\overline{a} \rangle|^2}{E_J - E_S} \neq 0 \tag{4.33b}$$

Both non-zero integrals are identical and through the Slater–Condon rules we arrive at the following second-order contribution of Φ_R and Φ_S to the diagonal elements $\langle \Phi_I | \hat{H}^{eff} | \Phi_I \rangle$ and $\langle \Phi_J | \hat{H}^{eff} | \Phi_J \rangle$

$$\frac{|\langle pp | \frac{1 - \hat{P}_{12}}{r_{12}} | hh \rangle|^2}{2\varepsilon_h - 2\varepsilon_p} \tag{4.34}$$

where the denominator is obtained by assuming the Møller–Plesset division for $\hat{H} = \hat{H}^{(0)} + \hat{V}$. In the general case the contribution of all the $2h$-$2p$ determinants to the diagonal elements is given by

$$\sum_{h,h'} \sum_{p,p'} \frac{|\langle pp' | \frac{1 - \hat{P}_{12}}{r_{12}} | hh' \rangle|^2}{\varepsilon_h + \varepsilon_{h'} - \varepsilon_p - \varepsilon_{p'}} \tag{4.35}$$

The summation only involves integrals that depend on the inactive (h, h') and virtual (p, p') orbitals, and hence, is exactly the same for all the diagonal elements in the model space. This uniform shift of the diagonal elements does not affect the energy differences of the eigenstates of the model space and in combination with the zero contribution to the off-diagonal elements, this shows that the $2h$-$2p$ determinants can be skipped in the CI expansion of the wave function. Note that this argument is based on second-order perturbation theory, the inclusion of higher-order interactions gives rise to small contributions and strictly speaking the mutual interaction between the $2h$-$2p$ determinants could affect the energy differences.

4.9 Consider a (non-degenerate) model space with neutral and ionic determinants. (a) Show that the $2h$-$1p$ determinant $\Phi_R = |a\overline{a}p\overline{b}|$ introduces non-zero off-diagonal elements between the ionic and neutral determinants of the model space. (b) Are the diagonal elements of the model space shifted uniformly by Φ_R?

The number of external determinants can even be more reduced when the model space is reduced to the neutral determinants $\Phi_I = \{|h\overline{h}a\overline{b}|, |h\overline{h}b\overline{a}|\}$. Under these circumstances, the list of determinants that do not affect the energy difference of the two states contained by the model space can be extended with the $2h$-$1p$ and $1h$-$2p$ classes. Taking as an example $\Phi_R = |a\bar{a}p\bar{b}|$ and $\Phi_S = |b\bar{b}p\bar{a}|$ ($2h$-$1p$ determinant generated from Φ_I and Φ_J, respectively), the same reasoning will be followed as above. In the first place it is easily seen that the second-order contribution to the off-diagonal elements is zero

$$\Phi_R: \quad \frac{\langle h\bar{h}ab|\hat{V}|a\bar{a}p\bar{b}\rangle\langle a\bar{a}p\bar{b}|\hat{V}|h\bar{h}b\bar{a}\rangle}{E_J - E_R} = 0 \tag{4.36a}$$

$$\Phi_S: \quad \frac{\langle h\bar{h}ab|\hat{V}|b\bar{b}p\bar{a}\rangle\langle b\bar{b}p\bar{a}|\hat{V}|h\bar{h}b\bar{a}\rangle}{E_J - E_S} = 0 \tag{4.36b}$$

The first integral in the numerator of the Φ_R contribution is non-zero because the determinants in the *bra* and the *ket* only differ by two columns, but $|a\bar{a}p\bar{b}|$ differs at three places from $|h\bar{h}b\bar{a}|$, and hence, leads to a zero contribution. The same holds for Φ_S. At first sight, the contribution to the diagonal elements of the model space may seem non-uniform:

$$\Phi_R: \quad \frac{|\langle h\bar{h}ab|\hat{V}|a\bar{a}p\bar{b}\rangle|^2}{E_I - E_R} \neq 0 \qquad \frac{|\langle h\bar{h}b\bar{a}|\hat{V}|a\bar{a}p\bar{b}\rangle|^2}{E_J - E_R} = 0 \tag{4.37a}$$

$$\Phi_S: \quad \frac{|\langle h\bar{h}ab|\hat{V}|b\bar{b}p\bar{a}\rangle|^2}{E_I - E_S} = 0 \qquad \frac{|\langle h\bar{h}b\bar{a}|\hat{V}|b\bar{b}p\bar{a}\rangle|^2}{E_J - E_S} \neq 0 \tag{4.37b}$$

At difference with the $2h$-$2p$ determinants discussed above, the two non-zero integrals are not necessarily equal in this case. However, the effect of $\Phi'_R = |a\bar{a}b\bar{p}|$ and $\Phi'_S = |b\bar{b}a\bar{p}|$ exactly compensates this disequilibrium:

$$\Phi'_R: \quad \frac{|\langle h\bar{h}ab|\hat{V}|a\bar{a}b\bar{p}\rangle|^2}{E_I - E'_R} = 0 \qquad \frac{|\langle h\bar{h}b\bar{a}|\hat{V}|a\bar{a}b\bar{p}\rangle|^2}{E_J - E'_R} \neq 0 \tag{4.38a}$$

$$\Phi'_S: \quad \frac{|\langle h\bar{h}ab|\hat{V}|b\bar{b}a\bar{p}\rangle|^2}{E_I - E'_S} \neq 0 \qquad \frac{|\langle h\bar{h}b\bar{a}|\hat{V}|b\bar{b}a\bar{p}\rangle|^2}{E_J - E'_S} = 0 \tag{4.38b}$$

The denominators in the non-zero contributions of Φ_R and Φ'_R are equal since $E_R = E'_R$, and $E_I = E_J$ in a degenerate model space. Furthermore, the integral $\langle h\bar{h}ab|\hat{V}|a\bar{a}p\bar{b}\rangle$ in Eq. 4.37a is exactly the same as the integral $\langle h\bar{h}b\bar{a}|\hat{V}|a\bar{a}b\bar{p}\rangle$ of Eq. 4.38a

$$\langle h\bar{h}ab|\hat{V}|a\bar{a}p\bar{b}\rangle = \langle h\bar{h}ab|\hat{H}|a\bar{a}p\bar{b}\rangle = -\langle h\bar{h}ab|\hat{H}|p\bar{a}ab\rangle =$$
$$- \langle h\bar{h}|\frac{1 - \hat{P}_{12}}{r_{12}}|p\bar{a}\rangle = \int \frac{h(1)h(2)p(1)a(2)}{r_{12}}d\tau_1 d\tau_2 \tag{4.39a}$$

$$\langle \overline{hh}b\overline{a}|\hat{V}|a\overline{a}b\overline{p}\rangle = \langle \overline{hh}b\overline{a}|\hat{H}|a\overline{a}b\overline{p}\rangle = -\langle \overline{hh}b\overline{a}|\hat{H}|a\overline{p}b\overline{a}\rangle =$$

$$- \langle \overline{hh}|\frac{1-\hat{P}_{12}}{r_{12}}|a\overline{p}\rangle = \int \frac{h(1)h(2)a(1)p(2)}{r_{12}}d\tau_1 d\tau_2 \quad (4.39b)$$

The same reasoning holds for Φ_S and Φ_S' showing that taking into account the $2h$-$1p$ and $1h$-$2p$ excitations only causes a uniform shift of the diagonal matrix elements, and hence, they can be left out of the calculation of the energy difference between the states of the model space.

This variant of the difference dedicated CI is commonly known as DDCI2 and gives reasonable energy differences for systems with a moderate importance of the ionic determinants. This is specially interesting for the treatment of organic biradicals or TM complexes with weakly coupled spin moments. However, the DDCI2 energy difference becomes increasingly more approximate when the CAS reference wave function contains non-negligible contributions from non-degenerate determinants. In these cases, one necessarily has to rely on the more expensive DDCI procedure. Finally, the external space is sometimes even further reduced by eliminating also the $2h$ and $2p$ determinants from the CI. The resulting CAS+S or DDCI1 method can be used to obtain a first impression of the relative size of the parameters, but does normally not provide accurate answers. Moreover, one should be aware of the practical problem that the implementations of this variant in different computer programs do not consider exactly the same list of determinants.

4.3.3 Multireference Perturbation Theory

As an alternative for the variational methods, one can also apply multireference perturbation theory (MRPT) to calculate magnetic interactions. In principle, this type of calculations makes it possible to treat larger systems with more unpaired electrons. Among the many different implementations, two schemes are especially popular in the field of magnetic interactions: CASPT2 [13] and NEVPT2 [14, 15], which will be shortly overviewed here.

The standard Møller–Plesset perturbation theory uses a single determinant reference wave function and defines the zeroth-order Hamiltonian as the sum of the Fock operators

$$\hat{H}_{MP}^{(0)} = \sum_i \hat{F}_i \quad (4.40)$$

By Koopmans' theorem, the eigenvalues of the Fock operator applied on occupied orbitals are proportional to ionization potentials and the eigenvalues corresponding to unoccupied orbitals are related to the electron affinities. CASPT2 extends the applicability to multireference cases by defining an effective one-electron Fock-type operator as zeroth order Hamiltonian in the following way

Fig. 4.11 Structure of the f-matrix of $\hat{H}^{(0)}$ in CASPT2. The *dark grey* blocks have non-zero values, the *white blocks* are zero and the *light grey blocks* are zero when the orbitals are optimized for the state under study. For zero active orbitals only the diagonal elements survive and the Møller–Plesset definition of $\hat{H}^{(0)}$ emerges

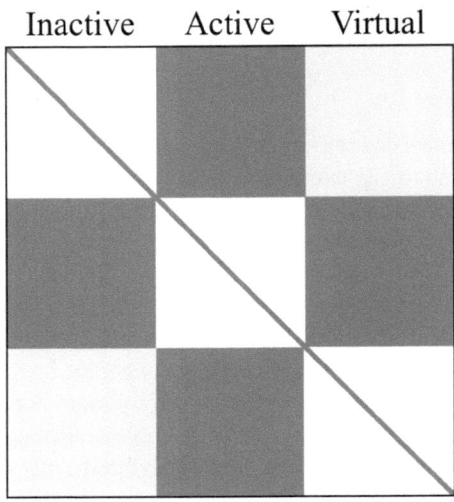

$$\hat{H}^{(0)} = \sum_{rs\sigma} f_{rs\sigma} \hat{E}_{rs}; \text{ with } f_{rs\sigma} = -\langle 0|[[\hat{H}, \hat{a}_{s\sigma}^{\dagger}], \hat{a}_{r\sigma}]_{+}|0\rangle \qquad (4.41)$$

where $|0\rangle$ is the CASSCF reference wave function and σ a general index for the spin coordinates. This definition may appear complicated at first sight, but a closer look on the f-matrix learns that it is in fact a rather straightforward expression that reduces to the Møller–Plesset zeroth-order Hamiltonian in the limit of zero active orbitals.

$$f_{rs} = \langle 0|\hat{a}_r \hat{H} \hat{a}_s^{\dagger}|0\rangle - \langle 0|\hat{a}_s^{\dagger} \hat{H} \hat{a}_r|0\rangle - \langle 0|\hat{a}_r \hat{a}_s^{\dagger} \hat{H}|0\rangle + \langle 0|\hat{H} \hat{a}_s^{\dagger} \hat{a}_r|0\rangle \qquad (4.42)$$

The structure of the matrix is schematically presented in Fig. 4.11. The inactive-virtual block of the matrix is given by

$$f_{hp} = \langle 0|\hat{a}_h \hat{H} \hat{a}_p^{\dagger}|0\rangle - \langle 0|\hat{a}_p^{\dagger} \hat{H} \hat{a}_h|0\rangle - \langle 0|\hat{a}_h \hat{a}_p^{\dagger} \hat{H}|0\rangle + \langle 0|\hat{H} \hat{a}_p^{\dagger} \hat{a}_h|0\rangle = 0 \quad (4.43)$$

where the first term is zero by $\langle 0|\hat{a}_h = \hat{a}_h^{\dagger}|0\rangle = 0$, since no particle can be created in an occupied orbital. The second and third term can be shown to be zero with an equivalent reasoning, while the fourth term is zero by the extended Brillouin theorem. The operator $\hat{a}_p^{\dagger} \hat{a}_h$ generates a singly excited configuration, which does not interact with the CASSCF wave function provided optimized orbitals are used. The diagonal elements of the inactive-inactive block are

$$f_{hh} = \langle 0|\hat{a}_h \hat{H} \hat{a}_h^{\dagger}|0\rangle - \langle 0|\hat{a}_h^{\dagger} \hat{H} \hat{a}_h|0\rangle - \langle 0|\hat{a}_h \hat{a}_h^{\dagger} \hat{H}|0\rangle + \langle 0|\hat{H} \hat{a}_h^{\dagger} \hat{a}_h|0\rangle = -IP_h \quad (4.44)$$

In this case, the first and third term are again zero for the same reason as exposed above. However, the operators in the second term annihilate an electron in the bra and the ket, which result in $\langle N-1|\hat{H}|N-1\rangle$, the energy of the ionized system.

The operator $\hat{a}_h^\dagger \hat{a}_h$ in the fourth term first annihilates an electron in orbital h and subsequently creates it again in the same orbital. This leads to $\langle 0|\hat{H}|0\rangle$, the CASSCF energy of the N-electron system. The off-diagonal terms $f_{hh'}$ are all zero. Finally, we consider the virtual-virtual diagonal elements of f:

$$f_{pp} = \langle 0|\hat{a}_p\hat{H}\hat{a}_p^\dagger|0\rangle - \langle 0|\hat{a}_p^\dagger\hat{H}\hat{a}_p|0\rangle - \langle 0|\hat{a}_p\hat{a}_p^\dagger\hat{H}|0\rangle + \langle 0|\hat{H}\hat{a}_p^\dagger\hat{a}_p|0\rangle = -EA_p \quad (4.45)$$

The action of $\hat{a}_p$ on $|0\rangle$ (annihilation of an electron in an empty orbital) results in zeros for the second and fourth terms. $\langle 0|\hat{a}_p$ and $\hat{a}_p^\dagger|0\rangle$ generate an electron in orbital p making the first term equal to the energy of the corresponding $(N+1)$-electron state. The third term is the energy of the CASSCF reference, and hence, the diagonal terms of the virtual-virtual block of f are electron affinities. The off-diagonal elements $f_{pp'}$ are zero. Then it is readily seen that the CASPT2 $\hat{H}^{(0)}$ reduces to the Møller–Plesset Hamiltonian in the limit of zero active orbitals.

It is important to realize that the one-electron nature of $\hat{H}^{(0)}$ makes that the expectation values of the excited configurations $E_R^{(0)}$ appearing in the denominator of the corrections to the energy and wave function do not coincide with the expectation values of the real Hamiltonian $\hat{H}$. In some specific cases, it can happen that $E_R^{(0)}$ is very close to (or even smaller than) the expectation value of the ground state. Such *intruder* states may cause a break-down of the perturbation theory. CASPT2 implementations provide a pragmatic solution to this problem by the so-called level-shift technique, in which near-degeneracies are removed by adding an extra term to the denominator.

Although this approach often resolves the intruder state problem very efficiently, a methodologically more satisfying route is taken in the *n*-electron valence state second-order perturbation theory (NEVPT2). By including two-electron interactions in $\hat{H}^{(0)}$, this perturbative scheme does not suffer from the intruder state problem, except in some pathological cases. The zeroth-order Hamiltonian proposed by Dyall [16] reads

$$\hat{H}_D^{(0)} = \hat{H}_i + \hat{H}_v + C \quad (4.46)$$

where $\hat{H}_i$ is of one-electron nature and acts on the inactive and virtual orbitals

$$\hat{H}_i = \sum_h \varepsilon_h \hat{E}_{hh} + \sum_p \varepsilon_p \hat{E}_{pp} \quad (4.47)$$

$\hat{H}_v$ is a two-electron operator but is restricted to the active space

$$\hat{H}_v = \sum_{a,b} h_{ab}^{eff} \hat{E}_{ab} + \frac{1}{2}\sum_{a,b,c,d} \langle ab|\frac{1-\hat{P}_{12}}{r_{12}}|cd\rangle\left(\hat{E}_{ac}\hat{E}_{bd} - \delta_{bc}\hat{E}_{ad}\right) \quad (4.48)$$

and C is an appropriate constant shift to ensure that $\hat{H}_D^{(0)}$ is equivalent to the full Hamiltonian in the active part.

Contracted versus uncontracted: The simplest way to define the first-order wave function is to apply single and double excitation operators on all the determinants (or CSFs) of the reference wave function.

$$\psi^{(1)} = \sum_I \sum_{rstu} c_{I,rstu} \hat{E}_{rs} \hat{E}_{tu} \Phi_I \tag{4.49}$$

The second-order correction to the energy is relatively straightforward to evaluate, but the number of terms in the summation rapidly becomes very large, especially for large reference wave functions. A second approach is to apply excitation operators not on the individual determinants of CSFs of the reference space, but on the reference wave function as a whole.

$$\psi^{(1)} = \sum_{rstu} c_{rstu} \hat{E}_{rs} \hat{E}_{tu} \psi^{(0)} \tag{4.50}$$

This approach generates much less terms since the external determinants appear as *contracted* sums in the first-order wave function. Moreover, the dimension of the external space does not grow as fast with the size of $\psi^{(1)}$ as in the uncontracted way of generating $\psi^{(1)}$. On the other hand, the calculation of the second-order correction to the energy relies on significantly more complicated expressions but once programmed this is just a minor issue compared to the limited length of $\psi^{(1)}$.

The differences between the contracted and uncontracted procedure are best illustrated by giving two examples with a very simple reference wave function: $\psi^{(0)} = \lambda|h\bar{h}a\bar{a}b| + \mu|h\bar{h}ab\bar{b}|$. In the first place, we will apply the single excitation operator involving the occupied orbital h and the unoccupied orbitals p and p'. The uncontracted wave function reads

$$\psi^{(1)} = c_1|\bar{h}a\bar{a}bp| + c_2|ha\bar{a}b\bar{p}| + c_3|\bar{h}a\bar{a}bp'| + c_4|ha\bar{a}b\bar{p}'|$$
$$+ c_5|\bar{h}ab\bar{b}p| + c_6|hab\bar{b}\bar{p}| + c_7|\bar{h}ab\bar{b}p'| + c_8|hab\bar{b}\bar{p}'| \tag{4.51}$$

and in the contracted formalism, the following function is generated

$$\psi^{(1)} = c_1\{\lambda|\bar{h}a\bar{a}bp| + \mu|\bar{h}ab\bar{b}p|\} + c_2\{\lambda|ha\bar{a}b\bar{p}| + \mu|hab\bar{b}\bar{p}|\}$$
$$+ c_3\{\lambda|\bar{h}a\bar{a}bp'| + \mu|\bar{h}ab\bar{b}p'|\} + c_4\{\lambda|ha\bar{a}b\bar{p}'| + \mu|hab\bar{b}\bar{p}'|\} \tag{4.52}$$

The uncontracted first-order correction has eight different coefficients to be determined, while the contracted variant generates the same determinants with only four different coefficients. The fact that the external determinants are weighted by the coefficients of the reference wave function does only slightly influence the final result. The second example applies the single excitation operator involving the active orbitals a and b, and the virtual orbital p. Again, the uncontracted algorithm generates a list of all five possible excited determinants

$$\psi^{(1)} = c_1|\overline{a}bp| + c_2|ab\overline{p}| + c_3|a\overline{a}p| + c_4|b\overline{b}p| + c_5|a\overline{b}p| \qquad (4.53)$$

where the $h\overline{h}$-part has been omitted for simplicity. The contracted wave function is shorter:

$$\psi^{(1)} = c_1\{\lambda|\overline{a}bp| + \mu|b\overline{b}p|\} + c_2\lambda|ab\overline{p}| + c_3\{\lambda|a\overline{a}p| + \mu|a\overline{b}p|\} + c_4\mu|ab\overline{p}| \quad (4.54)$$

Since the determinants of the second and fourth term are the same, the wave function presents a linear dependence, which should be removed and further reduces the number of coefficients.

The contraction written in Eq. 4.50 is used in CASPT2 and in the *partially-contracted* variant of NEVPT2. The latter method is also available in a *strongly-contracted* variant of NEVPT2, in which the contracted external functions are grouped together depending on the number of electrons added or removed from the active space. In this way a reduced set of orthogonal external functions is generated.

4.3.4 Spin Unrestricted Methods

The observation that except for the state of maximum multiplicity, spin states cannot be rigorously represented with a single determinant makes it very interesting to look at the possibility to study magnetic interactions in a spin unrestricted setting using a single determinant description of the spin states. We will start with the spatially symmetric 2-electron/2-orbital case and afterwards generalize for systems with more unpaired electrons.

The most widely applied approximation to extract J within a single determinant description of the spin states is the so-called *Broken Symmetry* approach which uses two determinants:

$$\Phi_{BS} = |\phi_1\overline{\phi}_2| \qquad \Phi_{HS} = |\phi_1\phi_2| \qquad (4.55)$$

where the closed-shell orbitals have been omitted for convenience. The $\hat{S}^2$ expectation value of Φ_{HS} is not exactly equal to 2 since the closed shell spin orbitals appear in pairs with slightly different spatial orbitals. However in most cases it is close to 2 and is generally considered as a good approximation to the triplet state obtained in a spin-restricted setting.

$$\Phi_{HS} \approx \Phi_T \qquad (4.56)$$

$$E_{HS} = \langle\Phi_{HS}|\hat{H}|\Phi_{HS}\rangle \approx E_T \qquad (4.57)$$

$$\langle\hat{S}^2\rangle_{HS} = \langle\Phi_{HS}|\hat{S}^2|\Phi_{HS}\rangle \approx 2 \qquad (4.58)$$

One the other hand, the $\hat{S}^2$ expectation value of Φ_{BS} is neither close to zero (singlet) nor to two (triplet), but rather somewhere in between. It is therefore a logical step to approximate the broken symmetry determinant as a linear combination of the spin-restricted singlet and triplet states [17]:

$$|\Phi_{BS}\rangle = \lambda|\Phi_S\rangle + \mu|\Phi_T\rangle \quad \text{with} \quad \lambda^2 + \mu^2 = 1 \quad (4.59)$$

with the following energy and $\hat{S}^2$ expectation value

$$E_{BS} = \langle \lambda\Phi_S + \mu\Phi_T|\hat{H}|\lambda\Phi_S + \mu\Phi_T\rangle = \lambda^2 E_S + \mu^2 E_T \quad (4.60)$$

$$\langle \hat{S}^2\rangle_{BS} = \lambda^2\langle\Phi_S|\hat{S}^2|\Phi_S\rangle + \mu^2\langle\Phi_T|\hat{S}^2|\Phi_T\rangle = 2\mu^2 \quad (4.61)$$

After substituting $\mu^2 = 1 - \lambda^2$ from the normalization condition in Eq. 4.61, we obtain

$$\lambda^2 = 1 - \frac{\langle\hat{S}^2\rangle_{BS}}{2} \quad \text{and} \quad \mu^2 = \frac{\langle\hat{S}^2\rangle_{BS}}{2} \quad (4.62)$$

which can be substituted in the energy expression of the BS determinant given in Eq. 4.60

$$E_{BS} = \left(1 - \frac{\langle\hat{S}^2\rangle_{BS}}{2}\right)E_S + \frac{\langle\hat{S}^2\rangle_{BS}}{2}E_T \quad (4.63)$$

The energy difference of the BS and HS determinants now reads

$$\begin{aligned} E_{BS} - E_{HS} &= E_S - \frac{\langle\hat{S}^2\rangle_{BS}}{2}(E_S - E_T) - E_T \\ &= \left(1 - \frac{\langle\hat{S}^2\rangle_{BS}}{2}\right)(E_S - E_T) \\ &= \frac{2 - \langle\hat{S}^2\rangle_{BS}}{2}(E_S - E_T) \end{aligned} \quad (4.64)$$

which leads to the final expression of the magnetic coupling parameter J as function of the energies of the spin-unrestricted HS and BS determinants

$$J = E_S - E_T = \frac{2(E_{BS} - E_{HS})}{2 - \langle\hat{S}^2\rangle_{BS}} \quad (4.65)$$

This is not the only expression used to relate the energy of the two determinants with the singlet-triplet energy difference. Under the assumption that the spin polarization in the closed shell orbitals is small enough to ensure their orthogonality, one can express the energy difference as function of the overlap of the magnetic orbitals. From Eq. 1.27 we can calculate $\langle \hat{S}^2 \rangle_{BS}$

$$\langle \phi_1 \overline{\phi}_2 | \hat{S}^2 | \phi_1 \overline{\phi}_2 \rangle = \langle \phi_1 \overline{\phi}_2 | \phi_1 \overline{\phi}_2 + \overline{\phi}_1 \phi_2 \rangle = \langle \phi_1 \overline{\phi}_2 | \phi_1 \overline{\phi}_2 - \phi_2 \overline{\phi}_1 \rangle$$
$$= 1 - \langle \phi_1 | \phi_2 \rangle^2 \qquad (4.66)$$

The substitution of this expression in Eq. 4.65 leads to

$$J = E_S - E_T = \frac{2(E_{BS} - E_{HS})}{1 + \langle \phi_1 | \phi_2 \rangle^2} \qquad (4.67)$$

which in the weak overlap limit evolves to

$$J = 2(E_{BS} - E_{HS}) \qquad (4.68)$$

and in the strong overlap limit to

$$J = E_{BS} - E_{HS} \qquad (4.69)$$

Note that in the latter case the overlap $\langle \phi_1 | \phi_2 \rangle$ tends to one, which means that ϕ_1 becomes equal to ϕ_2 and $\Phi_{BS} = |\phi_1 \overline{\phi}_1|$ represents a closed shell singlet state.

Expression 4.67 can be rewritten in terms of spin densities to avoid the less generally available overlap of the magnetic orbitals [18]. The simplest way to do this is to express the non-orthogonal magnetic orbitals ϕ_1 and ϕ_2 in the local orthogonal orbitals ψ_1 and ψ_2:

$$|\Phi_{BS}\rangle = |\phi_1 \overline{\phi}_2\rangle = |(\lambda \psi_1 + \mu \psi_2)(\mu \overline{\psi}_1 + \lambda \overline{\psi}_2)\rangle \qquad (4.70)$$

with $\langle \phi_1 | \phi_1 \rangle = \langle \phi_2 | \phi_2 \rangle = \lambda^2 + \mu^2 = 1$ and $\langle \phi_1 | \phi_2 \rangle = 2\lambda\mu$. The α and β spin densities arise from ϕ_1 and ϕ_2, respectively, and are equal to λ^2 and μ^2 for site 1. From this the total spin density can be obtained

$$\left. \begin{array}{l} \rho_1^\alpha = \lambda^2 \\ \rho_1^\beta = \mu^2 \end{array} \right\} \Rightarrow \rho_1^{\alpha-\beta} = \lambda^2 - \mu^2 \overset{2\lambda^2+\mu^2=1}{\Longrightarrow} \left\{ \begin{array}{l} 2\lambda^2 = 1 + \rho_1^{\alpha-\beta} \\ 2\mu^2 = 1 - \rho_1^{\alpha-\beta} \end{array} \right. \qquad (4.71)$$

Now the relation with $\langle \phi_1 | \phi_2 \rangle$ in Eq. 4.67 is easily made

$$\langle \phi_1 | \phi_2 \rangle^2 = 4\lambda^2 \mu^2 = (1 + \rho_1^{\alpha-\beta})(1 - \rho_1^{\alpha-\beta}) = 1 - (\rho_1^{\alpha-\beta})^2 = 1 - (\rho_1^{BS})^2 \quad (4.72)$$

Note that this equation assumes that all the spin density is localized on the magnetic centers.

With a slightly more elaborate derivation one can also handle cases with an important delocalization of the spin density onto the ligands [19]. The magnetic orbitals are written as a linear combination of three nonorthogonal basis functions

$$\phi_1 = \lambda\chi_1 + \mu\chi_2 + \nu\chi_3 \qquad \phi_2 = \mu\chi_1 + \lambda\chi_2 + \nu\chi_4 \tag{4.73}$$

with $\langle\chi_i|\chi_j\rangle \neq 0$, $\langle\chi_i|\chi_i\rangle = 1$ and $\lambda \gg \mu, \nu$. The basis functions χ_1 and χ_2 are centered on the magnetic site 1 and 2, respectively. The other two functions are ligand orbitals around site 1 (χ_3) and site 2 (χ_4). Furthermore, it holds that $\langle\chi_1|\chi_4\rangle = \langle\chi_2|\chi_3\rangle \ll \langle\chi_1|\chi_3\rangle = \langle\chi_2|\chi_4\rangle$ in a centro-symmetric system. The overlap of the two magnetic orbitals is

$$\begin{aligned}
\langle\phi_1|\phi_2\rangle = 2\lambda\mu &+ \nu^2\langle\chi_3|\chi_4\rangle + (\lambda^2 + \mu^2)\langle\chi_1|\chi_2\rangle \\
&+ 2\lambda\nu\langle\chi_1|\chi_4\rangle + 2\mu\nu\langle\chi_1|\chi_3\rangle
\end{aligned} \tag{4.74}$$

Many terms can be neglected in this expression. The terms with μ^2, ν^2 or $\mu\nu$ are small because these coefficients are much smaller than λ. Being located in different parts of the complex, the overlap integrals $\langle\chi_1|\chi_4\rangle$ and $\langle\chi_1|\chi_2\rangle$ are also expected to be small. This makes that the overlap of the magnetic orbitals can be roughly approximated by $2\lambda\mu$. The spin density on site 1 can be determined using the Mulliken population reasoning. The contribution due to ϕ_1 and ϕ_2 are

$$\phi_1 \text{ contribution:} \qquad \lambda^2 + \frac{1}{2}(2\lambda\mu\langle\chi_1|\chi_2\rangle + 2\lambda\nu\langle\chi_1|\chi_3\rangle) \tag{4.75}$$

$$\phi_2 \text{ contribution:} \qquad \mu^2 + \frac{1}{2}(2\lambda\mu\langle\chi_1|\chi_2\rangle + 2\mu\nu\langle\chi_1|\chi_4\rangle) \tag{4.76}$$

which reduce to λ^2 and μ^2 if we apply the same approximations as for the overlap of the magnetic orbitals. The spin density at the magnetic sites for Φ_{HS} and Φ_{BS} are given by

$$\rho_1^{HS} = \lambda^2 + \mu^2 \qquad \rho_1^{BS} = \lambda^2 - \mu^2 \tag{4.77}$$

Now it is easily derived that

$$(\rho_1^{HS})^2 - (\rho_1^{BS})^2 = 4\lambda^2\mu^2 = \langle\phi_1|\phi_2\rangle^2 \tag{4.78}$$

which can be used to replace the overlap integral in Eq. 4.67 with the more generally available spin populations. This expression is valid for centro-symmetric systems but improves the previous one by the fact that it is no longer implicit that the spin density in the HS state is entirely located on the magnetic center.

The extension of Eq. 4.65 to the general case of magnetic coupling between two centers with more than one unpaired electron is straightforward and follows the same logics. The spin-unrestricted HS determinant Φ_{HS} is assumed to be a good approximation to the spin eigenfunction of maximum multiplicity $\Phi_{S_{max}}$, and therefore,

$$\langle\hat{S}^2\rangle_{HS} = S_{max}(S_{max} + 1) \qquad E_{HS} = E_{S_{max}} \tag{4.79}$$

The broken symmetry determinant is written as a linear combination of the singlet Φ_S and the S_{max} spin eigenfunctions.[2]

$$|\Phi_{BS}\rangle = \lambda|\Phi_S\rangle + \mu|\Phi_{S_{max}}\rangle = \lambda|\Phi_S\rangle + \mu|\Phi_{HS}\rangle \tag{4.80}$$

From the $\hat{S}^2$ expectation value

$$\langle\hat{S}^2\rangle_{BS} = \lambda^2\langle\Phi_S|\hat{S}^2|\Phi_S\rangle + \mu^2\langle\Phi_{HS}|\hat{S}^2|\Phi_{HS}\rangle = \langle\hat{S}^2\rangle_{HS}\mu^2 \tag{4.81}$$

one arrives at

$$E_{BS} = \lambda^2 E_S + \mu^2 E_{HS} = \left(1 - \frac{\langle\hat{S}^2\rangle_{BS}}{\langle\hat{S}^2\rangle_{HS}}\right)E_S + \frac{\langle\hat{S}^2\rangle_{BS}}{\langle\hat{S}^2\rangle_{HS}}E_{HS} \tag{4.82}$$

Then, the energy difference between Φ_{BS} and Φ_{HS} is given by

$$E_{BS} - E_{HS} = E_S - \frac{\langle\hat{S}^2\rangle_{BS}}{\langle\hat{S}^2\rangle_{HS}}(E_S - E_{HS}) - E_{HS}$$

$$= \frac{\langle\hat{S}^2\rangle_{HS} - \langle\hat{S}^2\rangle_{BS}}{\langle\hat{S}^2\rangle_{HS}}(E_S - E_{HS}) \tag{4.83}$$

$$\Rightarrow E_S - E_{HS} = \frac{\langle\hat{S}^2\rangle_{HS}(E_{BS} - E_{HS})}{\langle\hat{S}^2\rangle_{HS} - \langle\hat{S}^2\rangle_{BS}} \tag{4.84}$$

from which the expression for J is directly derived

$$J = \frac{2(E_S - E_{HS})}{S_{max}(S_{max} + 1)} = \frac{2(E_S - E_{HS})}{\langle\hat{S}^2\rangle_{HS}} = \frac{2(E_{BS} - E_{HS})}{\langle\hat{S}^2\rangle_{HS} - \langle\hat{S}^2\rangle_{BS}} \tag{4.85}$$

This is the famous Yamaguchi relation originally derived in the framework of unrestricted Hartree–Fock calculations [20], but later also widely applied in DFT calculations. In the limit of zero overlap of the magnetic orbitals, $\langle\hat{S}^2\rangle_{BS}$ becomes equal to S_{max} and the following expression emerges

$$J = \frac{2(E_{BS} - E_{HS})}{S_{max}(S_{max} + 1) - S_{max}} = \frac{2(E_{BS} - E_{HS})}{S_{max}^2} \tag{4.86}$$

derived earlier by Noodleman [21] and which reduces to Eq. 4.68 for two magnetic centers with S=1/2. On the other hand, $\langle\hat{S}^2\rangle_{BS}$ is zero in the strong overlap limit and J relates to the energies of the HS and BS determinants as

---

[2]This is of course an approximation. There is no obvious reason to exclude the intermediate spin states from the linear combination.

$$J = \frac{2(E_{BS} - E_{HS})}{S_{max}(S_{max} + 1)} \tag{4.87}$$

which is the generalized form of Eq. 4.69. This expression is also used in DFT when the BS determinant is considered to be a good representation of the singlet (or lowest spin) state as proposed by Ruiz and co-workers [6, 19, 22]. These authors often replace the denominator by $2(2S_1S_2 + S_2)$ with $S_2 \geqslant S_1$ and $S_1 + S_2 = S_{max}$ to reflect situations with unequal spin moments on the two magnetic centers.

4.10 Calculate the expectation value of $\hat{S}^2$ for $\Phi_1 = |\phi_1\phi_2\overline{\phi}_3\overline{\phi}_4|$ in the zero overlap limit: $\langle\phi_i|\phi_j\rangle = \delta_{ij}$.

4.3.5 Alternatives to the Broken Symmetry Approach

The introduction of the broken symmetry determinant as representation of the low-spin coupled spin state not only provides (computational) chemists with a tool to calculate magnetic interactions with single determinant methods, it also makes a connection with the intuitive representations of spins with up- and downwards pointing arrows at each magnetic center. However, this representation does not lead to spin functions that are eigenfunctions of the total spin operator $\hat{S}^2$, as expected in a non-relativistic setting and explained in Chap. 1. From this point of view the broken symmetry approach is less satisfactory and there have been many attempts to design alternative approaches to calculate magnetic interactions with DFT to improve upon the shortcomings of the standard approach.

A natural starting point is to combine a multiconfigurational SCF approach to treat the static electron correlation[3] and DFT for the remaining (mainly dynamic) electron correlation. It is, however, not easy to design functionals that only take into account this latter part of the electron correlation and do not consider (part of) the static correlation. Despite many efforts, there seems no definitive solution to the double counting problem.

Restricted ensemble Kohn–Sham DFT Alternatively one can perform standard KS-DFT calculations on a collection of determinants with different occupations and take a weighted average of the individual energies to obtain an estimate of the multideterminantal situation. To avoid the independent calculation of several KS determinants, a generalization of this approach was proposed by Filatov and Shaik based on the coupling operator technique developed by Roothaan for restricted open-shell Hartree–Fock. This restricted open-shell Kohn–Sham (ROKS) approach was later extended to situations where fractional occupation numbers are not imposed by the

---

[3]In the case of magnetic interactions, the multideterminantal character of the N-electron states with $S < S_{max}$.

symmetry (as in atomic multiplets or ligand-field states in coordination complexes) but due to *accidental* (near-)degeneracies. This extended approach was named the restricted ensemble Kohn–Sham (REKS) method [23, 24]. An optimal set of Kohn–Sham orbitals and occupation numbers is obtained by a minimization procedure that always maintains the spin and spatial symmetry of the N-electron state under study.

The approach has been used to calculate the coupling of two localized spin moments in binuclear transition metal complexes and the singlet-triplet splitting in biradical systems, such as twisted ethylene. Rather reasonable values of the magnetic coupling parameters were obtained. In general, the couplings are slightly too small, which may be attributed to the lack of spin polarization. An important advantage of the method is the fact that geometries can be optimized for open-shell singlet states within the DFT framework.

Spin-flip time-dependent DFT The energy differences of the spin states involved in the magnetic interaction of two (or more) spin moments can be seen to some extent as vertical excitation energies, and hence, time-dependent DFT (or other linear response methods as equation of motions coupled cluster [25]) could in principle be used to determine the magnetic interaction between two spin moments. However, the standard implementation of TD-DFT only considers single, *spin-conserving* excitations, which prevents accounting for the multideterminantal character of the states with low-spin coupling [26]. Figure 4.12 shows the five determinants that are essential to describe the magnetic coupling in a two-electron/two-orbital problem. Using determinant Φ_1, Φ_2 or Φ_3 as reference will not generate all five determinants in standard TD-DFT, while Φ_4 and Φ_5 lead to the same spin contamination problems as in the BS approach discussed above. The spin-flip formalism (originally developed in the framework of Hartree–Fock and coupled cluster, and later implemented for TD-DFT) offers an interesting solution to this shortcoming. The determinant with maximum M_S-value is taken as reference (Φ_1 in Fig. 4.12) and all single excited determinants involving one spin-flip are generated from this. Within the space of the two-electron/two-orbital problem, this procedure generates the determinants $\Phi_2 \ldots \Phi_5$ of Fig. 4.12, and hence, gives access to the energy of the open-shell singlet within the TD-DFT framework without spin-contamination problems.

Constrained DFT The basic shortcomings of the BS approach can be summarized in two points. In the first place, the spin contamination, or the impossibility to represent the low-spin states with a single Kohn–Sham determinant. The second point is the fact that nearly all todays functionals tend to overestimate the delocalization of the spin density and overestimate the antiferromagnetic character of the coupling.

Fig. 4.12 The reference determinant Φ_1 and the four spin-flip determinants ($\Phi_2 \ldots \Phi_5$) generated in SF-TDDFT with a two-electron/two-orbital target space

Constrained DFT (C-DFT) remedies, at least partially, the latter by putting restrictions on the spatial distributions of the α and β electrons [27]. Two fragments p and q are defined such that both include one magnetic center and the atoms around it. Subsequently, the density is optimized under the restrictions that $N_\alpha^p - N_\beta^p = M_S^p$ and $N_\alpha^q - N_\beta^q = M_S^q$, where $N_{\alpha,\beta}^{p,q}$ are the summed spin populations of the atoms in the fragments and $M_S^{p,q}$ the prefixed excess of α or β electrons in each fragment. C-DFT results in less delocalized spin densities and therefore, in general, to smaller interaction parameters.

Problems

4.1 A master student wants to study the energy splitting $E_S - E_T$ in a planar $[Cu_2F_6]^{2-}$ model system, since experimental studies of similar di Cl-bridged Cu^{II} dimers suggested that $E_S - E_T$ depends strongly on the Cu–Cl–Cu angle θ. She performs RHF calculations on the triplet state in order to predict $E_S - E_T$ with the HTH model. She produces a Table of results, where the *gerade* and *ungerade* open shell orbitals are denoted 1 and 2, respectively.

θ	$\frac{J_{11}+J_{22}}{2} - J_{12}$ [K]	K_{12} [E$_h$]	$\varepsilon_1 - \varepsilon_2$ [E$_h$]
85°	24	0.4324	−0.0078
90°	20	0.4376	−0.0025
95°	22	0.4419	0.0034
100°	26	0.4456	0.0094
105°	32	0.4483	0.0150

Compute J (in K) for $\theta = 85° \ldots 105°$ using the HTH model. Do you observe a strong dependence of the coupling on the angle? Can the same conclusions be drawn when only considering the orbital energies?

4.2 Quantifying the counter-complementarity effect. Standard optimization of the molecular orbitals of a magnetic complex with two magnetic centers bridged by two different ligands normally leads to magnetic orbitals with contributions on both ligands (as φ_5 and φ_6 in Fig. 4.9). This makes it very hard to quantify the counter-complementary effect of the two ligands. Design a computational strategy to determine quantitatively the reduction of the magnetic coupling through ligand 1 by the counter-complementary effect of ligand 2. Hint: Many quantum chemical programs can divide the whole system into fragments.

4.3 Broken symmetry approach. The magnetic coupling of three binuclear TM complexes has been studied with DFT. The following results were obtained for the HS and BS determinants. (a) Calculate the magnetic coupling parameter J with the Yamaguchi equation (Eq. 4.85) and compare the outcomes to the alternative relations of Noodleman (Eq. 4.86) and Ruiz (Eq. 4.87).

TM	Energy [E_h]		$\langle \hat{S}^2 \rangle$	
	Φ_{HS}	Φ_{BS}	Φ_{HS}	Φ_{BS}
Cu^{2+}	-4061.7435920	-4061.7442381	2.0035	0.9957
Ni^{2+}	-3797.4742498	-3797.4767694	6.0083	1.9931
Mn^{2+}	-3082.7586297	-3082.7630415	30.0086	4.9936

(b) Calculate J in the Cu complex combining Eqs. 4.67 and 4.78 using $\rho^{\alpha-\beta}$ is 0.6864 and 0.6757 for HS and BS, respectively.

References

1. O. Kahn, B. Briat, J. Chem. Soc., Faraday Trans. **2**(72), 268 (1976)
2. P.J. Hay, J.C. Thibeault, R.J. Hoffmann, J. Am. Chem. Soc. **97**, 4884 (1975)
3. H.M. McConnell, J. Chem. Phys. **39**, 1910 (1963)
4. M. Deumal, J.J. Novoa, M.J. Bearpark, P. Celani, M. Olivucci, M.A. Robb, J. Phys. Chem. A **102**, 8404 (1998)
5. J.J. Novoa, M. Deumal, Struct. Bond. **100**, 33 (2001)
6. E. Ruiz, P. Alemany, S. Alvarez, J. Cano, J. Am. Chem. Soc. **119**, 1297 (1997)
7. O. Kahn, *Molecular Magnetism* (VCH Publishers, New York, 1993)
8. J.P. Launay, M. Verdaguer, *Electrons in Molecules: From Basic Principles to Molecular Electronics* (Oxford University Press, Oxford, 2014)
9. O. Kahn, J. Galy, Y. Journaux, J. Jaud, I. Morgenstern-Badarau, J. Am. Chem. Soc. **104**, 2165 (1982)
10. R. Costa, A. Garcia, J. Ribas, T. Mallah, Y. Journaux, J. Sletten, X. Solans, V. Rodríguez, Inorg. Chem. **32**, 3733 (1993)
11. T. Helgaker, P. Jørgensen, J. Olsen, *Molecular Electronic-structure Theory* (Wiley, Chichester, 2000)
12. J. Miralles, O. Castell, R. Caballol, J.P. Malrieu, Chem. Phys. **172**, 33 (1993)
13. K. Andersson, P.Å. Malmqvist, B.O. Roos, J. Chem. Phys. **96**, 1218 (1992)
14. C. Angeli, R. Cimiraglia, S. Evangelisti, T. Leininger, J.P. Malrieu, J. Chem. Phys. **114**, 10252 (2001)
15. C. Angeli, R. Cimiraglia, J.P. Malrieu, J. Chem. Phys. **117**(20), 9138 (2002)
16. K.G. Dyall, J. Chem. Phys. **102**, 4909 (1995)
17. J.P. Malrieu, R. Caballol, C.J. Calzado, C. de Graaf, N. Guihéry, Chem. Rev. **114**, 429 (2014)
18. R. Caballol, O. Castell, F. Illas, I. de P.R. Moreira, J.P. Malrieu, J. Phys. Chem. A **101**(42), 7860 (1997)
19. E. Ruiz, J. Cano, S. Alvarez, P. Alemany, J. Comput. Chem. **20**(13), 1391 (1999)
20. K. Yamaguchi, H. Fukui, T. Fueno, Chem. Lett. **15**, 625 (1986)
21. L. Noodleman, J. Chem. Phys. **74**, 5737 (1981)
22. J.P. Perdew, A. Savin, K. Burke, Phys. Rev. A **51**(6), 4531 (1995)
23. M. Filatov, S. Shaik, Chem. Phys. Lett. **288**, 689 (1998)
24. M. Filatov, S. Shaik, Chem. Phys. Lett. **304**, 429 (1999)
25. A.I. Krylov, Chem. Phys. Lett. **350**, 522 (2001)
26. Y.A. Bernard, Y. Shao, A.I. Krylov, J. Chem. Phys. **136**, 204103 (2012)
27. B. Kaduk, T. Kowalczyk, T. Van Voorhis, Chem. Rev. **112**, 321 (2012)

Chapter 5
Towards a Quantitative Understanding

Abstract Taking a binuclear copper complex as model system, the isotropic magnetic coupling is decomposed into different contributions. Perturbative expressions of the main contributions are derived and illustrated with numerical examples. An effective Hamiltonian is constructed that incorporates all important electron correlation effects and establishes a connection between the complex N-electron wave functions and the simpler qualitative methods discussed in the previous chapter. Subsequently an outline is given of the analysis of the coupling with a single determinant approach and the biquadratic and four-center interactions are decomposed. The chapter closes with the recently proposed method to extract DFT estimates for these complex interactions.

5.1 Decomposition of the Magnetic Coupling

The production of accurate electronic structure parameters is of course an important result for computational chemistry. However, it should not be the final goal and one has to go one step further on the road towards understanding. The qualitative valence methods described in the first sections of the previous chapter of this book are mainly focused on this understanding of the coupling, but here we discuss three approaches to analyse the results of the computational schemes that aim at a quantitative agreement with experiment. In this way quantitative accuracy can be combined with qualitative understanding.

The binuclear complex $[L_2Cu_2(\mu\text{-}1,3\text{-}N_3)_2]^{2+}$ $(L=N,N',N''\text{-trimethyl-1,4,7-}$ triaza-cyclononane) shows a large antiferromagnetic coupling with $J = -800\,$cm, nicely reproduced with a DDCI calculation using a CAS(2,2)SCF reference wave function on the model complex $[(NH_3)_6Cu_2(\mu\text{-}1,3\text{-}N_3)_2]^{2+}$ [1]. In this section, we will closely follow the work of Calzado and co-workers, decompose this 800 cm^{-1} into small pieces and ascribe each individual contribution to well defined physical mechanisms [2, 3].

© Springer International Publishing Switzerland 2016

C. Graaf and R. Broer, *Magnetic Interactions in Molecules and Solids*,
Theoretical Chemistry and Computational Modelling,
DOI 10.1007/978-3-319-22951-5_5

5.1.1 Valence Mechanisms

First we focus on the mechanisms that arise from interactions among the configurations in the active space, restricting ourselves to the role played by the two Cu^{2+} ions. For this centro-symmetric system, the active orbitals $g = (a + b)/\sqrt{2}$ and $u = (a - b)/\sqrt{2}$ shown in Fig. 5.1 define four different determinants

$$CAS = \{|g\bar{g}|, |u\bar{u}|, |g\bar{u}|, |u\bar{g}|\} \tag{5.1}$$

The diagonalization of the corresponding 4×4 matrix produces four eigenstates, three singlets and one triplet

$$
\begin{aligned}
S_g &= \lambda|g\bar{g}| - \mu|u\bar{u}| \\
S'_g &= \mu|g\bar{g}| + \lambda|u\bar{u}| \\
T_u &= (|g\bar{u}| - |u\bar{g}|)/\sqrt{2} \\
S_u &= (|g\bar{u}| + |u\bar{g}|)/\sqrt{2}
\end{aligned}
\tag{5.2}
$$

5.1 Demonstrate by substitution that S_g is dominated by the neutral determinants when $\lambda \approx \mu$. What situation is described for $\lambda \gg \mu$?

The energy difference of S_g and T_u defines J but the analysis is much easier in a representation with localized orbitals. Therefore, we rewrite the CAS in terms of the orthogonal localized Cu orbitals a and b, shown in Fig. 5.2. By defining the following electronic structure parameters

$$
\begin{aligned}
E_{ref} &= \langle a\bar{b}|\hat{H}|a\bar{b}\rangle = 0 \\
K_{ab} &= \langle a\bar{b}|\hat{H}|b\bar{a}\rangle \\
t_{ab} &= \langle a\bar{b}|\hat{H}|a\bar{a}\rangle \\
U &= \langle a\bar{a}|\hat{H}|a\bar{a}\rangle - \langle a\bar{b}|\hat{H}|a\bar{b}\rangle
\end{aligned}
\tag{5.3}
$$

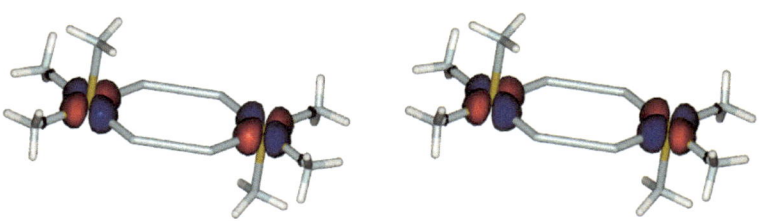

Fig. 5.1 Delocalized magnetic orbitals of *gerade* (*left*) and *ungerade* (*right*) symmetry

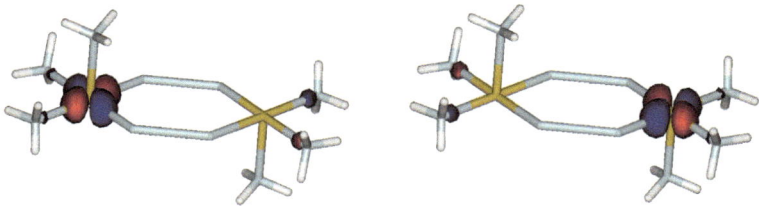

Fig. 5.2 *Left* and *right* localized magnetic orbitals

the CAS Hamiltonian matrix can be written as

$$
\begin{array}{c|cccc}
 & |a\bar{b}\rangle & |b\bar{a}\rangle & |a\bar{a}\rangle & |b\bar{b}\rangle \\
\hline
\langle a\bar{b}| & 0 & K_{ab} & t_{ab} & t_{ab} \\
\langle b\bar{a}| & K_{ab} & 0 & t_{ab} & t_{ab} \\
\langle a\bar{a}| & t_{ab} & t_{ab} & U & K_{ab} \\
\langle b\bar{b}| & t_{ab} & t_{ab} & K_{ab} & U
\end{array}
\tag{5.4}
$$

Here $E_{ref} = h_{aa} + h_{bb} + J_{ab}$ is the reference energy and has been subtracted from all diagonal matrix elements. K_{ab} is the direct exchange, U is the on-site repulsion parameter and t_{ab} is the hopping integral and gives a measure of the probability for the electron hopping from site a to b and *vice-versa*.

5.2 Express the energy of the ionic determinants $|a\bar{a}|$ and $|b\bar{b}|$ in terms of the one-electron integrals h and the Coulomb and exchange integrals J and K. What assumption has been made to reduce the diagonal element of these determinants to U?

The diagonalization of the 4×4 matrix gives four eigenvectors, equivalent to those listed above but expressed in local orbitals

$$
\begin{aligned}
S_g &= \lambda \left(|a\bar{b}| + |b\bar{a}|\right)/\sqrt{2} - \mu \left(|a\bar{a}| + |b\bar{b}|\right)/\sqrt{2} \\
S_g' &= \mu \left(|a\bar{b}| + |b\bar{a}|\right)/\sqrt{2} + \lambda \left(|a\bar{a}| + |b\bar{b}|\right)/\sqrt{2} \\
T_u &= \left(|a\bar{b}| - |b\bar{a}|\right)/\sqrt{2} \\
S_u &= \left(|a\bar{a}| - |b\bar{b}|\right)/\sqrt{2}
\end{aligned}
\tag{5.5}
$$

with energy eigenvalues

$$
E(S_g) = K_{ab} + \frac{U - \sqrt{U^2 + 16t_{ab}^2}}{2}
$$

$$E(S_g') = K_{ab} + \frac{U + \sqrt{U^2 + 16t_{ab}^2}}{2} \tag{5.6}$$

$$E(T_u) = -K_{ab}$$

$$E(S_u) = U - K_{ab}$$

This gives direct access to an analytical expression of J in terms of the previously defined electronic structure parameters

$$J = E(S_g) - E_T = 2K_{ab} + \frac{U - \sqrt{U^2 + 16t_{ab}^2}}{2} \tag{5.7}$$

To simplify this expression we use the Taylor expansion $\sqrt{p + q} = \sqrt{p} + \frac{1}{2}q/\sqrt{p} + \cdots$ with $p \gg q$.

$$J = 2K_{ab} - \frac{4t_{ab}^2}{U} \tag{5.8}$$

in which one can easily recognize the ferromagnetic ($2K_{ab}$) and antiferromagnetic ($4t_{ab}^2/U$) contributions of the qualitative Kahn–Briat and Hay–Thibeault–Hoffmann models.

5.3 Use the Taylor expansion $\sqrt{p + q} = \sqrt{p} + \frac{1}{2}q/\sqrt{p} + \cdots$ with $p = U^2$ and $q = 16t_{ab}^2$ to derive the simplified expression for J.

A pictorial understanding of this expression can be obtained within a QDPT reasoning using a model space limited to neutral determinants only. Figure 5.3 shows the two determinants $\Phi_I = |a\overline{b}|$ (left) and $\Phi_J = |\overline{a}b|$ on the right. The Heisenberg Hamiltonian matrix element of these two determinants is equal to $-\frac{1}{2}J$, see Eq. 3.34. The arrow connecting the determinants indicates the direct interaction between the determinants parametrized by the direct exchange K_{ab}.

$$\langle \Phi_I | \hat{H} | \Phi_J \rangle = -\langle a\overline{b} | \hat{H} | b\overline{a} \rangle = -K_{ab} \tag{5.9}$$

Fig. 5.3 Schematic representation of the interaction between the neutral determinants Φ_I and Φ_J by direct exchange and indirect interaction via ionic determinants

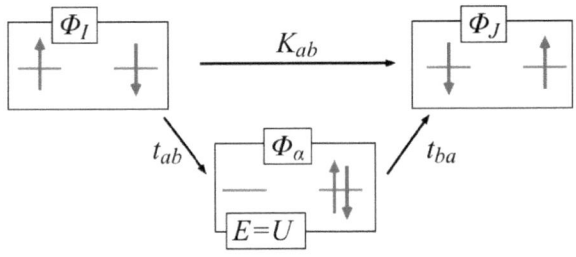

There is, however, also an indirect interaction between the two determinants via the ionic determinants $|a\bar{a}|$ and $|b\bar{b}|$ as shown in the lower part of the figure. Going from left to right, in the first step an electron is transferred from orbital a to orbital b to produce an ionic determinant at energy U with respect to the initial neutral determinant, and in the subsequent step the spin-down electron hops to orbital a to produce $|b\bar{a}|$. The interaction along this path is described with the second-order QDPT expression

$$\frac{\langle\Phi_I|\hat{H}|\Phi_\alpha\rangle\langle\Phi_\alpha|\hat{H}|\Phi_J\rangle}{E_J - E_\alpha} = \frac{\langle a\bar{b}|\hat{H}|b\bar{b}\rangle\langle b\bar{b}|\hat{H}|\bar{a}b\rangle}{0 - U}$$

$$= \frac{-\langle a\bar{b}|\hat{H}|b\bar{b}\rangle\langle b\bar{b}|\hat{H}|b\bar{a}\rangle}{-U} = \frac{t_{ab}\cdot t_{ba}}{U} = \frac{t_{ab}^2}{U} \qquad (5.10)$$

where Φ_I, Φ_α and Φ_J are defined in Fig. 5.3. Realizing that there is another indirect path connecting the neutral determinants and adding the direct interaction, we arrive at the following expression

$$\langle\Phi_I|\hat{H}^{eff}|\Phi_J\rangle = -K_{ab} + 2\frac{t_{ab}^2}{U} \qquad (5.11)$$

If we compare this to the matrix element of $|a\bar{b}|$ and $|\bar{a}b|$ of the Heisenberg Hamiltonian in Eq. 3.34, we obtain the same expression for J as derived from the diagonalization of the CAS given in Eq. 5.8.

5.4 The above described path can be denoted as $|a\bar{b}| \xrightarrow{t_{ab}} |b\bar{b}| \xrightarrow{t_{ba}} |b\bar{a}|$. Find the other path that connects the two neutral determinants via an ionic determinant.

As long as the variational space is restricted to the metal basis functions, the hopping parameter t_{ab} is extremely small due to the fact that orbitals a and b are strongly localized in different regions of space. Therefore, the antiferromagnetic contribution to J remains small and new mechanisms have to be introduced to describe the coupling. An important improvement is obtained when the role of the bridging ligand is taken into account as schematically represented in Fig. 5.4. Orbitals a and b are again the strongly localized metal orbitals and orbital h is localized on the bridge. In the first step, an electron is transferred from the ligand orbital h to site b, immediately followed by the movement of the electron on site a to the ligand, which creates the ionic determinant $|b\bar{b}|$. To arrive at the neutral determinant with inverted spins with respect to the initial determinant, a beta spin electron is transferred from the ligand to site a and the resulting hole is filled by the beta spin electron that resides on center b. This indirect interaction between the two neutral determinants involves three determinants outside the model space and hence the importance cannot be estimated by second-order QDPT. Instead, one has to apply fourth-order perturbation theory.

Fig. 5.4 Schematic representation of the interaction between the neutral determinants Φ_I and Φ_J through the bridging ligand

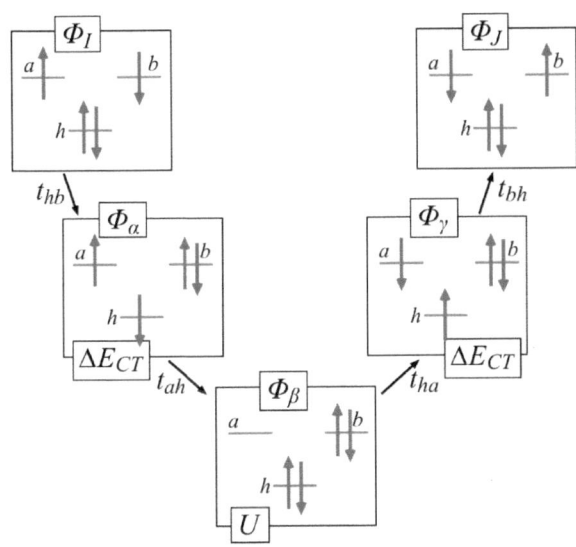

The full expression for the correction at fourth order is rather lengthy, but there is one term that exactly fits on the scheme of Fig. 5.4.

$$E^{(4)} = \cdots + \sum_{\alpha \notin S} \sum_{\beta \notin S} \sum_{\gamma \notin S} \frac{\langle \Phi_I | \hat{H} | \Phi_\alpha \rangle \langle \Phi_\alpha | \hat{H} | \Phi_\beta \rangle \langle \Phi_\beta | \hat{H} | \Phi_\gamma \rangle \langle \Phi_\gamma | \hat{H} | \Phi_J \rangle}{(E_J^{(0)} - E_\alpha^{(0)})(E_J^{(0)} - E_\beta^{(0)})(E_J^{(0)} - E_\gamma^{(0)})} + \cdots$$

$$(5.12)$$

After replacing the matrix elements in the numerator by the corresponding hopping parameters[1] and the relative energies of the external determinants in the denominator, we obtain a contribution that reads

$$\frac{t_{hb} \cdot t_{ah} \cdot t_{ha} \cdot -t_{bh}}{(0 - \Delta E_{CT})(0 - U)(0 - \Delta E_{CT})} = \frac{t_{ah}^2 t_{bh}^2}{\Delta E_{CT}^2 U} \qquad (5.13)$$

An identical expression if obtained for the pathway that starts with the electron hopping from h to a, and hence, the perturbative estimate has to be multiplied by two to obtain the QDPT expression of the matrix element between Φ_I and Φ_J with fourth-order corrections

$$\langle \Phi_I | \hat{H} | \Phi_J \rangle = -K_{ab} + 2\frac{t_{ab}^2}{U} + \frac{2 t_{ah}^2 t_{bh}^2}{\Delta E_{CT}^2 U} \qquad (5.14)$$

---

[1] $\Phi_I = |h\overline{h}a\overline{b}| \xrightarrow{t_{hb}} |b\overline{h}a\overline{b}| \xrightarrow{t_{ah}} |b\overline{h}h\overline{b}| \xrightarrow{t_{ha}} |b\overline{a}h\overline{b}| \xrightarrow{t_{bh}} |b\overline{a}h\overline{h}|$. A minus sign appears in the last step when the determinant is written as it appears in the model Hamiltonian $|h\overline{h}a\overline{b}| = \Phi_J$.

Now we replace the bare hopping matrix elements t_{ha} and t_{hb} by an effective parameter through

$$\frac{t_{ha}t_{hb}}{\Delta E_{CT}} = t_{ab}^{eff} \qquad (5.15)$$

and arrive at an analytical expression for J using Eq. 3.34

$$J = 2K_{ab} - \frac{4(t_{ab}^{eff})^2}{U} \qquad (5.16)$$

where the effect of the bare hopping parameter t_{ab} has been neglected being much smaller than t_{ab}^{eff}, which involves the bridging ligand(s). Note the similarity with the second-order expression of Eq. 5.11. The second, antiferromagnetic term is generally known as the kinetic exchange and is conceptually closely related to the superexchange of Anderson discussed at the end of Sect. 3.1.

> **5.5** Make a perturbative estimate of the contribution to J of the double LMCT configuration with an energy of ΔE_{2CT}

Putting these concepts to the numerical proof can be done by performing a CASCI calculation with triplet optimized orbitals. Instead of the strongly localized orbitals used in the conceptual reasoning, the optimal orbitals have important delocalization tails on the ligands, as shown in Fig. 5.5. These delocalization tails are just another representation of the through-ligand interaction discussed above, which is easily demonstrated by substituting the definition of the active orbitals with tails on the ligand

$$g = c_1(a+b) + c_2 h \qquad \text{and} \qquad u = c_3(a-b) + c_4 h' \qquad (5.17)$$

into the expression of the lowest singlet state given in Eq. 5.2.

$$\begin{aligned}
S_g = \lambda|(c_1(a+b) + c_2 h)(c_1(\overline{a}+\overline{b}) + c_2\overline{h})| \\
+ \mu|(c_3(a-b) + c_4 h')(c_3(\overline{a}-\overline{b}) + c_4\overline{h'})|
\end{aligned} \qquad (5.18)$$

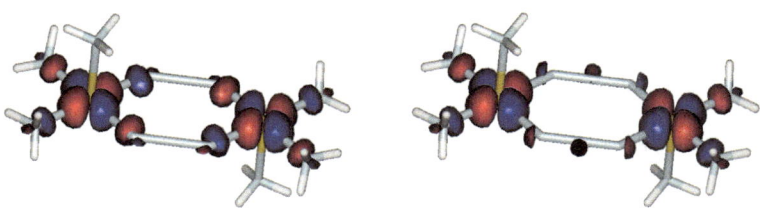

Fig. 5.5 Magnetic orbitals of *gerade* (*left*) and *ungerade* (*right*) symmetry with important delocalization tails on the ligands

Table 5.1 Decomposition of the CAS(2,2) magnetic coupling in the binuclear Cu^{2+} complex with a double azido bridge

Direct exchange	12
t_{ab}^{eff}	-2218
U	20.8×10^4 (=25.8 eV)
Kinetic exchange	-94
$J(CAS(2,2))$	$-82\ cm^{-1}$

Numbers are given in cm^{-1}

Apart from the previously seen neutral and ionic determinants $|a\overline{b}|$, $|b\overline{a}|$, $|a\overline{a}|$, $|b\overline{b}|$, other determinants such as such as $|h\overline{a}|$, $|b\overline{h}|$, etc. appear in the wave function involving ligand-to-metal charge transfer (LMCT) excitations that were shown to play an important role in the QDPT analysis of the coupling.

The results of the CAS calculation are listed in Table 5.1 and show how the kinetic exchange strongly dominates over the direct exchange, which is rather small as expected from the large distance between the Cu ions. The parameters are directly extracted by comparing the numerical values of the CASCI matrix with the symbolic representation given in Eq. 5.4. The choice for cm^{-1} as energy unit leads to big numbers for U, which is therefore often expressed in eV. The kinetic exchange contribution is calculated from t_{ab}^{eff} and U applying Eq. 5.16: $(-4 \cdot (-2218)^2/20800)$.

5.1.2 Beyond the Valence Space

It is obvious that this cannot be the whole story. The calculated magnetic coupling of the Cu^{2+} complex is just 10 % of the experimental and DDCI values. Hence, it is unavoidable to go beyond this valence-only description and incorporate more physical mechanisms in the description.

The 1h, 1p, 1h-1p excitations: In the first step towards the full DDCI result, we analyze the role of the $1h$, $1p$ and $1h$-$1p$ determinants as illustrated in Fig. 5.6. The determinants on the left are pure single excitations and those on the right are single excitations combined with an excitation within the CAS. Because of the Brillouin theorem the contributions of the pure single excitations are strictly zero for the spin state for which the orbitals have been optimized and tiny contributions are observed for the other spin states given that the optimal orbitals for the different spin states are in principle very similar.

The situation is quite different for the single excitations that are combined with electron replacements in the CAS. The $1h$-$1p$ excitation in the determinants marked as *spin polarization* not only excites one of the electrons from orbital h to orbital p but also changes the spin of the excited electron. These so-called triplet excitations have to be compensated by a simultaneous spin change in the active space to maintain the spin of the electronic state under consideration. This gives rise to a triplet coupled

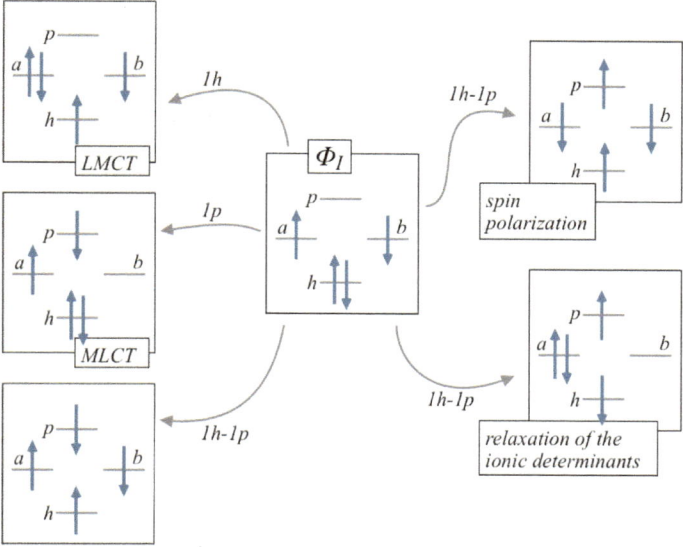

Fig. 5.6 Schematic representation of the 1*h*, 1*p* and 1*h*-1*p* determinants. The *left column* shows the pure single excitations, and the *right* the single excitations combined with a change in the occupation of the active orbitals. *h* and *p* are assumed to be ligand orbitals, LMCT = ligand-to-metal charge transfer, MLCT = metal-to-ligand charge transfer

electron pair *a-b* in the active space, from which triplet and singlet spin polarization determinants can be formed through the coupling with the *h-p* triplet coupled electron pair. The resulting determinants strongly interact both with S_g and T_u (see Eq. 5.5) via the neutral determinants of the reference wave functions, but the matrix element with the ionic determinants is zero since there are more than two differences in the orbital occupancies. The spin polarization introduces spin density on the ligand, which is opposite to the spin density on the metal centers. It can contribute both ferro- and antiferromagnetically depending on the structure of the complex, but it is general more important when the 1*h*-1*p* triplet excitation on the ligand is low in energy, as in conjugated bridges.

> **5.6** The *h-p* and the *a-b* electron pairs are triplet coupled (S = 1) in the determinants that cause spin polarization in the ligands. Which values can be assigned to the total spin by coupling the two S = 1 electron pairs? Are all spin states relevant to the binuclear Cu^{2+} system under study?

The second type of important 1*h*-1*p* determinants combines a spin-conserving *h* to *p* excitation with an electron replacement from *a* to *b* (or *vice versa*) in the active space. The resulting determinants can be considered as single excitations with respect to the ionic determinants, but Brillouin's theorem does not apply because the orbitals are not optimized for this ionic charge distribution but rather for the neutral situation. Hence, there is a strong interaction of these determinants with the $|a\bar{a}|$ and $|b\bar{b}|$ determinants, while the interaction with the neutral determinants is much weaker. Since, the ionic determinants are only present in the reference wave function of the S_g state, the addition to the wave function of these $1h$-$1p$ excitations leads to a significant stabilization of the singlet with respect to the triplet state, and consequently, an increase of the antiferromagnetic character of the coupling. Adding single excitations to a determinant that is not expressed in its optimal orbitals is a very efficient way to improve the orbitals. Therefore, this class of $1h$-$1p$ excitations is often interpreted as relaxing the ionic determinants in the wave function, lowering their energy with respect to the neutral determinants, that is, a decrease of U. In line with the expression for J given in Eq. 5.16, a smaller U makes the kinetic exchange more effective and J more antiferromagnetic.

The total effect of the single excitations is a large step in the right direction, both spin polarization and the relaxation of the ionic determinants cause antiferromagnetic contributions, but still the value of the coupling is only $\sim$50 % of the final value and other mechanisms have to be included.

> **5.7** Assuming that the $1h$-$1p$ excitations do not affect the hopping parameter t_{ab}^{eff}, calculate the energy lowering effect on U of the inclusion of the $1h$-$1p$ excitations combined with the electron replacement in the active space using the numerical data from Tables 5.1 and 5.2.

The last step: 2h, 2p, 2h-1p and 1h-2p excitations. The double excitations of the $2h$ and $2p$ class (shown in the left column of Fig. 5.7) only contribute very little to the magnetic coupling of the two Cu ions. They correspond to double ligand-to-metal or metal-to-ligand charge transfer excitations, respectively. The weak interaction is largely explained by the high energy of these determinants with respect to the neutral determinants. This energy difference enters the denominator of the perturbative expression of the effect of the external determinants, and hence, higher-lying determinants contribute less to J.

Table 5.2 Decomposition of the DDCI magnetic coupling in the binuclear Cu^{2+} complex with a double azido bridge

Direct exchange	12
Kinetic exchange	−94
Spin polarization	−59
Relaxation of the ionic determinants	−221
Double CT	−13
Relaxation of the CT determinants	−427
J	−802 cm^{-1}

Numbers are given in cm^{-1}

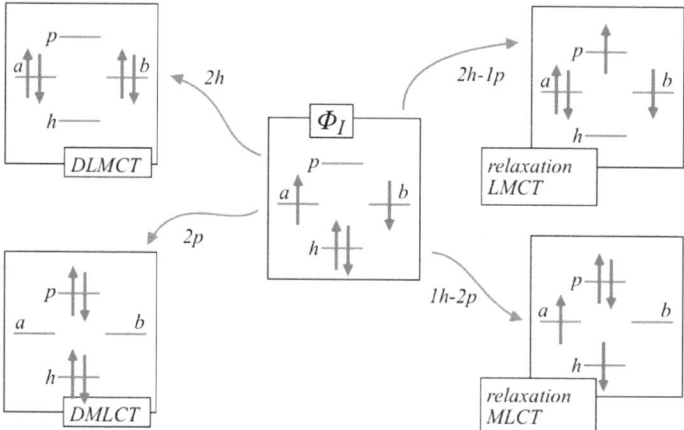

Fig. 5.7 Schematic representation of the $2h, 2p, 2h$-$1p$ and $1h$-$2p$ determinants. DLMCT = double ligand-to-metal charge transfer, DMLCT = double metal-to-ligand charge transfer

The single excitation connected to the ligand-to-metal charge transfer process does not interact with the reference wave function by Brillouin's theorem. However, the combination of a LMCT excitation with a single excitation from occupied to virtual orbitals ($1h + 1h$-$1p = 2h$-$1p$) gives rise to a relaxation process similar to the one described before for the ionic determinants, right column of Fig. 5.7. The addition of the relaxed LMCT determinants to the CI wave function has a large impact on the singlet-triplet splitting and virtually always favors the antiferromagnetic character of the coupling. Analogously, the $1h$-$2p$ excitations can be considered to introduce the relaxation of the metal-to-ligand charge transfer excitations. The addition of these determinants is ferromagnetic in most cases, but usually smaller than the effect of the $2h$-$1p$ determinants. Hence the net effect of these double excitations is a significant increase of the singlet-triplet gap as can be seen in Table 5.2. The $2h$ and $2p$ determinants give a tiny contribution to J, but the inclusion of the $2h$-$1p$ and $1h$-$2p$ determinants brings the computational estimate of J in close agreement with the experimental value.

Remember that inclusion of the largest group of determinants that can interact with the neutral determinants, the $2h$-$2p$ determinants, has a negligible effect on the energy difference and is not included in the DDCI wave function and not considered in the analysis of the mechanism of the coupling. This will be numerically shown for our example compound in the next section.

5.1.3 Decomposition with MRPT2

A similar exercise can be performed based on the results of a MRPT2 calculation with a CAS(2,2) reference wave function. Both NEVPT2 and CASPT2 distinguish the contributions to the second-order energy correction of the different excitation classes

Table 5.3 Decomposition of the CASPT2 contribution to the magnetic coupling in the binuclear Cu^{2+} complex with a double azido bridge

Excitation class	$E^{(2)}$(singlet)	$E^{(2)}$(triplet)	Difference
1h	0.000000	0.000000	0.0
1p	0.000000	0.000000	0.0
1h-1p	−0.017278	−0.018653	−301.8
2h	−0.000010	−0.000027	−3.9
2p	−0.000011	−0.000031	−4.4
2h-1p	−0.095270	−0.095938	−144.6
1h-2p	−0.213555	−0.213649	−20.8
2h-2p	−3.654102	−3.653976	27.5
CASPT2	−3.980233	−3.982275	−448.1[a]

Energies are given in Hartree, the difference in cm^{-1}
[a]The total magnetic coupling $J = J[CAS(2,2)] + CASPT2 = -101.4 + -448.1 = -549.5$ cm^{-1}

as exemplified in Table 4.1. The analysis is completely straightforward, one just has to subtract the energy contributions of the different spin states in a class-by-class manner to decompose the magnetic coupling. Table 5.3 shows the contribution of the different excitation classes in the example compound studied above with DDCI.

There are two major contributions to the energy difference of singlet and triplet. In the first place, the 1h-1p excitations, which cause spin polarization and relaxation of the ionic determinants. Unfortunately, it is not possible to separate the two contributions as in DDCI. The second large contribution arises from the 2h-1p excitations, which also enhances the singlet stability, as expected. The 2h and 2p excitations are nearly zero and the 1h-2p class also gives a rather small contribution for the present system. The total contribution of the 2h-2p class is by far the largest, it constitutes approximately 92 % of $E^{(2)}$, but the differential effect is very small. The non-zero contribution to the difference may seem surprising given the fact that the justification of DDCI is based on the zero contribution of these excitations at second-order perturbation theory. However, this reasoning is based on a common orbital basis for the spin states, which is not used in the CASPT2 calculation. It is common practice to optimize the orbitals for each spin state separately, contrary to MRCI where normally one set of orbitals is used. The use of state-specific orbitals also explains the strictly zero contribution of the 1h and 1p excitations for both states.

5.2 Mapping Back on a Valence-Only Model

The preceding section shows that a valence-only description of the coupling leads to rather poor predictions. Although the sign of the coupling is often (but not always) correctly reproduced, it can be stated that the strength of the coupling is underestimated by at least one order of magnitude. This is in sharp contrast with the success of the qualitative valence models discussed in Chap. 4, which are capable of explaining many magnetostructural correlations and rationalize the relative size of J in large families of compounds. These models seem to contain all the essential physics but their parametrization with *ab initio* calculations is deficient. Hence, it may be advantageous to construct a simple valence-only picture in which the values of the parameters t_{ab}, U and K_{ab} are replaced by effective values that absorb all the effects discussed above that go beyond the valence-only description. In fact, we have already seen how the bare hopping parameter t_{ab} was replaced by an effective t_{ab} due to the partial delocalization of the magnetic orbitals onto the ligands upon the change from strongly localized orbitals to self-consistently optimized molecular orbitals as schematically illustrated in Fig. 5.8.

A rigorous way to construct a valence-only model with *ab initio* methods is to make use of the effective Hamiltonian theory presented in Chap. 1. First, we define the basis of the model space as $\{|a\overline{b}|, |b\overline{a}|, |a\overline{a}|, |b\overline{b}|\}$ and use the matrix of Eq. 5.4 to represent the effective Hamiltonian. Then we replace the bare parameters obtained in a valence-only *ab initio* calculation with effective parameters that include all the effects that have been discussed in the previous section. This is done by selecting those four roots from the *ab initio* calculation that have the largest projection on the model space. After orthogonalization and normalization, Eq. 1.90 or 1.92 is used to construct a numerical Hamiltonian from which the new, effective parameters can be extracted by the comparison with Eq. 5.4.

To numerically illustrate the procedure, we will treat the magnetic coupling of two Cu^{2+} ions in the previously introduced $SrCu_2O_3$ compound with Cu_2O_3 layers separated by Sr^{2+} ions (see Sect. 3.4.2). We recall that the copper ions form a regular pattern that can best be described as a ladder structure as depicted in Fig. 5.9. Among

Fig. 5.8 *Top* direct (through space) hopping between two magnetic centers by t_{ab} with strongly localized atomic orbitals; *Bottom* effective (through ligand) hopping by t_{ab}^{eff} with self-consistently optimized magnetic orbitals, which have delocalization tails on the (bridging) ligands

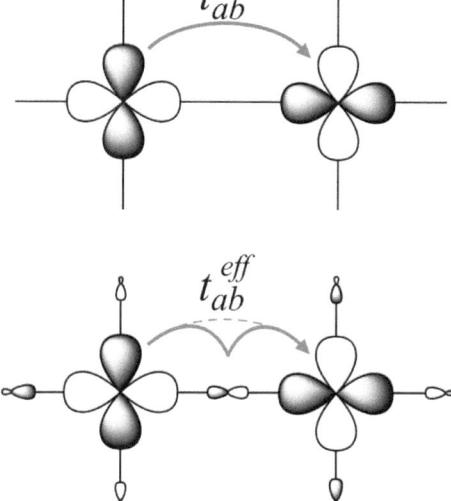

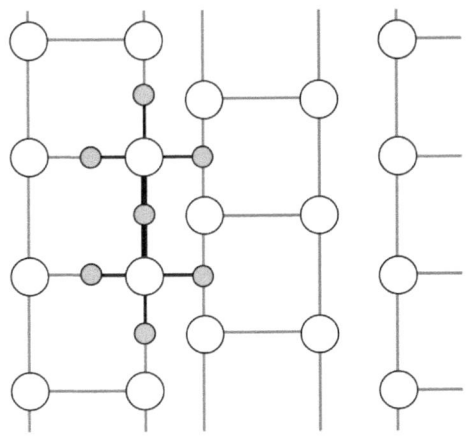

Fig. 5.9 Ladder-like structure of SrCu$_2$O$_3$. The *open circles* represent Cu^{2+} ions, the small *grey circles* are oxygens, only the oxygens belonging to the Cu$_2$O$_7$ cluster are shown. The Sr ions are below and above this plane

the sixteen lowest roots of a DDCI calculation on an embedded Cu$_2$O$_7$ cluster (see Sect. 6.3.1), the roots 1, 2, 7 and 8

$$\Psi_1 = -0.9224\big(|a\bar{b}| + |b\bar{a}|\big) - 0.1223\big(|a\bar{a}| + |b\bar{b}|\big) + \cdots$$
$$\Psi_2 = -0.6626\big(|a\bar{b}| - |b\bar{a}|\big) + \cdots \tag{5.19}$$
$$\Psi_7 = 0.4159\big(|a\bar{a}| - |b\bar{b}|\big) + \cdots$$
$$\Psi_8 = 0.1704\big(|a\bar{b}| + |b\bar{a}|\big) - 0.5324\big(|a\bar{a}| + |b\bar{b}|\big)$$

have the largest norm after projection on the model space. The normalized projections, denoted $\widetilde{\Psi}'_1 \ldots \widetilde{\Psi}'_4$, will be used for the construction of the effective Hamiltonian. The relative energies are 0.000, 0.158, 6.489 and 6.547 eV, respectively. The Bloch effective Hamiltonian uses biorthogonal vectors, which are obtained from Eq. 5.19 by

$$\widetilde{\Psi}'^{\dagger}_1 = \frac{1}{\sqrt{1-s^2}}\big(\widetilde{\Psi}'_1 - s\widetilde{\Psi}'_4\big) = -0.9524\big(|a\bar{b}| + |b\bar{a}|\big) - 0.3048\big(|a\bar{a}| + |b\bar{b}|\big)$$
$$\widetilde{\Psi}'^{\dagger}_2 = \widetilde{\Psi}'_2 = \big(|a\bar{b}| - |b\bar{a}|\big)/\sqrt{2} \tag{5.20}$$
$$\widetilde{\Psi}'^{\dagger}_3 = \widetilde{\Psi}'_3 = \big(|a\bar{a}| - |b\bar{b}|\big)/\sqrt{2}$$
$$\widetilde{\Psi}'^{\dagger}_4 = \frac{1}{\sqrt{1-s^2}}\big(-s\widetilde{\Psi}'_1 + \widetilde{\Psi}'_4\big) = 0.1315\big(|a\bar{b}| + |b\bar{a}|\big) - 0.9913\big(|a\bar{a}| + |b\bar{b}|\big)$$

where $s = \langle\widetilde{\Psi}'_1|\widetilde{\Psi}'_4\rangle$.

Table 5.4 Electronic structure parameters of a valence-only model for the magnetic interactions between two Cu^{2+} ions in $SrCu_2O_3$

	$\hat{H}^{eff}$	K_{ab}^{eff}	U^{eff}	t_{ab}^{eff}	J_{pert}
Valence-only	CASCI	16	24.6	−617	−30
Dressed model	Bloch	55	6.0	−1005/−417	−227
	Gram-Schmidt	−22	6.3	−427	−160

U is given in eV, the other parameters in meV. J_{pert} is calculated using Eq. 5.8. The magnetic coupling extracted from the DDCI energies of the lowest singlet and triplet states is −158 meV. The dressed models are extracted from DDCI calculations using triplet orbitals

> **5.8** Calculate the norm of the projections of Ψ_i on the model space and give the expressions of $\widetilde{\Psi}'_{1,4}$. Are all $\widetilde{\Psi}'_i$ mutually orthogonal? Check that the biorthogonal vectors $\widetilde{\Psi}_i^\dagger$ fulfill the orthogonality properties of Eq. 1.89.

Now we apply the Bloch formula (Eq. 1.90) to construct the effective Hamiltonian. The resulting parameters are given in the second line of Table 5.4 together with an estimate of J from the sum of the direct and kinetic exchange given in Eq. 5.8. As most particular results, we see that U is strongly reduced in comparison to the valence-only value (CASCI) and that the non-hermiticity is manifest in the two different values of the hopping parameter: $t_1 = \langle a\bar{b}|\hat{H}^{eff}|a\bar{a}\rangle$ and $t_2 = \langle a\bar{a}|\hat{H}^{eff}|a\bar{b}\rangle$.

The Gram–Schmidt procedure provides a simpler orthogonalization scheme that leads to a hermitian effective Hamiltonian. Since $\widetilde{\Psi}'_2$ and $\widetilde{\Psi}'_3$ are already orthogonal to the other projections, we only have to worry about $\widetilde{\Psi}'_1$ and $\widetilde{\Psi}'_4$. This means that the coefficients of $\widetilde{\Psi}_4^\perp$ are defined by $\widetilde{\Psi}'_1$, that is, if $\widetilde{\Psi}'_1 = \alpha\left(|a\bar{b}| + |b\bar{a}|\right) + \beta\left(|a\bar{a}| + |b\bar{b}|\right)$ then the orthogonal counterpart $\widetilde{\Psi}_4^\perp = -\beta\left(|a\bar{b}| + |b\bar{a}|\right) + \alpha\left(|a\bar{a}| + |b\bar{b}|\right)$, independent of the shape of Ψ_4 and only the energy of this state is used in the construction of $\hat{H}^{eff}$. The parameters extracted with the Gram-Schmidt orthogonalized vectors are listed in the third row of the table and reveal besides the expected large decrease of U, a negative effective direct exchange and, by construction, a hermitian form with only one estimate for t. The estimate of J based on Eq. 5.8 is in excellent agreement with the result of the full DDCI calculation.

The observed changes suffered by the parameters upon dressing them with the effects that go beyond the valence space can at least partially be rationalized by looking at the interaction of the model space determinants with those in the external space. The interaction of the spin-conserving $1h$-$1p$ excitations with the neutral determinants is (nearly) zero due to Brillouin's theorem. On the contrary, the interaction with the ionic determinants is strong (see the right part of Fig. 5.10). Hence, this class of external determinants largely decreases the on-site repulsion U as previously seen in Exercise 6.7 and confirmed here in the example.

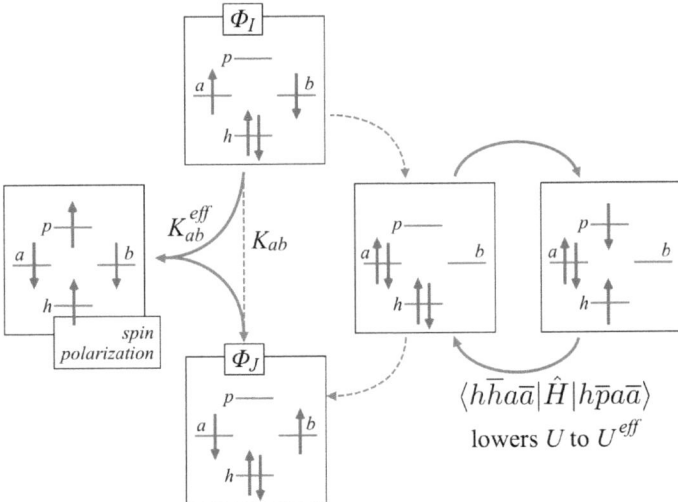

Fig. 5.10 Effect of the spin-conserving (*right*) and non spin-conserving (*left*) $1h$-$1p$ determinants on the on-site repulsion U and the direct exchange parameter K_{ab}. The interactions in a valence-only treatment are marked with *dashed lines*

The non spin-conserving excitations (the spin polarization) simultaneously interact with both neutral determinants:

$$\langle h\overline{h}ab|\hat{H}|h\overline{a}\,p\overline{b}\rangle\langle h\overline{a}\,p\overline{b}|\hat{H}|h\overline{h}b\overline{a}\rangle \neq 0 \qquad (5.21)$$

and hence, turns the direct exchange parameter $K_{ab} = \langle a\overline{b}|\hat{H}|b\overline{a}\rangle$ into an effective parameter that parametrizes the direct exchange dressed with spin polarization effects. Since this latter effect can be antiferromagnetic in nature, the apparent counterintuitive situation may arise that K_{ab}^{eff} turns out to be negative, whereas the bare K_{ab} is positive by definition.

Finally, t_{ab} is expected to be influenced by the $2h$-$1p$ excitations, which can be seen as a LMCT excitation ($1h$) coupled to a single excitation from an occupied to a virtual orbital ($1h$-$1p$) that relaxes the LMCT configuration. This should lead to an increase of the hopping between the magnetic centers. It is, however, difficult to isolate the effect of the $2h$-$1p$ excitations [4] due to the occurrence of interactions with other excited determinants and the interference of the $1h$-$2p$ excitations. Therefore, the hopping parameter t_{ab}^{eff} turns out to be somewhat smaller than the bare value. Experience has shown that t does not suffer dramatic changes when electron correlation is included in the valence-only models.

5.3 Analysis with Single Determinant Methods

The multiconfigurational wave function is a natural starting point for a decomposition of the magnetic coupling in polynuclear complexes. The relative magnitude of the coefficients of the determinants gives a straightforward strategy to analyze the importance of the different mechanisms. The disadvantage of this approach is of course the rather heavy computational burden, which limits the applicability to medium-sized systems. Alternatively, DFT can be used for larger systems but the analysis of the results cannot be made in the same way as discussed above, since there is only one Slater determinant that contains all the physics. Instead, one can decompose the coupling by a series of partial orbital optimizations of the high-spin and broken symmetry determinants [5]. The intermediate energy differences can be related to the direct and kinetic exchange, and the spin polarization as will be shown below.

The analysis starts with a restricted open-shell Kohn-Sham (ROKS) calculation on the HS state. If necessary, the magnetic orbitals are transformed to the representation with local orthogonal orbitals a and b, as shown in the first column of Fig. 5.11. Staying within the spin-restricted formalism makes that for each α orbital a β orbital can be found which has the same spatial part. In the first step, the direct exchange is estimated from the energy difference of the HS(ROKS) and a BS determinant in which only the spin of one of the unpaired electrons is inverted, but neither the *core* nor the magnetic orbitals are optimized.

$$E(\text{BS-ROKS}) = E(\text{HS-ROKS}) + K_{ab} \qquad (5.22)$$

With the help of Yamaguchi's expression (Eq. 4.65), one can define the direct exchange contribution (J_{DE}) to the total magnetic coupling as

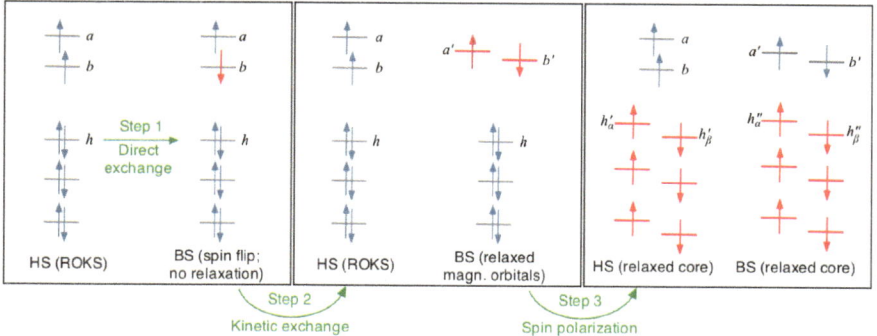

Fig. 5.11 Schematic representation of the decomposition of the magnetic coupling for computational schemes based on a single determinant. The orbitals and electrons marked in *red* are subject to changes with respect to the previous step, the rest is kept frozen

$$J_{DE} = \frac{2\big(E(\text{BS-ROKS}) - E(\text{HS-ROKS})\big)}{\langle \hat{S}^2 \rangle_{\text{HS-ROKS}} - \langle \hat{S}^2 \rangle_{\text{BS-ROKS}}} = \frac{2K_{ab}}{2-1} = 2K_{ab} \qquad (5.23)$$

Step 2 consists in the optimization of the magnetic orbitals of the BS determinant in the fixed field of the doubly occupied orbitals, a so-called frozen core (FC). The magnetic orbitals become more delocalized and, more specifically, gain some amplitude on the other magnetic center. This means that the unpaired electrons can move from one center to the other and activate the kinetic exchange mechanism. The contribution to J is

$$J_{KE} = \frac{2\big(E(\text{BS-FC}) - E(\text{HS-ROKS})\big)}{2 - \langle \hat{S}^2 \rangle_{\text{BS-FC}}} - J_{DE} \qquad (5.24)$$

To a very good approximation, the spatial part of the new magnetic BS orbitals a' and b' can be written as a weighted sum of the ROKS orbitals

$$a' = (\cos\alpha)a + (\sin\alpha)b \qquad b' = (\sin\alpha)a + (\cos\alpha)b \qquad (5.25)$$

5.9 Calculate the overlap of the spatial part of the relaxed magnetic orbitals for $\alpha = 0, \pi/60, \pi/20, \pi/4, \pi/2$.

The interaction with the virtual orbitals is very small and can be neglected for the present analysis purposes. Substituting these expressions in the BS determinant

$$\Phi_{BS} = (\cos\alpha)^2 |a\bar{b}| + (\sin\alpha)^2 |b\bar{a}| + (\sin\alpha\cos\alpha)\big(|a\bar{a}| + |b\bar{b}|\big) \qquad (5.26)$$

shows immediately that the relaxation activates the kinetic exchange by introducing the ionic determinants $|a\bar{a}|$ and $|b\bar{b}|$ in the BS determinant. The optimization of a and b makes that the $\hat{S}^2$ expectation value of the BS determinant is not exactly one as for the BS-ROKS determinant (see Problems).

The last step relaxes the core orbitals for the HS and BS determinants, keeping the magnetic orbitals fixed to what was obtained in step 2 (frozen magnetic orbitals: FM). Lifting the restrictions on the spin symmetry in the core orbitals introduces different α and β spin orbitals, and hence, accounts for the spin polarization of the core electrons in response to the parallel (HS) or antiparallel (BS) unpaired electrons. The energy difference between the BS-FM and HS-FM determinants gives access to the spin polarization contribution to J via

$$J_{SP} = \frac{2\big(E(\text{BS-FM}) - E(\text{HS-FM})\big)}{\langle \hat{S}^2 \rangle_{\text{HS-FM}} - \langle \hat{S}^2 \rangle_{\text{BS-FM}}} - J_{DE} - J_{KE} \qquad (5.27)$$

The sum of the three contributions is close to the magnetic coupling constant that is obtained in a standard calculation when the Kohn-Sham orbitals are optimized without imposing any restriction on the variational process.

$$J \approx J_{DE} + J_{KE} + J_{SP} \tag{5.28}$$

There is no *a priori* reason to determine the different contributions in this order. Alternatively, the spin polarization can be calculated before relaxing the magnetic orbitals (inverting step 2 and 3) or independently, both taking the orbitals of step 1 as starting point. However, the examples given in Ref. [5] show that the order chosen here gives the smallest deviation from the fully relaxed energy difference, and hence, includes the largest part of the physics.

5.4 Analysis of Complex Interactions

The analysis of the interaction between magnetic moments is not restricted to the isotropic bilinear exchange of two $S = 1/2$ centers but can also be applied to systems with higher spins and more magnetic centers. In this section, we will first decompose the magnetic coupling between two Ni^{2+} ($S = 1$) ions with a sizeable biquadratic exchange to pinpoint the origin of the deviations to the standard Heisenberg Hamiltonian. Secondly, we will focus attention on the four-spin cyclic exchange, and finally, we will describe how these interactions can be estimated within the DFT framework.

5.4.1 Decomposition of the Biquadratic Exchange

One of the central assumptions of the Heisenberg model Hamiltonian is that the local spin states are well separated in energy from excited spin states. We will demonstrate that non-Heisenberg behavior emerges as soon as this is no longer true. The biquadratic exchange is in general a rather small term in the total interaction of the spins in polynuclear TM-3d complexes. There are only a few examples where it is important to include them for obtaining an accurate description of the lowest-energy levels. Returning to the binuclear complexes with a double azido bridge, we will here analyze the magnetic coupling of the Ni^{2+} model complex shown in Fig. 5.12. The interaction of the spins strongly depends on the δ-angle and ranges from approximately 100 cm^{-1} for $\delta = 0°$ to nearly zero for $\delta = 45°$, which is accurately reproduced with DDCI [6]. More interestingly, the singlet, triplet and quintet DDCI energies do not strictly follow the expected Landé pattern, especially for small δ. Deviations up to 3 % are observed and we will use the corresponding wave functions to analyze the origin of the deviations to the standard Heisenberg spacing.

The magnetic orbitals are expressed again in orthogonal atomic-like orbitals, denoted φ_1, φ_2 for the two magnetic orbitals on site A and φ_3, φ_4 for the orbitals on

Fig. 5.12 $(NH_3)_3$–Ni–$(\mu$-$N_3)_2$–Ni–$NH_3)_3$ model complex and definition of the angle δ

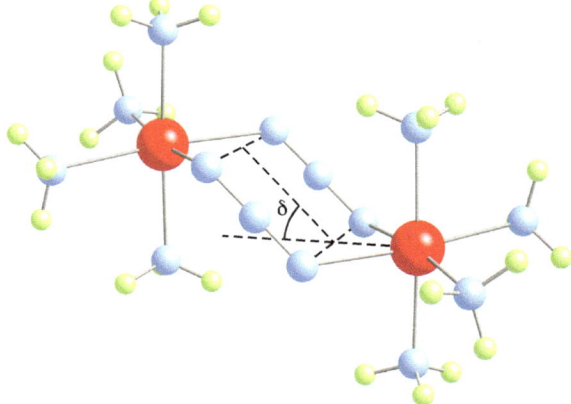

center B. The local ground state is a triplet denoted as $T_A^{1,0,-1}$ for the three degenerate M_S components on center A and $T_B^{1,0,-1}$ for the components of the triplet on center B.

$$
\begin{aligned}
T_A^+ &= |\varphi_1\varphi_2| & T_B^+ &= |\varphi_3\varphi_4| \\
T_A^- &= |\overline{\varphi}_1\overline{\varphi}_2| & T_B^- &= |\overline{\varphi}_3\overline{\varphi}_4| \\
T_A^0 &= (|\varphi_1\overline{\varphi}_2| - |\varphi_2\overline{\varphi}_1|)/\sqrt{2} & T_B^0 &= (|\varphi_3\overline{\varphi}_4| - |\varphi_4\overline{\varphi}_3|)/\sqrt{2}
\end{aligned}
\tag{5.29}
$$

Note that the superscript indicates the M_S-value of the function. In addition we also define a local singlet with the same orbital occupancy but a different spin coupling. This CSF dominates the lowest excited singlet state in octahedral Ni^{2+} complexes and will be named here a non-Hund state

$$
S_A^0 = (|\varphi_1\overline{\varphi}_2| + |\varphi_2\overline{\varphi}_1|)/\sqrt{2} \qquad S_B^0 = (|\varphi_3\overline{\varphi}_4| + |\varphi_4\overline{\varphi}_3|)/\sqrt{2}
\tag{5.30}
$$

The total wave functions of the binuclear complex can be constructed from the products of these local functions. In order to find any possible interactions between the lowest singlet, triplet and quintet states and the newly introduced non-Hund singlet, all products are written in their $M_S = 0$ variant.

$$
T_A^+ T_B^- = |\varphi_1\varphi_2\overline{\varphi}_3\overline{\varphi}_4| = T^+T^-
\tag{5.31}
$$

$$
T_A^- T_B^+ = |\overline{\varphi}_1\overline{\varphi}_2\varphi_3\varphi_4| = T^-T^+
\tag{5.32}
$$

$$
\begin{aligned}
T_A^0 T_B^0 &= \frac{1}{2}\big(|\varphi_1\overline{\varphi}_2| - |\varphi_2\overline{\varphi}_1|\big)\big(|\varphi_3\overline{\varphi}_4| - |\varphi_4\overline{\varphi}_3|\big) = \frac{1}{2}\big(|\varphi_1\overline{\varphi}_2\varphi_3\overline{\varphi}_4| \\
&\quad - |\varphi_1\overline{\varphi}_2\varphi_4\overline{\varphi}_3| - |\varphi_2\overline{\varphi}_1\varphi_3\overline{\varphi}_4| + |\varphi_2\overline{\varphi}_1\varphi_4\overline{\varphi}_3|\big) = T^0T^0
\end{aligned}
\tag{5.33}
$$

$$
\begin{aligned}
T_A^0 S_B^0 &= \frac{1}{2}\big(|\varphi_1\overline{\varphi}_2| + |\varphi_2\overline{\varphi}_1|\big)\big(|\varphi_3\overline{\varphi}_4| - |\varphi_4\overline{\varphi}_3|\big) = \frac{1}{2}\big(|\varphi_1\overline{\varphi}_2\varphi_3\overline{\varphi}_4| \\
&\quad + |\varphi_1\overline{\varphi}_2\varphi_4\overline{\varphi}_3| - |\varphi_2\overline{\varphi}_1\varphi_3\overline{\varphi}_4| - |\varphi_2\overline{\varphi}_1\varphi_4\overline{\varphi}_3|\big) = T^0S^0
\end{aligned}
\tag{5.34}
$$

$$
\begin{aligned}
S_A^0 T_B^0 &= \frac{1}{2}\left(|\varphi_1\overline{\varphi}_2| + |\varphi_2\overline{\varphi}_1|\right)\left(|\varphi_3\overline{\varphi}_4| - |\varphi_4\overline{\varphi}_3|\right) = \frac{1}{2}\left(|\varphi_1\overline{\varphi}_2\varphi_3\overline{\varphi}_4| \right. \\
&\quad \left. - |\varphi_1\overline{\varphi}_2\varphi_4\overline{\varphi}_3| + |\varphi_2\overline{\varphi}_1\varphi_3\overline{\varphi}_4| - |\varphi_2\overline{\varphi}_1\varphi_4\overline{\varphi}_3|\right) = S^0 T^0
\end{aligned}
\tag{5.35}
$$

$$
\begin{aligned}
S_A^0 S_B^0 &= \frac{1}{2}\left(|\varphi_1\overline{\varphi}_2| + |\varphi_2\overline{\varphi}_1|\right)\left(|\varphi_3\overline{\varphi}_4| + |\varphi_4\overline{\varphi}_3|\right) = \frac{1}{2}\left(|\varphi_1\overline{\varphi}_2\varphi_3\overline{\varphi}_4| \right. \\
&\quad \left. + |\varphi_1\overline{\varphi}_2\varphi_4\overline{\varphi}_3| + |\varphi_2\overline{\varphi}_1\varphi_3\overline{\varphi}_4| + |\varphi_2\overline{\varphi}_1\varphi_4\overline{\varphi}_3|\right) = S^0 S^0
\end{aligned}
\tag{5.36}
$$

These six CSFs can be combined to form spin eigenstates; three states with local triplet coupling on both magnetic centers and three more with at least one magnetic center in a locally excited (non-Hund) state.

$$
Q = \sqrt{\frac{2}{3}}\left(T^0 T^0 + \frac{1}{2}(T^+ T^-)\right)
\tag{5.37a}
$$

$$
T = \frac{1}{\sqrt{2}}\left(T^+ T^- - T^- T^+\right)
\tag{5.37b}
$$

$$
S = \frac{1}{\sqrt{3}}\left(T^0 T^0 - T^+ T^- - T^- T^+\right)
\tag{5.37c}
$$

$$
NH1 = \frac{1}{\sqrt{2}}\left(T^0 S^0 + S^0 T^0\right)
\tag{5.38a}
$$

$$
NH2 = \frac{1}{\sqrt{2}}\left(T^0 S^0 - S^0 T^0\right)
\tag{5.38b}
$$

$$
NH3 = S^0 S^0
\tag{5.38c}
$$

These six CSFs are the basis of the model space of the determinants of the four-electron/four-orbital CAS calculation with the restriction of one electron per orbital. The matrix representation of the model space is

	$|Q\rangle$	$|T\rangle$	$|S\rangle$	$|NH1\rangle$	$|NH2\rangle$	$|NH3\rangle$
$\langle Q|$	$E_0 - 2K - 2K'$					
$\langle T|$	0	$E_0 - 2K$				
$\langle S|$	0	0	$E_0 - 2K + K'$			
$\langle NH1|$	0	0	0	$E_0 + 2K'$		
$\langle NH2|$	0	$K_{24} - K_{13}$	0	0	$E_0 + 2K''$	
$\langle NH3|$	0	0	$\sqrt{3}(K'' - K')$	0	0	$E_0 + 2K - K'$

where E_0 contains all the one-electron terms and the Coulomb integrals. K is the average of the on-site exchange integrals $2K = K_{12} + K_{34}$. K' and K'' are sums of two-center exchange integrals, and hence, much smaller. $2K' = K_{13} + K_{24}$; $2K'' = K_{14} + K_{23}$. The different exchange integrals $K_{ij} = \langle\varphi_i\varphi_j|1/r_{ij}|\varphi_j\varphi_i\rangle$ are defined in Fig. 5.13.

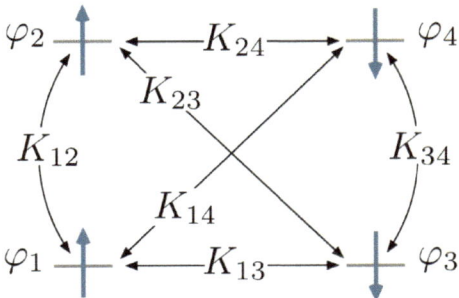

Fig. 5.13 Definition of the exchange integrals that appear in the matrix representation of the model space formed by the neutral determinants of the four-electron/four-orbital case. For the centrosymmetric case here considered $K_{12} = K_{34} \gg K_{13} \approx K_{24} > K_{14} = K_{23}$

In the first place, we recognize that without considering the non-Hund states, there is a strict regular order of the singlet, triplet and quintet states. The triplet-quintet splitting $(2K')$ is twice as large as the energy difference between triplet and singlet. Since this model space only considers the neutral determinants, it is not unexpected that the quintet spin coupling leads to the lowest energy. The non-Hund states with one local singlet (NH1 and NH2) lie at relative energies of approximately $2K$ and the double non-Hund state is found around $4K$ with respect to the Q, T, and S states.

The only non-zero off-diagonal terms in the matrix reveal small interactions between T and $NH2$ and between S and $NH3$. However, these matrix elements are usually very small. The exchange integrals involved are all two-center integrals and therefore rather small. Moreover, the different two-center integrals are similar in magnitude and tend to cancel each other. Obviously, the quintet state cannot have additional contributions from the non-Hund states, since singlet coupling on one of the magnetic centers cannot lead to a state with overall quintet coupling. Although these interactions are at the very origin of the deviations to the regular Landé spacing of the energies, a second ingredient is necessary to activate the contribution of the non-Hund states. The key to a sizeable non-Heisenberg behaviour lies in the interaction of the non-Hund states with the ionic determinants, which in turn interact with the singlet and triplet functions of Eq. 5.37. To illustrate this effect, the model space is enlarged with the eight ionic determinants that interact with the neutral ones defined in Eq. 5.31. The *plus* and *minus* combinations of the ionic determinants give rise to four singlet and four triplet CSFs with three electrons on one center and one electron on the other.

$$I_{1,2} = \left(|\varphi_1 \overline{\varphi}_1 \varphi_2 \overline{\varphi}_4| \pm |\varphi_1 \overline{\varphi}_1 \varphi_4 \overline{\varphi}_2| \right) / \sqrt{2} \qquad (5.39a)$$

$$I_{3,4} = \left(|\varphi_3 \overline{\varphi}_3 \varphi_4 \overline{\varphi}_2| \pm |\varphi_3 \overline{\varphi}_3 \varphi_2 \overline{\varphi}_4| \right) / \sqrt{2} \qquad (5.39b)$$

$$I_{5,6} = \left(|\varphi_2 \overline{\varphi}_2 \varphi_1 \overline{\varphi}_3| \pm |\varphi_2 \overline{\varphi}_2 \varphi_3 \overline{\varphi}_1| \right) / \sqrt{2} \qquad (5.39c)$$

$$I_{7,8} = \left(|\varphi_4 \overline{\varphi}_4 \varphi_1 \overline{\varphi}_3| \pm |\varphi_4 \overline{\varphi}_4 \varphi_2 \overline{\varphi}_1| \right) / \sqrt{2} \qquad (5.39d)$$

There are more ionic determinants, e.g. $|\phi_1\bar{\phi}_1\phi_2\bar{\phi}_3|$, but these do not interact with T or S when ϕ_1 and ϕ_3 belong to a different irreducible representation than ϕ_2 and ϕ_4. The use of spin symmetry adapted configurations allows us to write the full 15×15 matrix representation of the model space in three separate blocks. The first one is one-dimensional and only contains the neutral quintet CSF, the second one contains all the triplet CSFs: T, NH2 and the even-numbered I_i CSFs. The third sub-block of the total reference space is formed by the singlets: S, NH3 and the odd-numbered I_i CSFs. NH1 does not interact with any of the other CSFs due to symmetry. The triplet and singlet interaction matrices are

	$|T\rangle$	$|NH2\rangle$	$|I2\rangle$	$|I4\rangle$	$|I6\rangle$	$|I8\rangle$
$\langle T|$	$-2K$					
$\langle NH2|$	$K_{24} - K_{13}$	$2K''$				
$\langle I2|$	t_{13}	$-t_{13}$	$U - K_{24}$			
$\langle I4|$	t_{13}	$-t_{13}$	K_{13}	$U - K_{24}$		
$\langle I6|$	$-t_{24}$	$-t_{24}$	$\alpha - \beta$	$\beta - \gamma$	$U' - K_{13}$	
$\langle I8|$	$-t_{24}$	$-t_{24}$	$\beta - \gamma$	$\alpha - \beta$	K_{24}	$U' - K_{13}$

	$|S\rangle$	$|NH3\rangle$	$|I1\rangle$	$|I3\rangle$	$|I5\rangle$	$|I7\rangle$
$\langle S|$	$-2K + K'$					
$\langle NH3|$	$\sqrt{3}(K'' - K')$	$2K - K'$				
$\langle I1|$	$-3t_{13}/\sqrt{6}$	$t_{13}/\sqrt{2}$	$U + K_{24}$			
$\langle I3|$	$-3t_{13}/\sqrt{6}$	$t_{13}/\sqrt{2}$	K_{13}	$U + K_{24}$		
$\langle I5|$	$3t_{24}/\sqrt{6}$	$-t_{24}/\sqrt{2}$	$\alpha + \beta$	$\beta + \gamma$	$U' + K_{13}$	
$\langle I7|$	$3t_{24}/\sqrt{6}$	$-t_{24}/\sqrt{2}$	$\beta + \gamma$	$\alpha + \beta$	K_{24}	$U' + K_{13}$

with $\alpha = \langle\varphi_1\varphi_4|1/r_{12}|\varphi_2\varphi_3\rangle - \langle\varphi_1\varphi_4|1/r_{12}|\varphi_3\varphi_2\rangle$; $\beta = \langle\varphi_1\varphi_4|1/r_{12}|\varphi_2\varphi_3\rangle$ and $\gamma = \langle\varphi_1\varphi_2|1/r_{12}|\varphi_4\varphi_3\rangle - \langle\varphi_1\varphi_2|1/r_{12}|\varphi_3\varphi_4\rangle$. E_0 is omitted and the difference between the on-site repulsion integrals U and U' arises from the double occupancy of φ_1 or φ_3 in I_{1-4} versus φ_2 or φ_4 in I_{5-8}.

The interaction between the ionic states and the neutral states with Hund coupling cannot break the Landé pattern. This is very easily demonstrated by considering the effect of $I_{1,2}$ on S and T. The diagonalization of the two 2×2 matrices gives

$$E(T) = \frac{1}{2}\left(U \pm \sqrt{U^2 + 4t_{13}^2}\right)$$

$$E(S) = \frac{1}{2}\left(U \pm \sqrt{U^2 + 6t_{13}^2}\right) \tag{5.40}$$

To make it easier to see that these energies perfectly fit the energy differences described with the Heisenberg Hamiltonian, we simplify the expressions with the Taylor expansion used before in Eq. 5.8. The energies of the lowest two states are

$$E(T) = -\frac{t_{13}^2}{U}$$

$$E(S) = -\frac{3}{2}\frac{t_{13}^2}{U} \tag{5.41}$$

It is now trivial to see that $E(T) - E(Q) = 2\big(E(S) - E(T)\big)$. Note that taking into account the interaction of all the CSFs with ionic character leads to more elaborate expressions for the energies of S and T, but the principle is the same.

The simultaneous interaction of the ionic CSFs with S/T and $NH3/NH2$ makes that the non-Hund states gain some weight in the wave function of the lowest triplet and singlet states. This is exactly the same mechanism as in the configuration interaction of singles and doubles. The singles have no direct interaction with the Hartree-Fock determinant due the Brillouin theorem, but they appear in the CI wave function due to an indirect interaction via the doubles.

5.10 Rationalize the relative size for the estimates of J extracted from the singlet-triplet and from the triplet-quintet energy difference. Hint: compare the matrix elements of the ionic determinants with the non-Hund states and take into consideration the relative energy of the non-Hund states involved in the coupling.

To illustrate the above-discussed concepts, we first decompose the magnetic coupling of the Ni-azido complex with angle $\delta = 0$ in Table 5.5. The first column marked with K results from the diagonalization of the model space with only neutral determinants. The effect of the different exchange interactions makes the quintet the lowest state and no (measurable) deviations from the Heisenberg behaviour are observed. The inclusion of the spin polarization introduces important antiferromagnetic contributions but does not break the Landé pattern. By adding the ionic determinants to the CI wave functions of the three lowest spin states, the magnetic coupling further increases as expected. But, more interestingly, we observe a small difference in the estimates of J calculated from the singlet-triplet and the triplet-quintet energy difference. This non-Heisenberg behaviour becomes more pronounced in the CAS+S calculation when the ionic determinants are relaxed, leading to a stronger interaction with the Hund and non-Hund states. When all electron correlation effects are included, the calculated coupling is close to experiment and the deviations are yet a little larger.

It remains to establish which state, singlet or triplet, is most strongly affected by the interaction with the non-Hund states. For this purpose, we have decomposed the DDCI wave functions of the different spin states and listed the coefficients of

Table 5.5 Decomposition of the magnetic coupling of the binuclear Ni-azido complex with $\delta = 0$

	K	$K + SP$	CAS(4,4)	CAS(4,4)+S	DDCI
$(E_T - E_Q)/2$	2.41	-9.05	-12.62	-55.57	-104.48
$E_T - E_S$	2.41	-9.04	-12.58	-54.92	-101.47

Table 5.6 Decomposition of the DDCI wave function for the singlet, triplet and quintet state of the binuclear Ni-azido complex with three different values of δ

$\delta = 0°$	Quintet	Triplet	Singlet
Hund	0.97227	0.96552	0.94652
Non-Hund	–	0.00110	0.00002
Ionic	–	0.00163	0.00233
$(E_T - E_Q)/2$	−104.48		
$E_S - E_T$	−101.47		
$\delta = 22°$			
Hund	0.97028	0.96565	0.96545
Non-Hund	–	0.00079	$<10^{-5}$
Ionic	–	0.00117	0.00169
$(E_T - E_Q)/2$	−65.92		
$E_S - E_T$	−64.65		
$\delta = 45°$			
Hund	0.96578	0.96482	0.96633
Non-Hund	–	0.00026	$<10^{-5}$
Ionic	–	0.00037	0.00057
$(E_T - E_Q)/2$	−3.69		
$E_S - E_T$	−3.67		

the Hund, non-Hund and ionic CSFs in Table 5.6. If we first focus on the above-discussed case of $\delta = 0$, we see that the largest non-Hund contribution appears in the triplet function. This is in line with the larger matrix element of NH2 with the even-numbered ionic states I_i and the lower relative energy of NH2 (one atomic non-Hund state) with respect to NH3 (atomic non-Hund coupling on both magnetic centers). Hence, it is expected that the non-Hund states stabilize the triplet state more than the singlet, and hence, $E_S - E_T < (E_T - E_Q)/2$ as observed in the Ni-azido complexes.

Table 5.6 also shows that with increasing angle δ the magnetic coupling is strongly reduced, caused by the loss of efficiency of the kinetic exchange mechanism evidenced by the decrease of the coefficient of the ionic determinants in the wave function of the singlet and the triplet wave functions. At the same time, the deviations to the regular Heisenberg spacing are strongly suppressed as are the coefficients of the non-Hund states. This illustrates the role of the indirect coupling of the non-Hund states with the neutral determinants via the ionic ones; without significant contribution of the ionic determinants, i. e. no efficient kinetic exchange, all possible deviations from the Landé pattern of the energies of the lowest spin states are eliminated.

In summary, biquadratic exchange interactions are only expected in complexes with sizeable magnetic interaction, small on-site repulsion U, not too large on-site exchange interaction K and different inter-site interactions for the pairs of electrons on the different magnetic centers: $K_{13} \neq K_{24}$; $t_{13} \neq t_{24}$.

5.4.2 Decomposition of the Four-Center Interactions

The four-spin interaction as discussed in Chap. 3 (Sect. 3.4.2) is the effective matrix element between the determinants $\Phi_I = |a\bar{b}c\bar{d}|$ and $\Phi_J = |\bar{a}b\bar{c}d|$. Since the direct matrix element of the electronic Hamiltonian between them is zero (there are more than two different columns in the determinants), there must be other, indirect interactions that account for the non-zero value of this interaction. In analogy to the *normal* two-center magnetic interaction, we will review the role of the ionic determinants in the effective matrix elements. Figure 5.14 shows one of the pathways that connects Φ_I with Φ_J through three different ionic states. In the first step an electron hops from site A to B to form the ionic determinant Φ_α. The matrix element is

$$\langle \Phi_I | \hat{H} | \Phi_\alpha \rangle = \langle a\bar{b}c\bar{d} | \hat{H} | b\bar{b}c\bar{d} \rangle = \langle a | \hat{h} | b \rangle + \text{smaller}$$
$$\text{two-electron integrals} = t_{ab} = t \tag{5.42}$$

The two-electron integrals will be detailed in Chap. 6 but are here absorbed into the effective hopping parameter t_{ab}. The relative energy of Φ_α is U, the same parameter as in the analysis of the two-center interaction. The next step transfers an electron with β spin to center C to generate Φ_β with energy U. Assuming a square lattice, the matrix element with Φ_α equals t. The third and fourth step are similar and in total one gets the fourth-order perturbation contribution of this path to the effective matrix element between Φ_I and Φ_J by applying Eq. 5.12.

$$\frac{\langle \Phi_I | \hat{H} | \Phi_\alpha \rangle \langle \Phi_\alpha | \hat{H} | \Phi_\beta \rangle \langle \Phi_\beta | \hat{H} | \Phi_\gamma \rangle \langle \Phi_\gamma | \hat{H} | \Phi_J \rangle}{(E_\alpha - E_J)(E_\beta - E_J)(E_\gamma - E_J)} = \frac{t^4}{U^3} \tag{5.43}$$

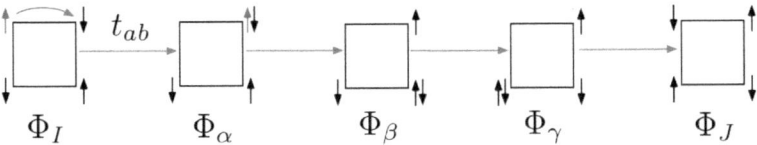

Fig. 5.14 Basic pathway to connect $\Phi_I = |a\bar{b}c\bar{d}|$ with $\Phi_J = |\bar{a}b\bar{c}d|$ via electron hopping among neighboring sites parametrized by $t = t_{ij}$. The relative energies of the intermediate determinants $\Phi_{\alpha,\beta,\gamma}$ is U

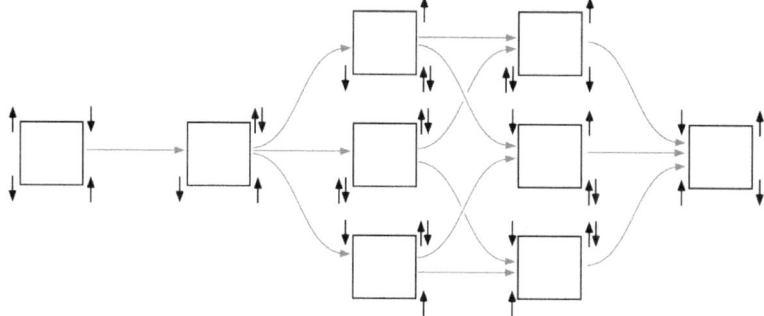

Fig. 5.15 The six pathways that connect $\Phi_I = |a\bar{b}c\bar{d}|$ with $\Phi_J = |\bar{a}b\bar{c}d|$ in a clockwise fashion. The relative energy of all intermediate determinants is U, expect the di-ionic determinant (*third column in the middle*), whose energy can be approximated by $2U$

5.11 Equation 5.43 assumes that the energies of Φ_α and Φ_β are strictly the same. Is this correct? Hint: Assume that A, B, C and D are uncharged in Φ_I and calculate the Coulomb interaction along the path in a simple point charge model.

The path in Fig. 5.14 is not the only possibility to go from Φ_I to Φ_J. In fact, there are 48 different pathways. Twelve of these start with an electron hopping from center A, six with a clockwise circulation and six go in an anti-clockwise fashion. Figure 5.15 shows the six clockwise electron circulations starting at center A, the upper path corresponds to the one that was discussed before. Putting the energy of the di-ionic state—third column in the middle of the figure—at $2U$, the total contribution of the six paths is

$$4 \times \frac{t^4}{U^3} + 2 \times \frac{t^4}{U \cdot 2U \cdot U} = 5\frac{t^4}{U^3} \tag{5.44}$$

The contribution of the anti-clockwise pathways is the same, increasing the prefactor to 10. By accounting for the pathways that start at the other magnetic centers, we can derive the total effective matrix element and a perturbative estimate of J_r

$$\langle \Phi_I | \hat{H}^{eff} | \Phi_J \rangle = 40\frac{t^4}{U^3} \Rightarrow J_r^{pt} = 80\frac{t^4}{U^3} \tag{5.45}$$

5.12 Write down an anti-clockwise path from Φ_I to Φ_J starting at center A that goes through a di-ionic determinant.

The comparison of this expression with the one derived for the ordinary two-center coupling $(4t^2/U)$ shows that J_r is expected to be significantly smaller than the interactions described in the standard Heisenberg Hamiltonian, dividing by U^3 instead of U makes the interaction much smaller. However, the very large prefactor in the perturbative estimate makes that the ring exchange is not necessarily negligible in all cases. As long as U is not too large and t sizeable, one can expect significant four-center interactions when the geometry of the system is square-like.

5.4.3 Complex Interactions with Single Determinant Approaches

Biquadratic exchange: The isotropic linear magnetic exchange can be calculated in a rather straightforward way with single determinant spin unrestricted methods. Assuming that the BS determinant is a linear combination of the spin states with lowest and highest possible spin moment, the Yamaguchi equation (Eq. 4.85) relates the energies of the BS and HS determinants with J in a straightforward way, independent of the number of unpaired electrons on the magnetic sites involved in the coupling. The biquadratic exchange can however not be addressed from energy differences only, simply because we have only access to one energy difference, obviously too few to determine two parameters.

Instead one can estimate the strength of the biquadratic exchange in an indirect way via the electronic structure parameters U, t and K. To derive the relevant equations we need to compare the expressions of the singlet and triplet states in terms of J and λ given in Eq. 3.75 with their fourth-order perturbation estimates using the matrix elements derived in Sect. 5.4.1. In the first place, we need a common zero of energy. This is easily achieved by putting the energy of the quintet state to zero. The expressions of singlet, triplet and quintet states in terms of J and λ then become

$$E(Q) = 0$$
$$E(T) = 2J$$
$$E(S) = 3J + 3\lambda \tag{5.46}$$

The fourth-order perturbation estimates give us expressions in terms of t, U and K, and hence, we can relate λ to these electronic structure parameters, which can be calculated with spin-unrestricted single determinant methods. To simplify the perturbation estimates we neglect α, β and γ, and all intersite exchange integrals in the interaction matrices derived in Sect. 5.4.1. The interaction matrices for triplet and singlet states then become

	$|T\rangle$	$|NH2\rangle$	$|I2\rangle$	$|I4\rangle$	$|I6\rangle$	$|I8\rangle$
$\langle T|$	0					
$\langle NH2|$	0	$2K$				
$\langle I2|$	t_{13}	$-t_{13}$	U			
$\langle I4|$	t_{13}	$-t_{13}$	0	U		
$\langle I6|$	$-t_{24}$	$-t_{24}$	0	0	U'	
$\langle I8|$	$-t_{24}$	$-t_{24}$	0	0	0	U'

	$|S\rangle$	$|NH3\rangle$	$|I1\rangle$	$|I3\rangle$	$|I5\rangle$	$|I7\rangle$
$\langle S|$	0					
$\langle NH3|$	0	$4K$				
$\langle I1|$	$-3t_{13}/\sqrt{6}$	$t_{13}/\sqrt{2}$	U			
$\langle I3|$	$-3t_{13}/\sqrt{6}$	$t_{13}/\sqrt{2}$	0	U		
$\langle I5|$	$3t_{24}/\sqrt{6}$	$-t_{24}/\sqrt{2}$	0	0	U'	
$\langle I7|$	$3t_{24}/\sqrt{6}$	$-t_{24}/\sqrt{2}$	0	0	0	U'

The second-order correction to the energy is obtained from the expression

$$E^{(2)} = \frac{\langle \Phi_I|\hat{H}|\Phi_\alpha\rangle\langle\Phi_\alpha|\hat{H}|\Phi_I\rangle}{E_I - E_\alpha} \tag{5.47}$$

which becomes

$$E_T^{(2)} = \sum_{i=2,4,6,8} \frac{\langle T|\hat{H}|Ii\rangle\langle Ii|\hat{H}|T\rangle}{-U_i} = -\frac{2t_{13}^2}{U} - \frac{2t_{24}^2}{U'} \tag{5.48}$$

for the triplet and

$$E_S^{(2)} = \sum_{i=1,3,5,7} \frac{\langle S|\hat{H}|Ii\rangle\langle Ii|\hat{H}|S\rangle}{-U_i} = -\frac{3t_{13}^2}{U} - \frac{3t_{24}^2}{U'} \tag{5.49}$$

for the singlet. From these equations we can define $J^{(2)}$, the second-order estimate of J, as $-\frac{t_{13}^2}{U} - \frac{t_{24}^2}{U'}$. The fourth-order contribution is determined with Eq. 5.12 and counts with much more terms, which are summarized below and illustrated for one of the cases in Fig. 5.16.

Fig. 5.16 Schematic representation of one of the fourth-order interactions contributing to the energy of the triplet state. The total contribution of this path equals $-t_{13}^2 t_{24}^2/UU'2K$

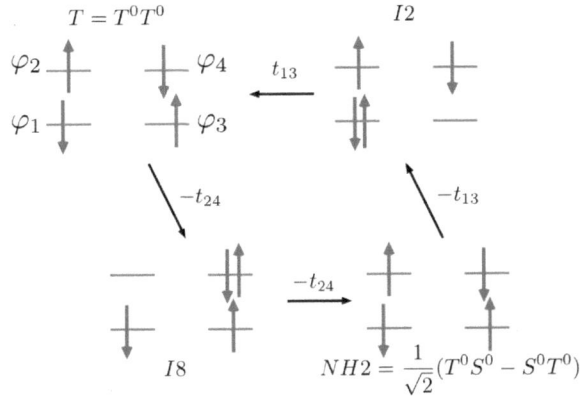

Φ_I	Φ_α	Φ_β	Φ_γ	Φ_I	
S	$I1$	$NH3$	$I1$	S	$-\frac{3}{4}t_{13}^4/U^24K$
			$I3$		$-\frac{3}{4}t_{13}^4/U^24K$
			$I5$		$-\frac{3}{4}t_{13}^2 t_{24}^2/UU'4K$
			$I7$		$-\frac{3}{4}t_{13}^2 t_{24}^2/UU'4K$
	$I3$		$I3$		$-\frac{3}{4}t_{13}^4/U^24K$
			$I5$		$-\frac{3}{4}t_{13}^2 t_{24}^2/UU'4K$
			$I7$		$-\frac{3}{4}t_{13}^2 t_{24}^2/UU'4K$
	$I5$		$I5$		$-\frac{3}{4}t_{24}^4/U'^24K$
			$I7$		$-\frac{3}{4}t_{24}^4/U'^24K$
	$I7$		$I7$		$-\frac{3}{4}t_{24}^4/U'^24K$

Φ_I	Φ_α	Φ_β	Φ_γ	Φ_I	
T	$I2$	$NH2$	$I2$	T	$-t_{13}^4/U^22K$
			$I4$		$-t_{13}^4/U^22K$
			$I6$		$t_{13}^2 t_{24}^2/UU'2K$
			$I8$		$t_{13}^2 t_{24}^2/UU'2K$
	$I4$		$I4$		$-t_{13}^4/U^22K$
			$I6$		$t_{13}^2 t_{24}^2/UU'2K$
			$I8$		$t_{13}^2 t_{24}^2/UU'2K$
	$I6$		$I6$		$-t_{24}^4/U'^22K$
			$I8$		t_{24}^4/U'^22K
	$I6$		$I6$		$-t_{24}^4/U'^22K$

The terms with the ionic states interchanged ($\Phi_\alpha = I3$, $\Phi_\gamma = I1$, etc.) should also be added to the perturbational estimate. This is easily done by multiplying all the terms by two, except the ones with $\Phi_\alpha = \Phi_\gamma$. Now the following corrections arise

$$E_T^{(4)} = \frac{2t_{13}^4}{U^2K} + \frac{2t_{24}^4}{U'^2K} - \frac{4t_{13}^2 t_{24}^2}{UU'K} = \frac{2B}{K} \tag{5.50}$$

$$E_S^{(4)} = -\frac{3}{4}\frac{t_{13}^4}{U^2K} - \frac{3}{4}\frac{t_{24}^4}{U'^2K} - \frac{3}{2}\frac{t_{13}^2 t_{24}^2}{UU'K} = -\frac{3J^{(2)2}}{4K} \tag{5.51}$$

with $B = t_{13}^2/U - t_{24}^2/U'$. Then the total perturbative estimate for the singlet and triplet energies are

$$E_T = E_T^{(2)} + E_T^{(4)} = -\frac{2t_{13}^2}{U} - \frac{2t_{24}^2}{U'} - \frac{2B^2}{K} = 2J^{(2)} - \frac{2B^2}{K} \tag{5.52}$$

$$E_S = E_S^{(2)} + E_S^{(4)} = -\frac{3t_{13}^2}{U} - \frac{3t_{24}^2}{U'} - \frac{3J^2}{4K} = 3J^{(2)} - \frac{3J^{(2)2}}{4K} \tag{5.53}$$

The comparison with the expression for the singlet and triplet energy eigenvalues of the Heisenberg Hamiltonian with biquadratic terms leads to a fourth-order estimate of J as

$$J = J^{(2)} - \frac{B^2}{K} \tag{5.54}$$

Then from the singlet energy we get

$$E_S = 3J + 3\lambda = 3\left(J^{(2)} - \frac{B^2}{K}\right) + 3\lambda = 3J^{(2)} - \frac{3J^{(2)2}}{4K} \tag{5.55}$$

from which the perturbative expression for λ in terms of t, U and K can be extracted

$$\lambda = \frac{B^2}{3K} - \frac{J^{(2)2}}{4K} \tag{5.56}$$

The next step concerns the calculation of the energy of a collection of spin unrestricted determinants with different occupations and relate their energies to the electronic structure parameters in order to calculate the λ parameter and in this way obtain a measure for the biquadratic interaction strength from methods like DFT (Fig. 5.17).

Assuming that $U = U'$, the energies of the following five determinants define the parameters that appear in the perturbative expressions of B and J, which in turn lead to an estimate of λ, the biquadratic exchange parameter.

Fig. 5.17 The five determinants that are needed to calculate the electronic structure parameters that define B and J in the perturbative expression of the biquadratic exchange strength

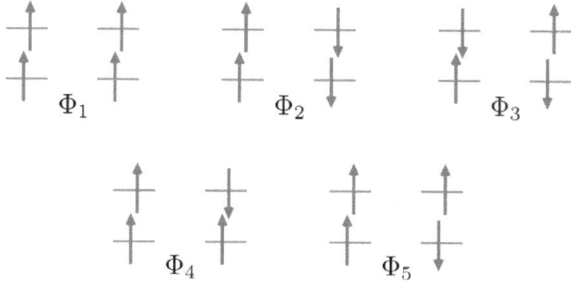

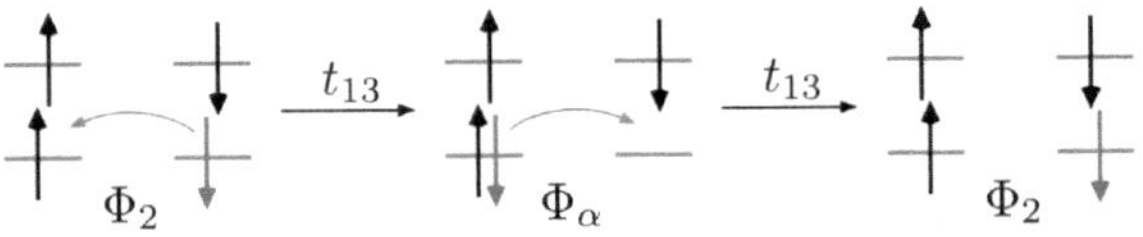

Fig. 5.18 One of the second-order contributions to the energy of Φ_2. This path contributes $t_{13}^2/-U$

$$M_S = 2: \quad \Phi_1 = |\varphi_1\varphi_2\varphi_3\varphi_4| \qquad E_1^{(2)} = 0 \qquad\qquad (5.57)$$

$$M_S = 0: \quad \Phi_2 = |\varphi_1\varphi_2\overline{\varphi}_3\overline{\varphi}_4| \qquad E_2^{(2)} = -\frac{2(t_{13}^2 + t_{24}^2)}{U} \qquad (5.58)$$

$$\Phi_3 = |\varphi_1\overline{\varphi}_2\overline{\varphi}_3\varphi_4| \qquad E_3^{(2)} = 2K - \frac{2(t_{13}^2 + t_{24}^2)}{U} \qquad (5.59)$$

$$M_S = 1: \quad \Phi_4 = |\varphi_1\varphi_2\varphi_3\overline{\varphi}_4| \qquad E_4^{(2)} = K - \frac{2t_{24}^2}{U} \qquad (5.60)$$

$$\Phi_5 = |\varphi_1\varphi_2\overline{\varphi}_3\varphi_4| \qquad E_5^{(2)} = K - \frac{2t_{13}^2}{U} \qquad (5.61)$$

The listed energies are the sum of the zeroth-order energies and second-order corrections. The latter are calculated by taking into account all possible interactions of these determinants with the ionic determinants, which are assumed to be degenerate with energy U relative to Φ_1. One example is given in Fig. 5.18, the rest is completely analogous. The zeroth-order energies $\langle \Phi_I | \hat{H} | \Phi_I \rangle$ only count the number of on-site exchange interactions K, all other terms are neglected or the same as in the reference energy $E_1^{(0)}$.

> **5.13** Write down the two ionic determinants that interact with Φ_4 and calculate the interaction matrix elements.

Four-center interactions: In Chap. 3, we have seen how the four-center interactions can be extracted using an effective Hamiltonian spanned by the six $M_S = 0$ determinants. To address the ring exchange within a spin unrestricted setting, this model space is no longer sufficient, but has to be extended with the $M_S = 2$ and the four $M_S = 1$ determinants. In any standard implementation of density functional theory, the main area of spin unrestricted methods, one has only access to the diagonal elements of this 11×11 model Hamiltonian; matrix elements between different determinants are not routinely calculated in most quantum chemistry packages. In addition, it should be realized that the four $M_S = 1$ determinants are all degenerate and as was shown in Eq. 3.84, the six $M_S = 0$ determinants are degenerate in pairs. Hence, one can count with at most five energies, i.e. four energy differences that can be used to determine four independent parameters. It is therefore intrinsically impossible to determine the interaction strength of the three cyclic permutations defined in Fig. 3.14, as can in principle be done with wave function based methods through the construction of a numerical effective Hamiltonian. However, in any practical case there is only one sizeable four-center interaction, namely the one with the $\hat{P}_{1234}$ operator associated to it.

The expectation values of the $M_S = 0$ determinants can be found in Eq. 3.84, but for the $M_S = 1$ and $M_S = 2$ determinants we will derive them here. Among the four degenerate $M_S = 1$ determinants we will focus on $|abc\bar{d}|$, or $|\alpha\alpha\alpha\beta|$ in a spin-only notation. With the following ingredients for the two-center interactions:

$$-J_1(\hat{S}_A\hat{S}_B + \hat{S}_C\hat{S}_D)\alpha\alpha\alpha\beta = -J_1\left(\frac{1}{4}\alpha\alpha\alpha\beta + \frac{1}{2}\alpha\alpha\beta\alpha - \frac{1}{4}\alpha\alpha\alpha\beta\right)$$

$$-J_2(\hat{S}_A\hat{S}_D + \hat{S}_B\hat{S}_C)\alpha\alpha\alpha\beta = -J_2\left(\frac{1}{2}\beta\alpha\alpha\alpha - \frac{1}{4}\alpha\alpha\alpha\beta + \frac{1}{4}\alpha\alpha\alpha\beta\right) \qquad (5.62)$$

$$-J_3(\hat{S}_A\hat{S}_D + \hat{S}_B\hat{S}_C)\alpha\alpha\alpha\beta = -J_3\left(\frac{1}{4}\alpha\alpha\alpha\beta + \frac{1}{2}\alpha\beta\alpha\beta - \frac{1}{4}\alpha\alpha\alpha\beta\right)$$

and for the four-center interaction:

$$J_r(\hat{S}_A\hat{S}_B)(\hat{S}_C\hat{S}_D)\alpha\alpha\alpha\beta = J_r(\hat{S}_A\hat{S}_B)\left(\frac{1}{2}\alpha\alpha\beta\alpha - \frac{1}{4}\alpha\alpha\alpha\beta\right)$$

$$= J_r\left(\frac{1}{8}\alpha\alpha\beta\alpha - \frac{1}{16}\alpha\alpha\alpha\beta\right)$$

$$J_r(\hat{S}_A\hat{S}_D)(\hat{S}_B\hat{S}_C)\alpha\alpha\alpha\beta = J_r(\hat{S}_A\hat{S}_D)\left(\frac{1}{4}\alpha\alpha\alpha\beta\right) \qquad (5.63)$$

$$= J_r\left(\frac{1}{8}\beta\alpha\alpha\alpha - \frac{1}{16}\alpha\alpha\alpha\beta\right)$$

$$-J_r(\hat{S}_A\hat{S}_C)(\hat{S}_B\hat{S}_D)\alpha\alpha\alpha\beta = -J_r(\hat{S}_A\hat{S}_C)\left(\frac{1}{2}\alpha\beta\alpha\alpha - \frac{1}{4}\alpha\alpha\alpha\beta\right)$$

$$= -J_r\left(\frac{1}{8}\alpha\beta\alpha\alpha - \frac{1}{16}\alpha\alpha\alpha\beta\right)$$

the matrix element becomes

$$\langle\alpha\alpha\alpha\beta|\hat{H}|\alpha\alpha\alpha\beta\rangle = -\frac{1}{16}J_r \qquad (5.64)$$

Applying the same procedure to $|\alpha\alpha\alpha\alpha|$ leads to

$$\langle\alpha\alpha\alpha\alpha|\hat{H}|\alpha\alpha\alpha\alpha\rangle = -\frac{1}{2}(J_1 + J_2 + J_3) + \frac{1}{16}J_r \qquad (5.65)$$

5.14 Check the matrix element of the $M_S = 2$ determinant of the spin Hamiltonian given in Eq. 3.83.

Taking the expectation value of the $M_S = 2$ determinant as zero of energy, the following relations emerge to determine the four parameters

$$
\begin{aligned}
E(|a\overline{b}c\overline{d}|) - E(|abcd|) &= J_1 + J_2 \\
E(|ab\overline{c}\overline{d}|) - E(|abcd|) &= J_1 + J_3 \\
E(|a\overline{b}\overline{c}d|) - E(|abcd|) &= J_2 + J_3 \\
E(|abc\overline{d}|) - E(|abcd|) &= \frac{1}{2}(J_1 + J_2 + J_3) - \frac{1}{8}J_r
\end{aligned}
\tag{5.66}
$$

Hence, the extraction of the four-spin cyclic exchange parameter within the spin-unrestricted setting of the DFT approach relies on obtaining converged solutions for the determinants with the required spin distributions, which is not always a trivial task.

Problems

5.1 Zeroth-order description. Write down the matrix of the model space that only considers neutral determinants expressed in local orbitals. Diagonalize the matrix and calculate the singlet-triplet energy difference. What is the state of lowest energy?

5.2 Construction of the CAS(2,2)CI matrix in the symmetry adapted CSF basis. The CASCI matrix given in Eq. 5.4 uses the four $M_S = 0$ determinants as basis. The matrix can be greatly simplified by a basis set change using symmetry adapted CSFs.

a. Write down the four symmetry adapted CSFs that arise from the linear combinations of the four $M_S = 0$ determinants. The expressions of the states after configuration interaction given in Eq. 5.5 may give a hint on the CSFs.
b. Calculate the energy expectation values of the four CSFs and place them on the diagonal of the matrix.
c. Identify the CSFs as singlet or triplet spin eigenfunctions and label them by *gerade/ungerade* spatial symmetry, assuming that the system has an inversion center. How many off-diagonal elements have non-zero value?
d. Calculate the remaining matrix elements to complete the CAS(2,2)CI matrix.

5.3 Spin contamination of the BS state. The relaxation of the magnetic orbitals of the BS determinant in the field of the frozen ROKS core orbitals introduces spin contamination. The amount of spin contamination can be determined analytically by rewriting Eq. 5.26 in terms of spin adapted CSFs instead of the neutral and ionic valence bond structures.

1. Which term in Eq. 5.26 is an eigenfunction of $\hat{S}^2$. Give the eigenvalue of this term.
2. The two other terms have to be written in the form of the singlet $(|a\overline{b}| + |b\overline{a}|)$ and triplet $(|a\overline{b}| - |b\overline{a}|)$ CSFs. Use the trigonometric relations $\sin^2 \phi + \cos^2 \phi = 1$ and

$\sin^2\phi - \cos^2\phi = \cos(2\phi)$. Hint: Add and subtract $\frac{1}{2}\cos^2\alpha|b\bar{a}| + \frac{1}{2}\sin^2\alpha|a\bar{b}|$ and split the first two terms of Eq. 5.26 in halfs. Then, order the terms to form the given trigonometric relations.

3. Calculate $\langle\Phi_{BS}|\hat{S}^2|\Phi_{BS}\rangle$ using the above derived expression $\Phi_{BS} = (|a\bar{b}| + |b\bar{a}|)/2 + (|a\bar{b}| - |b\bar{a}|)\cos(2\alpha)/2 + (|a\bar{a}| + |b\bar{b}|)\sin\alpha\cos\alpha$.

5.4 Kinetic exchange by second-order perturbation theory. Make a second-order estimate of the singlet energy taking into account the interaction between $S = \frac{1}{\sqrt{2}}(|a\bar{b}| + |b\bar{a}|)$ and the ionic states $I_1 = \frac{1}{\sqrt{2}}(|a\bar{a}| + |b\bar{b}|)$ and $I_2 = \frac{1}{\sqrt{2}}(|a\bar{a}| - |b\bar{b}|)$. The energy of the triplet, $T = \frac{1}{\sqrt{2}}(|a\bar{b}| - |b\bar{a}|)$, equals $E_{ref} - K_{ab}$ and should be taken as reference.

5.5 Biquadratic exchange versus t_{13}/t_{24}. Express the perturbative estimate of λ (Eq. 5.56) in terms of t_{13} and t_{24} with a common denominator for the two terms. Determine for which values of t_{13} and t_{24} the biquadratic exchange vanishes and for which values it can be expected to be maximal.

5.6 Estimating J_r. Accurate calculations on a polynuclear paramagnetic compound with four $S = \frac{1}{2}$ magnetic centers indicate that the only significant interactions are the following bilinear isotropic interactions: $J_{12} = J_{34} = -25.1$ meV, and $J_{23} = J_{14} = -39.5$ meV. Nevertheless, the experimental temperature dependence of the magnetic susceptibility could not be fitted satisfactorily with these values. Provide a perturbational estimate for the four-center cyclic exchange to improve the fitting (extra data: $U = 5.3$ eV).

References

1. J. Cabrero, C. de Graaf, E. Bordas, R. Caballol, J.P. Malrieu, Chem. Eur. J. **9**, 2307 (2003)
2. C.J. Calzado, J. Cabrero, J.P. Malrieu, R. Caballol, J. Chem. Phys. **116**(7), 2728 (2002)
3. C.J. Calzado, J. Cabrero, J.P. Malrieu, R. Caballol, J. Chem. Phys. **116**(10), 3985 (2002)
4. C.J. Calzado, C. Angeli, D. Taratiel, R. Caballol, J.P. Malrieu, J. Chem. Phys. **131**, 044327 (2009)
5. E. Coulaud, J.P. Malrieu, N. Guihéry, N. Ferré, J. Chem. Theory Comput. **9**, 3429 (2013)
6. R. Bastardis, N. Guihéry, C. de Graaf, J. Chem. Phys. **129**, 104102 (2008)

Chapter 6
Magnetism and Conduction

Abstract After the description of the electron hopping in systems where not all the magnetic centers have the same number of unpaired electrons, a short account is given of the double exchange mechanism in mixed-valence systems. Although this phenomenon can certainly be found in transition metal complexes, it is more common to happen in doped systems in the solid state. Therefore, the second part of this chapter introduces the basics of the quantum chemical approach to magnetic interactions in extended systems. The embedded cluster approach will be contrasted against band structure calculations. Thereafter, some concepts will be introduced that are widely used in the condensed matter physics community. We do not give a full description of all the magnetic phenomena in solid state compounds but rather help the reader with a quantum chemical background to find its way in the rich literature on this topic.

6.1 Electron Hopping

In all the magnetic systems described so far the number of magnetic orbitals was equal to the number of unpaired electrons. These systems are generally known as half-filled systems and the price (in terms of energy) to move an electron from one site to another is proportional to the on-site repulsion parameter U. Since this parameter is in general huge in comparison to the magnetic interactions, the electrons are considered to be immobile or in other words, trapped on the magnetic sites. The situation changes drastically when the number of electrons in the magnetic orbitals is no longer equal to the number of magnetic orbitals, that is when the system is doped with electrons (more electrons than magnetic orbitals) or doped with holes (less electrons than magnetic orbitals). In these systems, the electron is no longer necessarily trapped and can move from site to site under certain circumstances that will be described below. A commonly used classification of magnetic compounds by the degree of electron mobility was given by Robin and Day [1], who divided the so-called mixed valence compounds into three groups. Class I contains all the compounds where the magnetic centers have different oxidation states but the electrons are nevertheless trapped. Class III is quite the opposite; the magnetic centers have formally a distinct oxidation state

© Springer International Publishing Switzerland 2016
C. Graaf and R. Broer, *Magnetic Interactions in Molecules and Solids*,
Theoretical Chemistry and Computational Modelling,
DOI 10.1007/978-3-319-22951-5_6

Fig. 6.1 Initial and final
states of electron hopping
processes in hole-doped,
neutral and electron-doped
magnetic systems

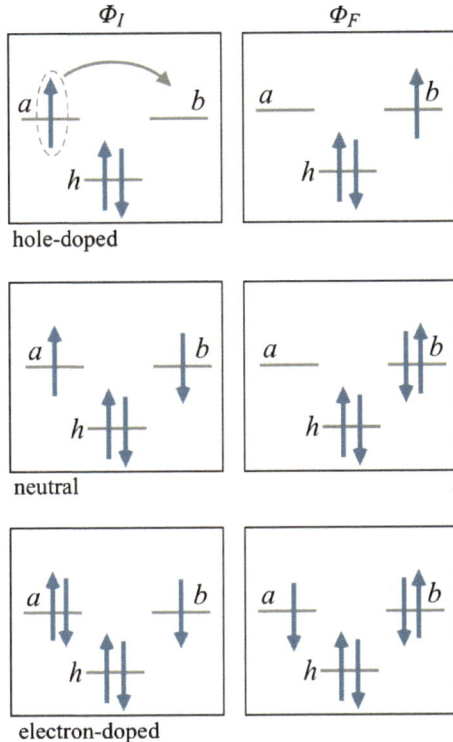

but the electrons are completely delocalized and in practice all magnetic centers share the same average oxidation, often a non-integer number. In between, one finds the probably most interesting case of Class II compounds. There is a certain degree of localization but the hopping of an electron from one site to a neighbouring one has a low energy barrier and occurs frequently.

In the background of a collection of inactive doubly occupied orbitals h, three different scenarios can be envisaged to describe electron hopping processes. Figure 6.1 illustrates these scenarios and from top to bottom we recognize the hopping process from a singly occupied orbital to an empty orbital; from a singly occupied to another singly occupied orbital; and from a doubly occupied (filled) to a singly occupied orbital. Taking the system in the middle as reference *neutral* system, the upper part of the figure is indicative for electron hopping in a hole-doped (or electron-ionized) system, while the bottom illustrates the hopping in an electron-doped system. In this case the process is often interpreted in terms of hole mobility, where the figure illustrates how a hole on site B moves to site A.

The probability for these hopping processes is normally condensed into a single parameter referred to as t_{ab}, but V_{ab} and β (Hückel theory) are also used. Intuitively one would say that the hopping parameter is the same for all three processes, since one electron moves from a to b, while the rest of the occupations stay the same in all cases. But the calculation of the $\langle \Phi_I | \hat{H} | \Phi_F \rangle$ matrix element shows that this is not exactly the case. The interaction matrix elements of the initial and final states are easily determined with the Slater–Condon rules. In the first case, the hopping of an electron to an empty orbital is defined by

$$\Phi_I = |h\bar{h}a| \qquad \Phi_F = |h\bar{h}b|$$

$$t_{ab}^+ = \langle \Phi_I | \hat{H} | \Phi_F \rangle = \sum_h \langle ah | \frac{1}{r_{12}} | bh \rangle + \langle a | \hat{h} | b \rangle \tag{6.1a}$$

where a and b are (orthogonal) atomic-like orbitals centered on the centers A and B and h is one of the inactive doubly occupied orbitals. The sum runs over all the inactive orbitals. In the second scenario, the initial and final states and their matrix element are

$$\Phi_I = |h\bar{h}a\bar{b}| \qquad \Phi_F = |h\bar{h}b\bar{b}|$$

$$t_{ab}^0 = \langle \Phi_I | \hat{H} | \Phi_F \rangle = \sum_h \langle ah | \frac{1}{r_{12}} | bh \rangle + \langle a | \hat{h} | b \rangle + \langle ab | \frac{1}{r_{12}} | bb \rangle \tag{6.1b}$$

and finally, the process on the bottom of the figure from doubly to singly occupied is described by

$$\Phi_I = |h\bar{h}a\bar{a}b| \qquad \Phi_F = |h\bar{h}ab\bar{b}|$$

$$t_{ab}^- = \langle \Phi_I | \hat{H} | \Phi_F \rangle = \sum_h \langle ah | \frac{1}{r_{12}} | bh \rangle + \langle a | \hat{h} | b \rangle + \langle ab | \frac{1}{r_{12}} | bb \rangle + \langle aa | \frac{1}{r_{12}} | ba \rangle$$

$$\tag{6.1c}$$

The contribution of the inactive doubly occupied orbitals is the same in the three cases as is the one-electron term h_{ab}. However, the appearance of two-electron integrals for those cases with more than one electron in the magnetic orbitals introduces differences in the interaction matrix elements.

Numerical estimates of the hopping parameter are relatively easy to obtain with the different computational schemes discussed in Chap. 4. Starting with t_{ab}^+ in a centrosymmetric two-site system, two electronic states can be defined with doublet spin coupling

$$D_1 = |h\bar{h}g| \qquad D_2 = |h\bar{h}u| \tag{6.2}$$

with g and u the bonding and anti-bonding combinations of the local orbitals a and b. The energy of the two doublets is

$$E(D_1) = \langle h\bar{h}g|\hat{H}|h\bar{h}g\rangle = \frac{1}{2}\langle h\bar{h}(a+b)|\hat{H}|h\bar{h}(a+b)\rangle$$

$$= \frac{1}{2}\Big(\langle h\bar{h}a|\hat{H}|h\bar{h}a\rangle + 2\langle h\bar{h}a|\hat{H}|h\bar{h}b\rangle + \langle h\bar{h}b|\hat{H}|h\bar{h}b\rangle\Big) \qquad (6.3)$$

$$E(D_2) = \langle h\bar{h}u|\hat{H}|h\bar{h}u\rangle = \frac{1}{2}\langle h\bar{h}(a-b)|\hat{H}|h\bar{h}(a-b)\rangle$$

$$= \frac{1}{2}\Big(\langle h\bar{h}a|\hat{H}|h\bar{h}a\rangle - 2\langle h\bar{h}a|\hat{H}|h\bar{h}b\rangle + \langle h\bar{h}b|\hat{H}|h\bar{h}b\rangle\Big) \qquad (6.4)$$

Combining the energy difference of the two doublets

$$\Delta E_{12} = E(D_1) - E(D_2) = 2\langle h\bar{h}a|\hat{H}|h\bar{h}b\rangle \qquad (6.5)$$

with the definition given in Eq. 6.1a, the hopping parameter can be calculated by

$$t_{ab}^+ = \frac{1}{2}\Delta E_{12} \qquad (6.6)$$

In practice, an effective hopping parameter can be obtained from accurate *ab initio* energies for the two doublets. For non-centrosymmetric systems, the calculation is slightly more involved. The two doublets are now defined as

$$D_1 = c_1|h\bar{h}a| + c_2|h\bar{h}b| \qquad D_2 = c_2|h\bar{h}a| - c_1|h\bar{h}b| \qquad (6.7)$$

and the energy difference is

$$\Delta E_{12} = (c_1^2 - c_2^2)(H_{aa} - H_{bb}) + 4c_1c_2H_{ab} \qquad (6.8)$$

with $H_{ij} = \langle h\bar{h}i|\hat{H}|h\bar{h}j\rangle$. This leads to the following expression for t_{ab}^+

$$t_{ab}^+ = \frac{\Delta E_{12} - (c_1^2 - c_2^2)(H_{aa} - H_{bb})}{4c_1c_2} \qquad (6.9)$$

To determine t, the energy difference is no longer sufficient and information is required from the wave function. The magnetic orbitals have to be expressed in orthogonal atomic-like orbitals and the wave functions projected on the model space $\{|h\bar{h}a|, |h\bar{h}b|\}$. After orthonormalization, a numerical 2×2 effective Hamiltonian can be constructed

$\hat{H}^{eff}$	$\|h\bar{h}a\rangle$	$\|h\bar{h}b\rangle$
$\langle h\bar{h}a\|$	H_{aa}	t_{ab}^+
$\langle h\bar{h}b\|$	t_{ab}^+	H_{bb}

and the hopping parameter can be directly determined from the off-diagonal matrix element.

6.1 Show that the expression of t_{ab}^{+} for the non-centrosymmetric case reduces to $\Delta E_{12}/2$ for a centrosymmetric system.

The determination of t_{ab}^{0} has already been discussed in Sect. 5.2. It requires the construction of a 4×4 effective Hamiltonian with a basis of two neutral and two ionic determinants. The hopping integral is defined as the matrix element between neutral and ionic determinants. The calculation of t_{ab}^{-} is analogous to the procedure for estimating t_{ab}^{+}. The two doublets that can be defined in a centrosymmetric complex with three electrons in the two magnetic orbitals g and u (omitting $h\bar{h}$ for simplicity)

$$D_1 = |gu\bar{u}| \qquad D_2 = |g\bar{g}u| \qquad (6.10)$$

are re-expressed in the orthogonal atomic-like orbitals a and b

$$D_1 = \frac{1}{2\sqrt{2}}|(a+b)(a-b)(\bar{a}-\bar{b})| = \frac{1}{2\sqrt{2}}|-ab\bar{a}+ab\bar{b}+ba\bar{a}-ba\bar{b}|$$
$$= \frac{1}{\sqrt{2}}(|ab\bar{b}|+|a\bar{a}b|) \qquad (6.11a)$$

$$D_2 = \frac{1}{2\sqrt{2}}|(a+b)(\bar{a}+\bar{b})(a-b)| = \frac{1}{2\sqrt{2}}|-a\bar{a}b-a\bar{b}b+b\bar{a}a+b\bar{b}a|$$
$$= \frac{1}{\sqrt{2}}(|ab\bar{b}|-|a\bar{a}b|) \qquad (6.11b)$$

The energies of the two states are

$$E_1 = \frac{1}{2}\langle ab\bar{b}+a\bar{a}b|\hat{H}|ab\bar{b}+a\bar{a}b\rangle$$
$$= \frac{1}{2}\left[\langle ab\bar{b}|\hat{H}|ab\bar{b}\rangle + 2\langle ab\bar{b}|\hat{H}|a\bar{a}b\rangle + \langle a\bar{a}b|\hat{H}|a\bar{a}b\rangle\right] \qquad (6.12a)$$

$$E_2 = \frac{1}{2}\langle ab\bar{b}-a\bar{a}b|\hat{H}|ab\bar{b}-a\bar{a}b\rangle$$
$$= \frac{1}{2}\left[\langle ab\bar{b}|\hat{H}|ab\bar{b}\rangle - 2\langle ab\bar{b}|\hat{H}|a\bar{a}b\rangle + \langle a\bar{a}b|\hat{H}|a\bar{a}b\rangle\right] \qquad (6.12b)$$

and t_{ab}^- can again be calculated from the energy difference of the two doublets, see Eq. 6.10:

$$t_{ab}^- = \frac{1}{2} \Delta E_{12} = \langle ab\bar{b}|\hat{H}|a\bar{a}b \rangle \tag{6.13}$$

6.2 Double Exchange

The concept of *double exchange* was introduced by Zener in the 1950s to explain the sudden drop in the resistance of certain manganites when an external magnetic field was applied [2]. The manganese ions in these compounds have either three or four unpaired electrons in the $3d$-shell; three electrons in t_{2g}-like orbitals and the fourth electron in an e_g-like orbital. Electron hopping in such compounds takes place in the presence of other unpaired electrons and depending on the inter-site spin coupling of the t_{2g} electrons, the e_g electron has more or less probability to hop to a neighboring site. Assume that the antiferromagnetic coupling dominates in the absence of an external magnetic field. As shown in the upper part of Fig. 6.2, the electron hopping leads to a state that does not have the maximum spin on the center that receives the electron, whereas maximum spin coupling at one site is preferred as stated by Hund's rule. By applying a sufficiently large external magnetic field, the spins on all centers can be forced in a ferromagnetic alignment. In such a case, see the lower part of the figure, the electron hopping creates a high-spin state on the receiving center, which corresponds to the local ground state and this is more favorable for electron mobility, i.e. an external magnetic field can drastically lower the resistance to electric conductance.

The effectiveness of the electron hopping between two metal centers separated by a closed-shell ion, typically O^{2-}, inspired Zener to introduce the concept of *double exchange* illustrated in Fig. 6.3. The simultaneous hopping process of an electron with α-spin from the first metal center to the O^{2-} ion and from this O^{2-} ion to the second metal center was held responsible for the hopping. Contrary to the superexchange described in Fig. 5.4, the double exchange only involves electrons of the same spin. Therefore, the intuitive picture is that since the electron transfers can take place simultaneously, in contrast to the superexchange, the double exchange hopping is very efficient.

This simple electron hopping explanation has later been revised to incorporate the strong electron-phonon coupling caused by the Jahn-Teller splitting of the Mn^{3+} ions. The conduction is due to the hopping of a *magnetic polaron* rather than a bare electron [3].

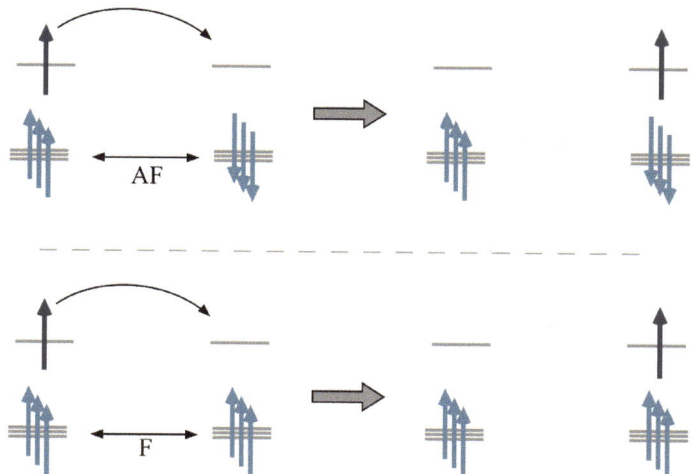

Fig. 6.2 Schematic explanation for the spin dependence of the hopping probability of the extra electron in a background of unpaired electrons

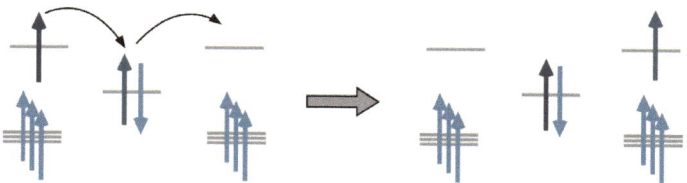

Fig. 6.3 Double exchange mechanism proposed by Zener consisting in the simultaneous hopping of an electron from center 1 to the bridge and from the bridge to metal 2

Fig. 6.4 Simple model to describe electron hopping in the background of unpaired electrons, the orbitals are mutually orthogonal

A rigorous description of electron hopping in the presence of other unpaired electrons can be written down for the simple model system defined in Fig. 6.4. The two electrons in the a_1 and b_1 orbitals provide a static background of unpaired electrons for the mobile electron in the a_2 b_2 channel. Choosing $M_S = \frac{1}{2}$, there are 24 ways to distribute the three electrons over the four orbitals, but only six of them are essential to the basic description of the hopping process of the electron between the a_2 and b_2 orbitals. The six determinants are

$$\Phi_1 = |a_1 b_1 \bar{a}_2| \qquad\qquad \Phi_4 = |a_1 b_1 \bar{b}_2|$$

$$\Phi_2 = |a_1 \bar{b}_1 a_2| \qquad\qquad \Phi_5 = |\bar{a}_1 b_1 b_2| \qquad (6.14)$$

$$\Phi_3 = |\bar{a}_1 b_1 a_2| \qquad\qquad \Phi_6 = |a_1 \bar{b}_1 b_2|$$

6.2 Find the four determinants (or linear combinations of determinants) that represent a triplet spin coupling of the electrons on center a or b. What is the spin coupling of the other two (linear combinations) of determinants? What do you expect for the relative energies of the two groups?

The interaction matrix elements are readily written down using the Slater–Condon rules given in Chap. 1. We will work out three examples and leave the others as exercise to the reader.

$$\langle \Phi_1 | \hat{H} | \Phi_1 \rangle = \langle a_1 b_1 \bar{a}_2 | \hat{H} | a_1 b_1 \bar{a}_2 \rangle = \langle a_1 | \hat{h} | a_1 \rangle + \langle b_1 | \hat{h} | b_1 \rangle + \langle a_2 | \hat{h} | a_2 \rangle$$

$$+ \langle a_1 b_1 | \frac{1 - \hat{P}_{12}}{r_{12}} | a_1 b_1 \rangle + \langle a_1 \bar{a}_2 | \frac{1 - \hat{P}_{12}}{r_{12}} | a_1 \bar{a}_2 \rangle + \langle b_1 \bar{a}_2 | \frac{1 - \hat{P}_{12}}{r_{12}} | b_1 \bar{a}_2 \rangle$$

$$(6.15)$$

The two-electron part becomes

$$\langle a_1 b_1 | \frac{1}{r_{12}} | a_1 b_1 \rangle - \langle a_1 b_1 | \frac{1}{r_{12}} | b_1 a_1 \rangle + \langle a_1 a_2 | \frac{1}{r_{12}} | a_1 a_2 \rangle + \langle b_1 a_2 | \frac{1}{r_{12}} | b_1 a_2 \rangle$$

$$= J_{a_1 b_1} + J_{a_1 a_2} + J_{b_1 a_2} - K_{a_1 b_1} \qquad (6.16)$$

The one-electron part and the Coulomb integrals J_{xy} are common to all diagonal matrix elements and the sum of these terms can be taken as the zero of energy. Then, the matrix element $\langle \Phi_1 | \hat{H} | \Phi_1 \rangle$ reduces to the exchange integral $-K_{a_1 b_1}$. Likewise, the diagonal elements involving Φ_2 and Φ_3 reduce to $K_{a_1 a_2}$ and to 0, respectively. The off-diagonal matrix element between Φ_2 and Φ_3 is relatively simple

$$\langle \Phi_2 | \hat{H} | \Phi_3 \rangle = \langle a_1 \bar{b}_1 a_2 | \hat{H} | \bar{a}_1 b_1 a_2 \rangle = \langle a_1 \bar{b}_1 | \frac{1 - \hat{P}_{12}}{r_{12}} | \bar{a}_1 b_1 \rangle$$

$$= -\langle a_1 b_1 | \frac{1}{r_{12}} | b_1 a_1 \rangle = -K_{a_1 b_1} \qquad (6.17)$$

but the matrix element between Φ_1 and Φ_4 is slightly more involved

$$\langle \Phi_1|\hat{H}|\Phi_4\rangle = \langle a_1b_1\bar{a}_2|\hat{H}|a_1b_1\bar{b}_2\rangle = \langle a_2|\hat{h}|b_2\rangle + \langle a_1\bar{a}_2|\frac{1-\hat{P}_{12}}{r_{12}}|a_1\bar{b}_2\rangle$$

$$+ \langle b_1\bar{a}_2|\frac{1-\hat{P}_{12}}{r_{12}}|b_1\bar{b}_2\rangle = \langle a_2|\hat{h}|b_2\rangle + \langle a_1a_2|\frac{1}{r_{12}}|a_1b_2\rangle + \langle b_1a_2|\frac{1}{r_{12}}|b_1b_2\rangle$$

$$(6.18)$$

where the two-electron integrals cannot be written as Coulomb or exchange integrals. The sum of the three terms can be considered as the hopping parameter t, similar to the expressions given in Eqs. 6.1b and 6.1c. The complete interaction matrix is

$\hat{H}$	$|\Phi_1\rangle$	$|\Phi_2\rangle$	$|\Phi_3\rangle$	$|\Phi_4\rangle$	$|\Phi_5\rangle$	$|\Phi_6\rangle$
$\langle\Phi_1|$	$-K'$	0	$-K$	t	0	0
$\langle\Phi_2|$	0	$-K$	$-K'$	0	0	t
$\langle\Phi_3|$	$-K$	$-K'$	0	0	t	0
$\langle\Phi_4|$	t	0	0	$-K'$	0	$-K$
$\langle\Phi_5|$	0	0	t	0	$-K$	$-K'$
$\langle\Phi_6|$	0	t	0	$-K$	$-K'$	0

$K = K_{a_1a_2} = K_{b_1b_2}$ is the on-site exchange interaction and $K' = K_{a_1b_1}$ is the intersite exchange. Two approximations have been made to obtain this matrix. In the first place, it is assumed that $K_{a_ib_j}$ with $i \neq j$ can be neglected. Furthermore, we assume that the effect of the so-called singlet displacement operator is small enough to be omitted. The action of this operator is illustrated in Fig. 6.5 and transforms Φ_1 into Φ_5 or Φ_6, and Φ_4 into Φ_2 or Φ_3.

6.3 Show that the other zeros in the matrix are real zeros and not due to any additional approximation. Explain why in the Hamiltonian matrix $H_{11} = H_{44}$, $H_{22} = H_{55}$, $H_{33} = H_{66}$, and $H_{13} = H_{46}$.

The diagonalization of the matrix yields six eigenstates, two quartets and four doublets with the following energies after shifting all states with $K + K'$ to let the zero of energy coincide with the average energy of the quartet states

Fig. 6.5 Action of the singlet displacement operator

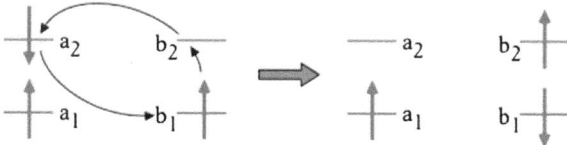

$$E(Q_{1,2}) = \pm t \tag{6.19}$$

$$E(D_{1,2}) = K + K' - \sqrt{K^2 + t(t \pm K) + K'^2 - K'(K \pm 2t)} \tag{6.20}$$

$$E(D_{3,4}) = K + K' + \sqrt{K^2 + t(t \pm K) + K'^2 - K'(K \pm 2t)} \tag{6.21}$$

The first two doublets are dominated by the CSFs with triplet coupling on center a or b, and hence, much lower in energy than the third and fourth doublets with local singlet coupling. The latter states are similar to the non-Hund states invoked to explain the deviations to the regular Heisenberg pattern in Sect. 5.4. The quartet states are in-phase and out-of-phase linear combinations of the high spin coupled determinants with the extra electron on center a or center b, which is most conveniently seen in the $M_S = 3/2$ components of these states.

$$Q_1(M_S = 3/2) = \frac{1}{\sqrt{2}}(a_1 b_1 a_2 + a_1 b_1 b_2) \tag{6.22}$$

$$Q_2(M_S = 3/2) = \frac{1}{\sqrt{2}}(a_1 b_1 a_2 - a_1 b_1 b_2) \tag{6.23}$$

In the simplest description of the hopping, the on-site exchange integral is assumed to be so large in comparison to the other parameters that the doublet states dominated by the non-Hund determinants are not relevant and the energy of the lower doublet states can be simplified to

$$E(D_{1,2}) = K - \sqrt{K^2} + \frac{1}{2}\frac{t(t \pm K) + K'^2 - K'(K \pm 2t)}{\sqrt{K^2}} = \pm\frac{1}{2}t + \frac{3}{2}K' \quad (6.24)$$

using the Taylor expansion $\sqrt{p + q} = \sqrt{p} + \frac{1}{2}q/\sqrt{p} + \dots$ and neglecting all terms proportional to K^{-1} because K is very large in comparison to t and K'. A general expression for any number of unpaired electrons within this approximation is

$$E(S) = \pm t\frac{S + 1/2}{S_{max} + 1/2} + \frac{1}{2}\big(S_{max}(S_{max} + 1) - S(S + 1)\big)K' \tag{6.25}$$

Recalling that the exchange parameters $K_{a_1 b_2}$ and $K_{a_2 b_1}$ have been neglected, the K' parameter plays exactly the same role as the Heisenberg J in the description of Girerd, who included the magnetic coupling between the two sites with S_A and S_B spin moments in the description of the double exchange. Then the equation can also be written in a more familiar form [4].

$$E(S) = \pm t\frac{S + 1/2}{S_{max} + 1/2} + \frac{1}{2}J\big(S_{max}(S_{max} + 1) - S(S + 1)\big) \tag{6.26}$$

It is interesting to see that even when the intersite interaction is completely neglected by putting J (or K') to zero, the hopping process forces the system into the ferromagnetic state. Only when J is very strongly antiferromagnetic and t relatively small, one may expect a low-spin ground state. In the more common case that t dominates, we see that the transfer integral is reduced by the factor $(S + 1/2)/(S_{max} + 1/2)$.

An important aspect of the physics of double exchange compounds is the interaction between the electron distribution and the movement of the nuclei by vibronic coupling in complexes or electron-phonon interaction in extended systems. This goes beyond the scope of this book and we refer the interested reader to Ref. [5] for further reading.

Semi-classical description of the double exchange: The first description of the double exchange by Zener [2] gave a simple (yet convincing) explanation of the strong dependence of the electric resistivity on the strength of the external magnetic field. The model only considers the hopping parameter t and assumes that the intra-atomic exchange integral is infinitely large, which makes that the electron can only move through the material when all spins at the magnetic sites are ferromagnetically aligned. A more detailed description was given by Anderson and Hasegawa [6], who derived the first right-hand-side term of Eq. 6.26. Here, we will review the semi-classical description of these authors to illustrate the concept of spin dependent hopping which is the basis of the Goodenough–Kanamori rules treated in the Sect. 6.4.

The Anderson–Hasegawa model describes the electron transfer from site A to B in the field of the spin moments $\mathbf{S}_A$ and $\mathbf{S}_B$, which are described as classical vectors. The spin moments are not necessarily co-linear but have an angle θ. The justification for this semi-classical description is that for large spin moments the quantum mechanical description converges with the classical one. Being applied to describe the electron hopping in manganites, this approximation is not as severe due to the relatively large spin moment on the manganese ions. Since the magnetic axes frames on site A and B do not have the same orientation, the basis of spin functions of site A (α and β) has to be expressed in terms of the basis of spin functions of site B (α' and β').

$$\alpha = \cos(\theta/2)\alpha' + \sin(\theta/2)\beta' \tag{6.27a}$$
$$\beta = -\sin(\theta/2)\alpha' + \cos(\theta/2)\beta' \tag{6.27b}$$

The basis functions of this semi-classical model are $\phi_1 = a\alpha$, $\phi_2 = a\beta$, $\phi_3 = b\alpha'$ and $\phi_4 = b\beta'$, where a and b define the spatial part of the orbitals that carry the mobile electron. The following definition of the interaction for $\theta = 0$ is used

$$\langle a\alpha|\hat{H}|b\alpha\rangle = \tau \tag{6.28a}$$
$$\langle a\alpha|\hat{H}|b\beta\rangle = 0 \tag{6.28b}$$

and for $\theta \neq 0$, slightly more elaborated expressions are obtained

$$\langle a\alpha|\hat{H}|b\alpha'\rangle = \langle a\big(\cos(\theta/2)\alpha' + \sin(\theta/2)\beta'\big)|\hat{H}|b\alpha'\rangle = \tau\cos(\theta/2) \qquad (6.28c)$$

$$\langle a\alpha|\hat{H}|b\beta'\rangle = \tau\sin(\theta/2) \qquad (6.28d)$$

$$\langle a\beta|\hat{H}|b\alpha'\rangle = -\tau\sin(\theta/2) \qquad (6.28e)$$

$$\langle a\beta|\hat{H}|b\beta'\rangle = \tau\cos(\theta/2) \qquad (6.28f)$$

The spin moment of $a\alpha$ and $b\alpha'$ is parallel to $\mathbf{S}_A$ and $\mathbf{S}_B$, respectively. In the simplest description, the energy of these states with respect to the ones with antiparallel alignment ($a\beta$ and $b\beta'$) is given by the number of exchange interactions between the extra electron and the electrons that give rise to the background spin moments S_A and S_B.

$$E_{1,3} = -K \cdot 2S_{A,B} \qquad E_{2,4} = 0 \qquad (6.29)$$

However, in a formalism with correct spin eigenfunctions the energies become

$$E_{1,3} = -K(S_{A,B} + 1) \qquad E_{2,4} = +KS_{A,B} \qquad (6.30)$$

6.4 Consider a magnetic site with a $S = 1$ background spin moment (triplet coupled electrons in φ_1 and φ_2) and an electron in φ_3 that can hop to neighboring centers. Calculate the energies of $|\varphi_1\varphi_2\varphi_3|$, $|\varphi_1\varphi_2\overline{\varphi}_3|$ and the CSF for the doublet with triplet coupling for φ_1 and φ_2 and compare to the Eqs. 6.29 and 6.30.

With these ingredients the matrix representation of the model Hamiltonian can be constructed

	$|a\alpha\rangle$	$|a\beta\rangle$	$|b\alpha'\rangle$	$|b\beta'\rangle$
$\langle a\alpha|$	$-K(S_A + 1)$	0	$\tau\cos(\theta/2)$	$\tau\sin(\theta/2)$
$\langle b\beta|$	0	KS_A	$-\tau\sin(\theta/2)$	$\tau\cos(\theta/2)$
$\langle a\alpha'|$	$\tau\cos(\theta/2)$	$-\tau\sin(\theta/2)$	$-K(S_B + 1)$	0
$\langle b\beta'|$	$\tau\sin(\theta/2)$	$\tau\cos(\theta/2)$	0	KS_B

$$(6.31)$$

In the case of $S_A = S_B = S$, the diagonalization of the matrix leads to four eigenvalues which read as follows:

$$E = \frac{1}{2}K \pm \sqrt{\big(K(S + {}^1\!/_2) \pm \tau\cos(\theta/2)\big)^2 + \tau^2\sin^2(\theta/2)} \qquad (6.32)$$

Figure 6.6 gives a clue on how to simplify this expression. In the first place, we see that $\cos(\theta/2)$ can be written as $S_0/2S$ with $S_0 = |\mathbf{S}_A + \mathbf{S}_B|$. In a classical description, that is in the limit of infinitely large S_0, the alignment of the spin moment of the extra electron to S_0 is irrelevant and the total spin moment of the system S_T is

Fig. 6.6 Definition of S_0 as $|\mathbf{S}_A + \mathbf{S}_B|$ and $\cos(\theta/2)$ as $S_0/2S$

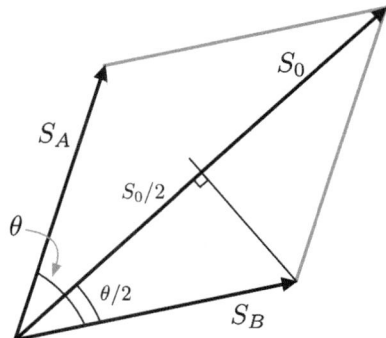

directly equal to S_0. Applying the same correction for the quantum nature of the spin moments as done in Eq. 6.30, we obtain $\cos(\theta/2) = (S_0 + 1/2)(2S_{max} + 1)$, where $S_{max} = |\mathbf{S}_A + \mathbf{S}_B| + \frac{1}{2}$, corresponding to the maximum spin moment that can be realized by all the unpaired electrons. The second simplification arises from the fact that $\tau \ll K$ and justifies the neglect of the term quadratic in τ in the square root. The expression for the energy now becomes

$$E = \frac{1}{2}K \pm \big(K(S + 1/2) \pm \tau \cos(\theta/2)\big) = \frac{1}{2}K \pm \left(KS + \frac{1}{2}K\right) \pm \tau \frac{S_0 + 1/2}{2S_{max} + 1} \quad (6.33)$$

By choosing the reference energy equal to $-KS$, the expression reduces to

$$E_- = \pm \tau \frac{S_0 + 1/2}{2S_{max} + 1} \quad (6.34)$$

$$E_+ = K + 2KS \pm \tau \frac{S_0 + 1/2}{2S_{max} + 1} \quad (6.35)$$

Considering the quantum correction due to the use of spin eigenfunctions, τ can be replaced by $2t$, which turns the expression for E_- into the first term of Eq. 6.26 and describes the energies of the low-lying states with the spin moment of the extra electron parallel aligned with $\mathbf{S}_A$ or $\mathbf{S}_B$. E_+ applies to the states with anti-parallel alignment, and hence, lie at much higher energy.

6.3 A Quantum Chemical Approach to Magnetic Interactions in the Solid State

Many of the macroscopic manifestations of the interaction between localized, delocalized or itinerant unpaired electrons in solid state compounds require a description that goes far beyond the possibilities of the computational schemes that are routinely applied in molecular quantum chemistry. The theoretical treatment of the long-range magnetic ordering, Kondo effect, domain formation, superconductivity, metal-insulator transitions, etc. belongs typically to the field of condensed matter physics and several excellent books have been published on this topic, see for example Refs. [7–10]. This does however not mean that quantum chemistry cannot contribute to the understanding of magnetic phenomena in solid state compounds. We have already seen in Sect. 3.3 how the calculation of the magnetic interaction parameters can serve as the basis for the calculation of the magnetic susceptibility and the determination of the magnetic structure, or more precise the magnetic unit cell. In fact, a large part of the parameters that typically appear in the model Hamiltonians of condensed matter physics can be calculated accurately through quantum chemical calculations provided that one can establish an accurate finite representation of the relevant part of the crystal. In the case of molecular crystals, this issue is nearly trivially answered: taking one or several discrete units as model often suffices to calculate the desired microscopic electronic structure parameters. The situation becomes more complicated when dealing with ionic lattices (oxides, pnictides among others) and is even worse for crystals with only covalently bonded atoms (e.g. silicon or graphene doped with holes). There are however several well-established approaches to extract reliable information at least for the ionic crystals. Also in the more difficult case of (partly) covalent lattices quantum chemical strategies can offer interesting insights in the electronic structure related to magnetic interactions.

6.3.1 Embedded Cluster Approach

The intrinsic local nature of the interaction between two localized spin moments suggests the possibility to study the magnetic interactions in solids with a cluster model. In this approach, a small yet relevant piece is cut from the crystal and treated like a molecule. These bare clusters are only a reasonable choice in the case of molecular crystals, but otherwise nearly always too crude a representation. Therefore it is necessary to account for the effect of the rest of the crystal especially when dealing with ionic or covalent lattices. Here, we will shortly review a few representative examples of the different approaches for improving the bare cluster model that find their basis in the theory of electron separability of McWeeny, the subsystem formulation of DFT of Cartona or the incremental scheme of Fulde and Stoll.

Electrostatic embedding: In the case of ionic compounds, the largest contribution to the potential exerted by the rest of the crystal on the (central region of the) cluster is due to long-range electrostatic interactions. These are accurately represented by the point charge approximation, that is, the Madelung potential:

$$V_M(\kappa \in \mathbb{K}) = \sum_{\lambda \in \mathbb{L}} \frac{q_\kappa q_\lambda}{r_{\kappa\lambda}} \tag{6.36}$$

where $\mathbb{K}$ is the set of ions that belong to the cluster and $\mathbb{L}$ contains all other ions, q corresponds to the formal ionic charge of each center. This interaction is easily included in the calculation by placing an array of point charges around the cluster at the lattice sites. Their value is either taken as the formal ionic charge (with fractional charges on the edge of the array to ensure charge neutrality) or fitted in such a way that a relatively small set of point charges reproduces the electrostatic effect of the whole crystal.

The presence of point charges at lattice sites in the immediate neighbourhood of the cluster often artificially polarizes the electron density of the cluster. This polarization is especially large when the cluster has anions on the outside and the first shell of charges contains positive charges. Actually, this is the common situation when magnetic interactions in ionic transition metal compounds are studied. The usual cluster has two (or sometimes more) transition metals and the anions (O^{2-}, F^-, etc.) of the first coordination sphere. The first shell around this cluster is formed by either the transition metal ions, ternary cations or a combination of these, depending on the crystal structure. In any case, positive point charges are located directly around the highly polarizable anions causing important distortions of the electronic structure not only in the border regions of the cluster, but also in the central part. To improve the description a border region is created between the point charges and the cluster. In this intermediate region the lattice sites are occupied by potentials that model the Coulomb and exchange interactions between the electron density of the cluster and the ions in the intermediate region [11]. Figure 6.7 shows how the potentials separate the cluster from the bare point charges and avoid the artificial polarization of the cluster electron density.

Density-based embeddings: This approach starts with a calculation on the whole system to construct an approximate yet accurate representation of the total density ρ_{tot} by performing a periodic DFT calculation. Then, a guess density of the cluster is constructed from a calculation on the isolated unit or using some simple embedding scheme as described above. The total density is now divided in two parts $\rho_{tot} = \rho_1 + \rho_2$ and the one-electron embedding potential is constructed from the functional derivative of interaction energy with respect to the cluster density ρ_1.

$$E^{int} = T_s^{int} + E_{ne}^{int} + E_{xc}^{int} + E_H^{int} + E_{nn}^{int}$$
$$v_{emb}(r) = \frac{\partial E_{int}}{\partial \rho_1} \tag{6.37}$$

where the interaction energy is written as a sum of the kinetic, electron-nuclear, exchange correlation, Coulomb repulsion, and nuclear repulsion energy. This embedding potential is added to the standard Kohn-Sham equation for the cluster and a new energy and density ρ_1 are calculated. Since the embedding potential depends on the density of the cluster, v_{emb} is updated and the Kohn-Sham equations of the cluster are solved again. This process is repeated until a self-consistent description is obtained. In addition to the here sketched DFT in DFT (cluster in embedding) procedure, the variants with wave function (WF) based methods have also been described. The

Fig. 6.7 Two-dimensional impression of a typical embedded cluster model to calculate the interaction strength between two spin moments. The atoms in the *shaded area* constitute the cluster, the *small spheres* on the outside constitute the first shell of positive and negative bare point charges (the rest is not shown), and the spheres with the *dotted outline* in the intermediate region are model potentials that separate the cluster from the point charges

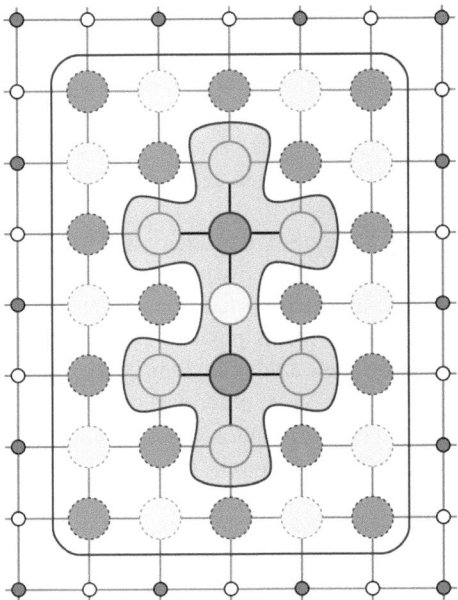

WF in DFT approach is especially interesting for the application to systems with unpaired electrons because the multideterminental nature of the wave function can be rigorously treated while the embedding can be generated with DFT.

Induced dipoles: There are also embedding schemes that go beyond the static representation of the cluster environment and model the polarization of the electron density in response to changes in the electronic structure of the cluster, for example ionizations or electron excitation processes. In the so-called shell model, the bare point charges are split in a positive point charge (the nucleus) and a negative shell (the electron cloud of the ion) connected through a harmonic potential. The shells interact with a Buckingham potential and the total energy of the system (cluster + shell environment) is minimized not only with respect to the electron distribution in the cluster region but also with respect to the position of the shells. Another scheme places a set of polarizable dipoles in the environment and the values of the induced dipoles are optimized in a self-consistent procedure along with the electron density of the cluster.

Once, a convenient embedded cluster model is constructed, one can apply all the regular methods from molecular quantum chemistry to evaluate the electronic structure parameters of interest, hopping parameters, magnetic coupling strength, local anisotropy, biquadratic exchange, etc. The validity of the embedded cluster model has been established in many applications either by comparing the results to periodic calculations or by checking the stability of the results against the size of the cluster.

6.3.2 Periodic Calculations

Magnetic interactions in extended systems can also be studied without creating the more or less approximate representation of the material with an embedded cluster. The approach based on the translational symmetry in the crystal naturally leads to the well-known band structures of the Bloch functions, periodic one-electron functions.

$$\psi_k(r) = \sum_{r'} e^{i\mathbf{k}\cdot\mathbf{r}'} \phi(r') \tag{6.38}$$

The difficulty of constructing spin eigenfunctions with $S < S_{max}$ for extended systems with unpaired electrons makes that most of the periodic calculations are performed within a single determinant method and no restrictions on the spatial part of the spin orbitals. The results are then necessarily interpreted with the Ising model Hamiltonian described in Sect. 3.2.2. In practice, the total energy of the magnetic unit cell (not necessarily of the same size as the structural unit cell) is calculated for different spin orientations (that is, different M_S values) and the relative energies are compared to the matrix elements of the Ising Hamiltonian to determine the magnetic coupling strength between the ions in the crystal. This is not necessarily limited to isotropic bilinear coupling but can also be used to extract estimates for biquadratic and four-center interactions.

To illustrate the procedure of extracting magnetic coupling parameters by periodic calculations, we will focus on the perovskite structure Sr_2CuO_3, related to the previously used spin ladder compound $SrCu_2O_3$, although the structural motif here is formed by CuO_4 units arranged in linear chains along the b-axis of the unit cell. Figure 6.8 illustrates the structure of this oxide and indicates the unit cell with a dashed box. The unit cell has two symmetry inequivalent Cu^{2+} ions with an $S = \frac{1}{2}$ spin moment each.

In the first place, we calculate the energy per unit cell with all spins aligned ferromagnetically as schematically depicted in the left panel of Fig. 6.9. This calculation can be done within any spin unrestricted periodic computationally scheme, either HF or DFT and it gives us $E_F(a, b, c)$. Subsequently, this energy has to be expressed as an Ising energy. The Ising Hamiltonian for this compound is defined as

$$\hat{H} = -J_a \sum_{i,j} \hat{S}_{z,i}\hat{S}_{z,j} - J_b \sum_{k,l} \hat{S}_{z,k}\hat{S}_{z,l} - J_d \sum_{m,n} \hat{S}_{z,m}\hat{S}_{z,n} \tag{6.39}$$

where the interaction along the c-direction is neglected and J_d is the interaction along the body diagonal of the unit cell. As can be seen in the left panel of Fig. 6.9, the unit cell contains 8 times the interaction along the body diagonal. The four vertices along a and b represent the J_a and J_b interactions, but each of these have to be counted only for 1/4 since the vertices are shared by four unit cells. Furthermore, the copper ion in the center of the unit cell interact with the copper ions in the adjacent unit cells and this contributes two times $1/2J_a$ and two times $1/2J_b$ to the Ising energy. In total the energy expression becomes

Fig. 6.8 Graphical
representation of the
perovskite Sr_2CuO_3. The
unit cell is given as a *dashed
box*. Cu ions are represented
as *large white spheres* and
oxygen as *smaller gray
spheres*, Sr ions are not
depicted for clarity

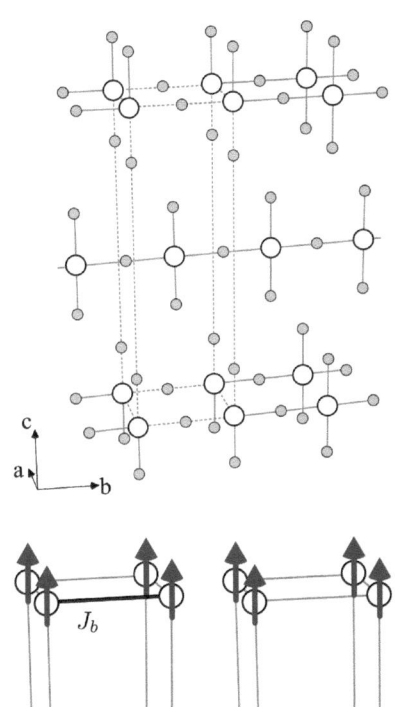

Fig. 6.9 Ferromagnetic
(*left*) and antiferromagnetic
(*right*) spin settings in the
unit cell of Sr_2CuO_3. The
spin on symmetry equivalent
copper ions are marked with
the same *gray scale*

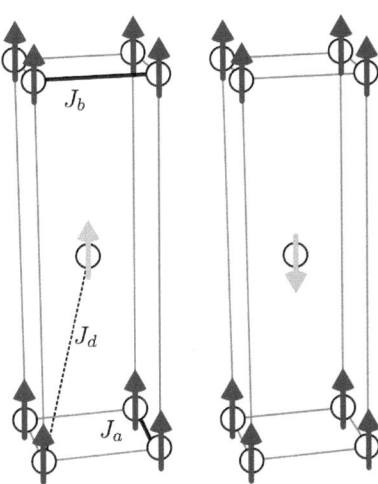

$$E_F(a, b, c) = -J_a \left(4 \cdot \frac{1}{2} \cdot \frac{1}{2}/4 + 2 \cdot \frac{1}{2} \cdot \frac{1}{2}/2 \right) - J_b \left(4 \cdot \frac{1}{2} \cdot \frac{1}{2}/4 + 2 \cdot \frac{1}{2} \cdot \frac{1}{2}/2 \right)$$

$$- J_d \cdot 8 \cdot \frac{1}{2} \cdot \frac{1}{2} = -2J_d - \frac{1}{2}J_a - \frac{1}{2}J_b \qquad (6.40)$$

The second step consists of the calculation of the energy per unit cell after flipping the spin on the central copper (right panel of Fig. 6.9) and relating it to the
energy expression obtained with the Ising Hamiltonian. As far as the interactions
along a and b are concerned, the situation remains unchanged with respect to the
ferromagnetic alignment. The expression only changes for the interaction along the

diagonal, since $\hat{S}_{z,m}$ and $\hat{S}_{z,n}$ result now in $\frac{1}{2}$ and $-\frac{1}{2}$. With this, the expression is readily written down as

$$E_{AF}(a, b, c) = 2J_d - \frac{1}{2}J_a - \frac{1}{2}J_b \qquad (6.41)$$

and from the energy difference of the two calculations, we can determine J_d

$$E_{AF}(a, b, c) - E_F(a, b, c) = 4J_d \qquad (6.42)$$

The calculation of J_b (and J_a) cannot be done with the simple unit cell, no other spin flips can be made. Therefore, we double the unit cell in the b direction to obtain a new *magnetic* unit cell, the super cell $(a, 2b, c)$, represented in Fig. 6.10. The energy of the fully ferromagnetic supercell is in principle exactly twice $E_F(a, b, c)$, but it is highly recommendable to repeat the HF or DFT calculation for this double unit cell due to numerical precision issues. Subsequently, we flip the spins on the copper ions in the middle of the cell to obtain a spin arrangement with antiferromagnetic ordering along the b-axis. A careful analysis of the interactions contained in these two magnetic unit cells gives the energy expressions of the Ising Hamiltonian

$$E_F(a, 2b, c) = 2E_F(a, b, c) = -4J_d - J_a - J_b \qquad (6.43)$$
$$E_{AF}(a, 2b, c) = -J_a + J_b \qquad (6.44)$$

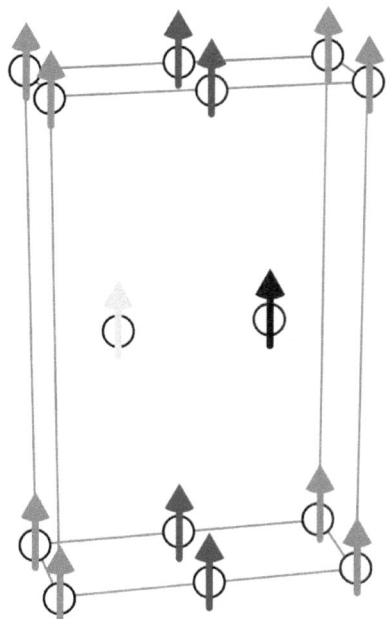

Fig. 6.10 Magnetic unit cell obtained by doubling the simple unit cell along the b direction. Symmetry equivalent copper ions have spins with the same *gray scale*. The antiferromagnetic unit cell is obtained by flipping the *dark gray* spins at the lattice positions $(0,1,0)$ and $(\frac{1}{2}, \frac{3}{2}, \frac{1}{2})$

and the energy difference of the two spin arrangements becomes

$$E_{AF}(a, 2b, c) - E_F(a, 2b, c) = 4J_d + 2J_b \qquad (6.45)$$

which allows us to extract J_b, given that J_d is already determined in the simple unit cell calculations. To calculate J_a one should double the unit cell along the a direction and follow the same strategy as for the $(a, 2b, c)$ super cell.

Method of increments: By taking the appropriate linear combinations of the delocalized Bloch functions one can construct orbitals that are localized on an atom or a small group of atoms of the crystal. These so called Wannier orbitals form the basis of the method of increments for calculating the cohesive energy of an extended solid [12, 13] and other related properties such as lattice constants, bulk modulus, absorption energies, among others. Excited state properties can also be studied and from there one has access to the band structure. The method was originally formulated for closed shell systems, but recently variants have been developed to treat compounds with unpaired electrons. Hence, the method can in principle also be used for the study of magnetic interactions in solids.

In its most basic formulation, the procedure starts with a periodic Hartree-Fock calculation. The correlation energy is calculated by *increments*. The unit cell is divided in m subunits A_i, either individual atoms or small clusters of atoms. The Bloch functions optimized in the periodic HF calculation are transformed to Wannier functions that are localized on the different subunits and the local correlation energy $E_i^{corr} = E_i^{tot} - E_i^{HF}$ is calculated for each subunit A_i with a standard (size-extensive) post-HF method. This is not the final estimate because all non-additive terms in the correlation energy are still missing. Therefore one subsequently calculates the two-center corrections through calculations on subunits A_i–A_j: $E_{ij}^{corr} = E_{ij}^{tot} - E_{ij}^{HF} - E_i^{corr} - E_j^{corr}$. The index i runs over all groups in the unit cell, but j can in principle be any atom (group of atoms) in the system. Fortunately, the size of the increment decays rapidly with the distance between the groups and hence the number of terms to be calculated remains relatively small. This can be repeated with three-center corrections and higher order increments. The total correlation energy is then determined

$$E^{corr} = \sum_i E_i^{corr} + \frac{1}{2} \sum_{i \neq j} E_{ij}^{corr} + \frac{1}{6} \sum_{i \neq j \neq k} E_{ijk}^{corr} + \cdots \qquad (6.46)$$

and added to the Hartree-Fock energy of the periodic calculation.

Correlated band structures: Periodic single determinant approaches are well suited to give qualitative answers or to serve as benchmark for checking the validity of embedded cluster results. On the other hand, the accurate treatment of (strong) electron correlation effects in crystalline materials, for example to predict the subtle interplay of magnetism and electrical conductance, requires an accurate, balanced description of all states involved, and this is still a challenge.

Examples of strongly correlated systems are transition metal and rare-earth metal compounds. In these materials on-site Coulomb repulsion between the metal valence electrons dominates the width of the corresponding one-electron energy bands. Widely used independent electron methods such as DFT in the local density approximation (LDA) are not suited to study the magnetic properties of such systems. Therefore, a correction term U to the LD functional has been introduced that accounts for the strong on-site Coulomb interactions between d (or f) electrons on the metal ions, giving rise to the LDA+U method [14]. Although LDA+U was introduced as a method without adjustable parameters, the values used for U vary significantly in different studies on the same compound.

Within Green function theory, many-electron effects can be introduced through a non-local and energy-dependent self-energy operator [15]. Since the self-energy is hard to calculate, various approximations are introduced and among the simplest ones is the so-called GW approximation, which is derived from many-body perturbation theory. Although the GW approximation offers in principle a sophisticated account of the electron correlation effects, practical realizations are commonly also based on the LDA method.

Finally, algorithms have been developed which incorporate electron correlation effects explicitly in wave function based band theory for crystalline solids [16, 17]. These algorithms construct the many-electron Hamiltonian matrix for a periodic system by extracting the matrix elements from calculations on finite embedded clusters. In this way the incorporation of correlation effects leads to many-electron energy bands, not only associated with hole states and added-electron states but also with excited states. More recently, Pisani and co-workers [18] introduced a post-Hartree-Fock program based on periodic local second order Møller-Plesset perturbation theory.

A word of warning is in place when these techniques are employed for the study of magnetic interactions. The tiny energy differences associated with these interactions demand that the procedure is capable to deliver not only an accurate but also a balanced treatment of the various states involved. This means that approximate computation and cut-offs of integrals etc. have to be exactly the same for all states.

6.4   Goodenough–Kanamori Rules

The Goodenough–Kanamori (GK) rules have evolved from the studies to explain the magnetism in manganese oxides in the 1950s and have been applied ever since mostly in the field of ionic insulators; often oxides of one or several third-row transition metal ions. Studies of the magnetic interactions in these compounds commonly reduce to a three center problem with two metals that carry a spin moment and a non-magnetic anion in between. Before explaining the rules, which are sometimes (incorrectly) referred to as the Goodenough–Kanamori–Anderson rules, we need to introduce some concepts related with the electron hopping involving the magnetic sites and the ligand that connects them. Goodenough defined *superexchange* as the virtual electron transfer between two atoms with a net spin moment and *semi covalent exchange* as the virtual electron transfer between the anion and the two magnetic centers. Note that Goodenough's definition of superexchange differs from the one given by Kramers (see Sect. 3.1).

To understand the use of the term *virtual* in these definitions it is best to contrast it against the electron hopping discussed in the previous section for doped or mixed-valence systems. These processes represent a real electron movement in which the formal oxidation state of the metals changes by ± 1. There are three different cases as defined in Eq. 6.1, which correspond to electron movement from half-filled to an empty orbitals, from half-filled to half-filled and from filled to half-filled orbitals. On the contrary a virtual electron transfer process does not cause changes in oxidation state and the initial electron count per atom is always restored. It can very much be compared to the perturbative interaction paths introduced in Chap. 5, the super-exchange is comparable to the mechanism shown in Fig. 5.3 and the semi covalent exchange is strongly related to the one depicted in Fig. 5.4. The direct exchange K_{ab} is normally not considered in the GK reasonings to explain magnetic interactions in solid state compounds.

The Goodenough–Kanamori rules state that superexchange and semi covalent exchange give an antiferromagnetic contribution to the coupling of the spin moments on site A and B when the virtual electron transfer is between overlapping orbitals that are half-filled. A ferromagnetic contribution arises when the virtual transfer is from half-filled to empty orbitals or from filled to half-filled orbitals. Moreover it is taken for granted that the electron transfer can only take place between overlapping orbitals. For orthogonal orbitals, the hopping is zero and Hund's rule prevails leading to a ferromagnetic contribution (Fig. 6.11).

Figure 6.12 illustrates the prototypical case of the magnetic interactions in the CuO_2 layers of the parent compounds of the high T_c superconductors. The Cu^{2+} ions have a $3d^9$ electronic configuration with one unpaired electron in the $3d_{x^2-y^2}$ orbital.

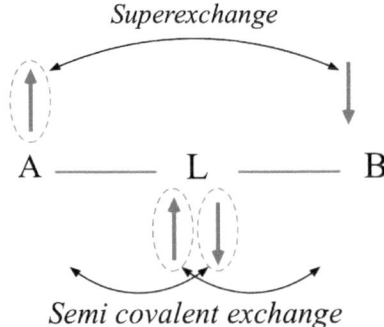

Fig. 6.11 Representation of the two *virtual* electron exchange mechanisms that are the basis of the Goodenough–Kanamori rules. *Above* the virtual exchange between two atoms (A and B) with non-zero spin moment, known as *superexchange* and *below* the virtual exchange between two atoms and a shared anion, the so-called *semi covalent exchange*

Fig. 6.12 Virtual electron superexchange between two overlapping half-filled orbitals leading to antiferromagnetic coupling

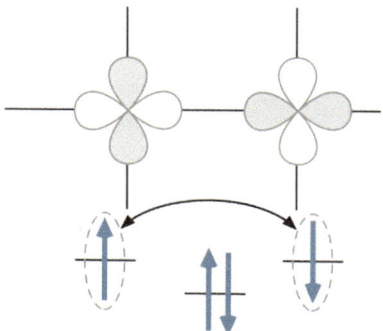

The virtual electron transfer between these two copper atoms, i.e. the superexchange, involves half-filled orbitals and hence contributes in an antiferromagnetic manner to the coupling. The semi covalent exchange contributes in the same direction, since it also involves the half-filled orbitals on the metals. Together the two effects give a qualitative explanation of the strong antiferromagnetic interactions between the Cu^{2+} ions.

The second example concerns $LaMnO_3$, which upon hole-doping shows a spectacular drop in the electrical resistivity when an external magnetic field is applied, the so-called colossal magnetoresistance effect. The electronic configuration of the Mn^{3+} ions is $3d^4$, with three unpaired electrons in the t_{2g} and one in the e_g orbitals assuming an octahedral coordination of the Mn cations. However, this configuration is Jahn-Teller active and induces displacements of the oxygen anions as indicated by the arrows in the left part of Fig. 6.13. In consequence, the occupied $3d$-orbitals of e_g symmetry are rotated by 90° at each magnetic center. This is called *orbital ordering* in the literature. Now, the superexchange between half-filled e_g orbitals cannot take place because they are orthogonal as shown in the left panel of Fig. 6.13. The only overlapping e_g orbitals are the half-filled on the left and the empty orbital on the right, see the right side of Fig. 6.13. The GK rules state that this superexchange (and the semi covalent exchange as well) is ferromagnetic in nature. The total interaction between the two magnetic centers is therefore expected to be ferromagnetic, although attenuated by the superexchange interactions in the weakly overlapping half-filled $3d(t_{2g})$ orbitals.

In previous chapters we have considered the magnetic interactions in the spin ladder compound $SrCu_2O_3$. There we focused on the interactions along the legs and the rungs, which share the common feature of a linear Cu–O–Cu linkage. However, taking a closer look at the structure (see Fig. 5.9) it becomes immediately clear that these copper ions are not nearest neighbours. Instead, the distance to the copper ion on the next ladder is shorter and one could naively think that the interactions of such pairs are also important. We have already seen in Sect. 4.2 that the interaction between two magnetic centers connected by a (double) bridge making an angle of around 90° is in general ferromagnetic and rather weak. A qualitative picture of the weak ferromagnetic interaction can also be obtained by applying the GK rules. The

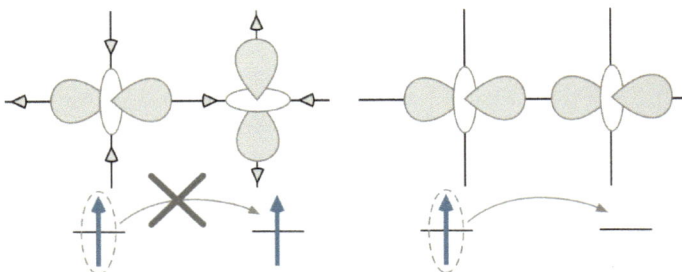

Fig. 6.13 The virtual hopping on the *left* is not permitted due to the orthogonality of the (occupied) orbitals on the two centers. The superexchange on the *right* is due to a virtual electron hopping from a half-filled to an empty orbital and hence ferromagnetic in character

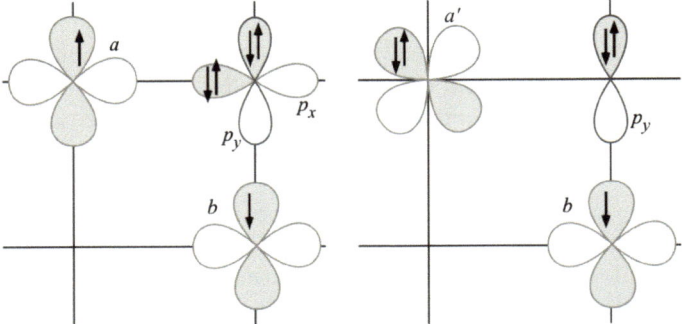

Fig. 6.14 *Left* the superexchange between a and b gives a small antiferromagnetic contribution. The semi covalent exchange with a and b is inoperative. *Right* Ferromagnetic contribution to the coupling by semi covalent exchange with a half-filled and a filled orbital

half-filled orbitals a and b shown on the left side of Fig. 6.14 overlap, and hence, the superexchange mechanism gives a antiferromagnetic contribution, albeit rather small since the overlap is not very strong. The semi covalent exchange involving a and b is inoperative because these two half-filled orbitals do not overlap with the same orbital on the anion. To activate the semi covalent exchange we have to consider one of the filled $3d$ orbitals that overlaps with one of the ligand orbitals, which in turn has a non-zero overlap with the half-filled $3d$ orbital on the other cation. Such situation is outlined in the right panel of Fig. 6.14, where a' is one of the doubly occupied $3d(t_{2g})$ orbitals (to be more precise the $3d_{xy}$ in this case) and b the half-filled $3d_{x^2-y^2}$ orbital. Whereas the overlap of a' and b is zero (no superexchange) both orbitals overlap with the O-$2p_y$ orbital and the semi covalent exchange becomes active. Since the virtual electron transfer is between a filled and a half-filled orbital, the contribution is ferromagnetic.

A pictorial explanation for the ferromagnetic nature of the semicovalent exchange between filled and half-filled orbitals is given in Fig. 6.15. In the upper part, we can see how the subsequent electron transfer from p_y to b and from a' to p_y leads to a

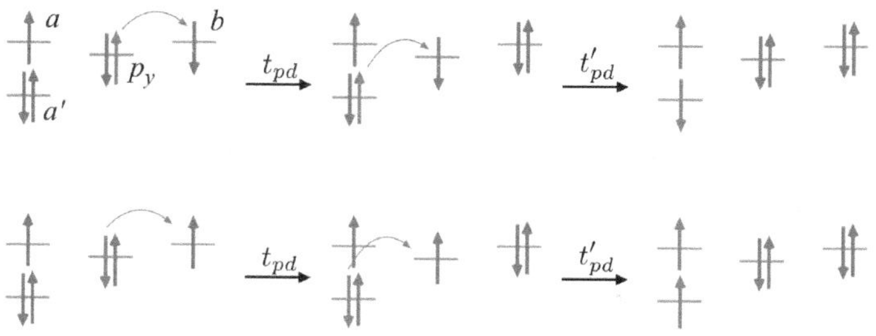

Fig. 6.15 Semi covalent exchange between filled and half-filled orbitals as operative in the magnetic coupling between two copper ions with an oxygen bridge forming an angle of 90°. *Upper part* for antiferromagnetic coupling, and *below* for ferromagnetic interaction

non-Hund determinant with two anti-parallel electrons on center A. This unfavorable electronic configuration is avoided in the case of a ferromagnetically coupled initial state shown in the lower part of the figure.

Estimation with perturbation theory: The ferromagnetic nature of the interaction between two magnetic centers with $S = \frac{1}{2}$ through a ligand under an angle of 90° can also be rationalized with perturbation theory in the same way as discussed in Sect. 5.1.1 for the magnetic interactions in a linear geometry or in Sect. 5.4.3 to derive the equations to estimate the magnitude of the four-center interactions with single determinant methods. The derivation is similar to the one presented by Koch in Ref. [19] with this difference that we here work within a spin restricted formalism.

There are four orbitals involved in the coupling as can be seen in Fig. 6.14. The six electrons can be distributed in sixteen different ways over the orbitals under the restriction of $M_S = 0$:

$$\Phi_1 = |a_1 p_x \bar{p}_x p_y \bar{p}_y \bar{b}_2| \quad \Phi_2 = |\bar{a}_1 p_x \bar{p}_x p_y \bar{p}_y b_2| \quad \Phi_3 = |a_1 \bar{p}_x p_y \bar{p}_y b_2 \bar{b}_2|$$

$$\Phi_4 = |\bar{a}_1 p_x p_y \bar{p}_y b_2 \bar{b}_2| \quad \Phi_5 = |a_1 \bar{a}_1 p_x \bar{p}_x p_y \bar{p}_y| \quad \Phi_6 = |a_1 \bar{a}_1 p_x \bar{p}_x p_y \bar{b}_2|$$

$$\Phi_7 = |a_1 \bar{a}_1 p_x p_y \bar{p}_y \bar{b}_2| \quad \Phi_8 = |a_1 \bar{a}_1 \bar{p}_x p_y \bar{p}_y b_1| \quad \Phi_9 = |a_1 p_x \bar{p}_x \bar{p}_y b_2 \bar{b}_2|$$

$$\Phi_{10} = |\bar{a}_1 p_x \bar{p}_x p_y b_2 \bar{b}_2| \quad \Phi_{11} = |a_1 \bar{a}_1 p_x \bar{p}_y b_2 \bar{b}_2| \quad \Phi_{12} = |a_1 \bar{a}_1 \bar{p}_x p_y b_2 \bar{b}_2|$$

$$\Phi_{13} = |a_1 \bar{a}_1 p_x \bar{p}_x p_y \bar{p}_y| \quad \Phi_{14} = |p_x \bar{p}_x p_y \bar{p}_y b_2 \bar{b}_2| \quad \Phi_{15} = |a_1 \bar{a}_1 p_x \bar{p}_x b_2 \bar{b}_2|$$

$$\Phi_{16} = |a_1 \bar{a}_1 p_y \bar{p}_y b_2 \bar{b}_2|$$

Taking the energy of Φ_1 and Φ_2 as reference, the other determinants lie at ΔE_{CT} ($\Phi_3 \ldots \Phi_6$), $\Delta E'_{CT}$ ($\Phi_7 \ldots \Phi_{10}$), ΔE_{2CT} (Φ_{11} and Φ_{12}), U_d (Φ_{13} and Φ_{14}) and $\Delta E_{2CT} + U_p$ (Φ_{15} and Φ_{16}). The last two determinants are high in energy and will be neglected. The subscript p refers to the O-$2p_x$ or $2p_y$ orbital and the subscript d to the a or b Cu-$3d$ orbital.

To simplify the derivation of the singlet and triplet energy, we first construct spin-symmetry adapted CSFs by forming linear combination of the above-listed determinants with unpaired electrons:

$$\Psi_{1,2} = \frac{1}{\sqrt{2}}(\Phi_1 \pm \Phi_2) \qquad \Psi_{3,4} = \frac{1}{\sqrt{2}}(\Phi_1 \pm \Phi_2) \qquad \Psi_{5,6} = \frac{1}{\sqrt{2}}(\Phi_5 \pm \Phi_6)$$

$$\Psi_{7,8} = \frac{1}{\sqrt{2}}(\Phi_7 \pm \Phi_8) \qquad \Psi_{9,10} = \frac{1}{\sqrt{2}}(\Phi_9 \pm \Phi_{10}) \qquad \Psi_{11,12} = \frac{1}{\sqrt{2}}(\Phi_{11} \pm \Phi_{12})$$

$$\Psi_{13,14} = \frac{1}{\sqrt{2}}(\Phi_{13} \pm \Phi_{14})$$

The plus (minus) combinations are triplet (singlet) functions, except for the combination of closed-shell determinants Ψ_{13} and Ψ_{14}, which are both singlets. Now we can construct the 6×6 configuration interaction matrix for the triplet functions and an 8×8 matrix for the singlet and then determine the energy either by diagonalizing the matrices or (simpler) with perturbation theory.

$S=1$	$\lvert\Psi_1\rangle$	$\lvert\Psi_3\rangle$	$\lvert\Psi_5\rangle$	$\lvert\Psi_7\rangle$	$\lvert\Psi_9\rangle$	$\lvert\Psi_{11}\rangle$
$\langle\Psi_1\rvert$	0	0	0	$-t_{pd}$	$-t_{pd}$	0
$\langle\Psi_3\rvert$	0	ΔE_{CT}	0	t_{ab}	0	0
$\langle\Psi_5\rvert$	0	0	ΔE_{CT}	0	t_{ab}	0
$\langle\Psi_7\rvert$	$-t_{pd}$	t_{ab}	0	$\Delta E'_{CT}$	0	$-t_{pd}$
$\langle\Psi_9\rvert$	$-t_{pd}$	0	t_{ab}	0	$\Delta E'_{CT}$	$-t_{pd}$
$\langle\Psi_{11}\rvert$	0	0	0	$-t_{pd}$	$-t_{pd}$	$\Delta E_{2CT} - K_{xy}$

$S=0$	$\lvert\Psi_2\rangle$	$\lvert\Psi_4\rangle$	$\lvert\Psi_6\rangle$	$\lvert\Psi_8\rangle$	$\lvert\Psi_{10}\rangle$	$\lvert\Psi_{12}\rangle$	$\lvert\Psi_{13}\rangle$	$\lvert\Psi_{14}\rangle$
$\langle\Psi_2\rvert$	0	0	0	$-t_{pd}$	$-t_{pd}$	0	$2t_{ab}$	0
$\langle\Psi_4\rvert$	0	ΔE_{CT}	0	$-t_{ab}$	0	0	t_{pd}	$-t_{pd}$
$\langle\Psi_6\rvert$	0	0	ΔE_{CT}	0	$-t_{ab}$	0	t_{pd}	t_{pd}
$\langle\Psi_8\rvert$	$-t_{pd}$	$-t_{ab}$	0	$\Delta E'_{CT}$	0	$-t_{pd}$	0	0
$\langle\Psi_{10}\rvert$	$-t_{pd}$	0	$-t_{ab}$	0	$\Delta E'_{CT}$	$-t_{pd}$	0	0
$\langle\Psi_{12}\rvert$	0	0	0	$-t_{pd}$	$-t_{pd}$	$\Delta E_{2CT} + K_{xy}$	0	0
$\langle\Psi_{13}\rvert$	$2t_{ab}$	t_{pd}	t_{pd}	0	0	0	U_d	0
$\langle\Psi_{14}\rvert$	0	$-t_{pd}$	t_{pd}	0	0	0	0	U_d

In these matrices we have neglected the intersite exchange integrals K_{ab} and K_{pd}. The hopping parameter t_{ab} parametrizes the electron transfer from cation to cation and t_{pd} the transfer from O-$2p_x$ to a and from p_y to b, which are strictly the same. The hopping from p_y to a is zero by symmetry. This is most easily seen in Fig. 6.14 (left). The symmetry behavior under $180°$ rotation around the x-axis is different for p_y (changes sign) and for a (no sign change). Because the Hamiltonian is totally symmetric, the integral $\langle p_y \lvert \hat{h} \rvert a \rangle$ is zero. A similar reasoning shows that the hopping from p_x to b is zero.

In an order-by-order perturbational approach we will derive the singlet-triplet energy gap to estimate the character and size of the magnetic coupling of the two Cu^{2+} ions bridged by an oxygen anion under $90°$. Up to first order, the energies are zero for the lowest singlet and triplet functions: $\langle \Psi_1|\hat{H}|\Psi_1\rangle = \langle \Psi_2|\hat{H}|\Psi_2\rangle = 0$. Remember that the intersite exchange interactions have been neglected, otherwise the zeroth-order triplet-singlet gap would be $2K_{ab}$. The second-order correction to the energy of the triplet and singlet are

$$\text{Triplet: } E_T^{(2)} = \frac{|\langle \Psi_1|\hat{H}|\Psi_7\rangle|^2}{E_0 - E_7} + \frac{|\langle \Psi_1|\hat{H}|\Psi_9\rangle|^2}{E_0 - E_9} = \frac{-2t_{pd}^2}{\Delta E_{CT}'} \tag{6.47a}$$

$$\text{Singlet: } E_S^{(2)} = \frac{-2t_{pd}^2}{\Delta E_{CT}'} - \frac{4t_{ab}^2}{U_d} \tag{6.47b}$$

The singlet is lower in energy by $4t_{ab}^2/U_d$, which corresponds to the antiferromagnetic superexchange by the direct electron transfer between the half-filled orbitals. The energy lowering is however small since t_{ab} is in general very small as long as there is no delocalization onto the ligand, conform the discussion of the valence mechanisms in Sect. 5.1.1. Note that the electron transfer from ligand to metal does give a significant energy lowering but that the differential effect is zero, the contribution to both states is the same.

6.5 Demonstrate that the energies up to second-order are given by the expressions in Eq. 6.47.

At fourth-order perturbation, there are many more contributions, but a large part is again identical for singlet and triplet. Only the contributions that involve Ψ_{11}, Ψ_{12} and Ψ_{13} have a differential effect. These can be separated in two contributions. First, the singlet-only contribution involving Ψ_{13} and second with either Ψ_{11} (singlet) or Ψ_{12} (triplet). The singlet-only contribution is given by

$$\sum_{i=4,6} \frac{\langle \Psi_2|\hat{H}|\Psi_{13}\rangle \langle \Psi_{13}|\hat{H}|\Psi_i\rangle \langle \Psi_i|\hat{H}|\Psi_{13}\rangle \langle \Psi_{13}|\hat{H}|\Psi_2\rangle}{-E_i \cdot E_{13}^2} = -\frac{8t_{ab}^2 t_{pd}^2}{\Delta E_{CT} U_d^2} \tag{6.48}$$

The fourth-order differential contribution to the triplet is

$$\sum_{i,j=7,9} \frac{\langle \Psi_1|\hat{H}|\Psi_i\rangle \langle \Psi_i|\hat{H}|\Psi_{11}\rangle \langle \Psi_{11}|\hat{H}|\Psi_j\rangle \langle \Psi_j|\hat{H}|\Psi_1\rangle}{-E_i \cdot E_j \cdot E_{11}}$$

$$= -\frac{16t_{pd}^4}{(\Delta E_{CT}')^2(\Delta E_{2CT} - K_{xy})} \tag{6.49a}$$

and the analogous contribution to the singlet is

$$\sum_{i,j=8,10} \frac{\langle\Psi_2|\hat{H}|\Psi_i\rangle\langle\Psi_i|\hat{H}|\Psi_{12}\rangle\langle\Psi_{12}|\hat{H}|\Psi_j\rangle\langle\Psi_j|\hat{H}|\Psi_2\rangle}{-E_i\cdot E_j\cdot E_{12}}$$

$$= -\frac{16t_{pd}^4}{(\Delta E'_{CT})^2(\Delta E_{2CT} + K_{xy})} \tag{6.49b}$$

The singlet-only term is small because of the presence of t_{ab} in the numerator, and hence, the other contribution is expected to be the dominant one. This second contribution is identical for both states except for the energy of the intermediate state with two unpaired electrons on the oxygen: Ψ_{11} and Ψ_{12} for triplet and singlet, respectively. Recalling Hund's rule for the tendency towards maximum spin multiplicity of unpaired electrons on one atom makes clear that the energy of Ψ_{11} will be significantly lower than the intermediate state on the singlet path (by $2K_{xy}$ to be precise), and hence, the fourth-order correction to the energies favors the ferromagnetic alignment of the spin moments on the cations.

6.5 Spin Waves for Ferromagnets

The last part of this chapter leaves behind the local viewpoint of the electronic structure and explores the description of magnetic interactions from a periodic perspective. Let us consider a lattice with N sites. Each site has a spin angular moment of S and all spins are aligned along the principal magnetization axis ($M_S = S$), corresponding to the ground state of a set of ferromagnetically coupled centers. The Heisenberg Hamiltonian for such a lattice reads

$$\hat{H} = -\sum_{i<j} J_{ij} \left[\frac{1}{2}\left(\hat{S}^+(i)\hat{S}^-(j) + \hat{S}^-(i)\hat{S}^+(j)\right) + \hat{S}_z(i)\hat{S}_z(j)\right] \tag{6.50}$$

and the wave function is characterized by the M_S value at each lattice site

$$\Phi_0 = |S_1, S_2, \ldots, S_i, S_j, \ldots, S_N\rangle \tag{6.51}$$

To calculate the energy of Φ_0 we evaluate the effect of the different terms of the Hamiltonian separately and then add them up to obtain the energy.

$$\hat{S}^+(i)\hat{S}^-(j)|S_1, S_2, \ldots, S_i, S_j, \ldots, S_N\rangle = 0$$

$$\hat{S}^-(i)\hat{S}^+(j)|S_1, S_2, \ldots, S_i, S_j, \ldots, S_N\rangle = 0 \tag{6.52}$$

$$\hat{S}_z(i)\hat{S}_z(j)|S_1, S_2, \ldots, S_i, S_j, \ldots, S_N\rangle = S^2|S_1, S_2, \ldots, S_i, S_j, \ldots, S_N\rangle \tag{6.53}$$

The zero's in the first two contributions are due to the fact that a spin with maximum M_S-value cannot climb further on the ladder by $\hat{S}^+$. From this, we confirm that Φ_0 is an eigenfunction with eigenvalue

$$E_0 = -S^2 \sum_{i<j} J_{ij} \tag{6.54}$$

Next, we study the low-lying excitations of the ferromagnet, following Kaxiras [20]. To generate an excited state, the M_S value at one of the sites is lowered from $M_S = S$ to $S - 1$ by applying $\hat{S}^-(i)$ on Φ_0, the smallest change that can be imagined. To ensure that the excited state is also an eigenfunction of the Heisenberg Hamiltonian two determinants are needed

$$\Phi_1 = |S_1, S_2, \ldots, S_i - 1, S_j, \ldots, S_N\rangle$$
$$\Phi_2 = |S_1, S_2, \ldots, S_i, S_j - 1, \ldots, S_N\rangle \tag{6.55}$$

The action of the ladder operators on such functions is defined in Eq. 1.23 and results in

$$\hat{S}^+(i)\hat{S}^-(j)\Phi_1 = \hat{S}^+(i)\sqrt{(S+S)(S+1-S)}|S_1, S_2, \ldots, S_i - 1, S_j - 1, \ldots, S_N\rangle$$
$$= \sqrt{2S}\sqrt{(S-S+1)(S+1+S-1)}|S_1, S_2, \ldots, S_i, S_j - 1, \ldots, S_N\rangle = 2S\Phi_2 \tag{6.56}$$

and, similarly,

$$\hat{S}^-(i)\hat{S}^+(j)\Phi_1 = 0$$
$$\hat{S}^+(i)\hat{S}^-(j)\Phi_2 = 0$$
$$\hat{S}^-(i)\hat{S}^+(j)\Phi_2 = 2S\Phi_1 \tag{6.57}$$
$$\hat{S}_z(i)\hat{S}_z(j)\Phi_1 = (S-1)S\Phi_1$$
$$\hat{S}_z(i)\hat{S}_z(j)\Phi_2 = S(S-1)\Phi_2$$

By defining $\Psi_\pm = (\Phi_1 \pm \Phi_2)/\sqrt{2}$, eigenfunctions of the Heisenberg Hamiltonian are obtained with the following eigenvalues

$$\Psi_+ : \quad \hat{H}(\Phi_1 + \Phi_2)/\sqrt{2} = -\sum_{i<j} J_{ij}\big(S\Phi_1 + S\Phi_2 + (S-1)S(\Phi_1 + \Phi_2)\big)/\sqrt{2}$$
$$= -\sum_{i<j} J_{ij}(S + S(S-1))(\Phi_1 + \Phi_2)/\sqrt{2} = -\sum_{i<j} J_{ij}S^2\Psi_+ \tag{6.58}$$

$$\Psi_- : \quad \hat{H}(\Phi_1 - \Phi_2)/\sqrt{2} = -\sum_{i<j} J_{ij}\left(-S\Phi_1 + S\Phi_2 + (S-1)S(\Phi_1 - \Phi_2)\right)/\sqrt{2}$$

$$= -\sum_{i<j} J_{ij}(-S + S^2 - S)(\Phi_1 - \Phi_2)/\sqrt{2} = -\sum_{i<j} J_{ij}(S^2 - 2S)\Psi_-$$

$$(6.59)$$

E_+ is identical to the ground state value and the corresponding wave function has the same spin multiplicity as Φ_0 but the total M_S value is lowered by one. The second energy, E_-, is higher than E_0 (remember that the J_{ij} are positive for a ferromagnetic system) and describes a state where the total spin moment is no longer equal to the maximum value.

6.6 Consider a system with two $S = 1$ magnetic sites. The ferromagnetic solution is $\Phi_0 = \alpha\alpha\alpha\alpha$. Check that $\Psi_\pm = (M_{S,max}, M_{S,max}-1) \pm (M_{S,max}-1, M_{S,max})$ are indeed eigenfunctions of the Heisenberg Hamiltonian and that the *plus* combination corresponds to a quintet and the *minus* combination to a triplet.

This description of the excited state does however not respect the translational symmetry of the crystal and an extra step has to be taken to obtain a more complete description. First, we change from discrete point indexation $(1, 2, \ldots i, j, \ldots, N)$ to a more convenient representation based on the distance between two lattice sites. Figure 6.16 shows how the discrete labeling of lattice sites can be replaced by a representation based on the distance r between these through the vectors **r**. Although slightly more abstract, this choice is more versatile for an extended system with, in principle, infinite lattice sites and translational symmetry.

The Heisenberg Hamiltonian of Eq. 6.50 remains the same except that the indices i and j are replaced by $\mathbf{r}'$ and $\mathbf{r}''$.

Fig. 6.16 Definition of **r**, **r'** and **r''** used in the derivation of the spin wave representation of the excited states of an Heisenberg ferromagnetic extended system

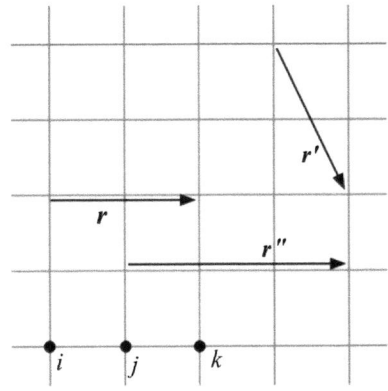

$$\hat{H} = -\sum_{\mathbf{r'}<\mathbf{r''}} J(\mathbf{r'} - \mathbf{r''}) \left[\frac{1}{2} \left(\hat{S}^+(\mathbf{r'}) \cdot \hat{S}^-(\mathbf{r''}) + \hat{S}^-(\mathbf{r'}) \cdot \hat{S}^+(\mathbf{r''}) \right) + \hat{S}_z(\mathbf{r'})\hat{S}_z(\mathbf{r''}) \right]$$

(6.60)

The general expression for the excited state that respects the translational symmetry is obtained by applying the same procedure that is followed to construct the well-known Bloch functions to represent the single-particle wave functions in a crystal.

$$|\Phi_k\rangle = \sum_{\mathbf{r}} e^{i\mathbf{k}\cdot\mathbf{r}} \hat{S}^-(\mathbf{r})|\Phi_0\rangle$$

(6.61)

where Φ_0 is the ferromagnetic ground state with maximum M_S-value (all spin moments aligned along the principal magnetic axis) and Φ_k a state with $M_S = M_{S,max} - 1$. We will follow the same strategy as above to determine the energy of this extended wave function by letting the Hamiltonian act on it. First, the action of $\hat{S}_z(\mathbf{r'})\hat{S}_z(\mathbf{r''})$ on the spin dependent part of $|\Phi_k\rangle$:

$$\hat{S}_z(\mathbf{r'})\hat{S}_z(\mathbf{r''})\hat{S}^-(\mathbf{r})|\Phi_0\rangle = [S(S-1)(\delta_{\mathbf{r'r}} + \delta_{\mathbf{r''r}}) + S^2(1 - \delta_{\mathbf{r'r}} - \delta_{\mathbf{r''r}})]\hat{S}^-(\mathbf{r})|\Phi_0\rangle$$
$$= (S^2 - S\delta_{\mathbf{r'r}} - S\delta_{\mathbf{r''r}})\hat{S}^-(\mathbf{r})|\Phi_0\rangle$$

(6.62)

This expression shows that the product of two $\hat{S}_z$ operators results nearly always in S^2, except when $\mathbf{r}$ coincides with $\mathbf{r'}$ or $\mathbf{r''}$ where it acts on a spin function with an M_S-value lowered by 1, resulting in $S(S-1)$. This is conveniently represented with the Kronecker delta functions in the expression. This results leads us directly to the expression that reflects the action of the last term of the Hamiltonian on Φ_k

$$-\sum_{\mathbf{r'}<\mathbf{r''}} J(\mathbf{r'} - \mathbf{r''})(S^2 - S\delta_{\mathbf{r'r}} - S\delta_{\mathbf{r''r}}) \sum_{\mathbf{r}} e^{i\mathbf{k}\cdot\mathbf{r}}\hat{S}^-(\mathbf{r})|\Phi_0\rangle$$

$$= \left[-S^2 \sum_{\mathbf{r'}<\mathbf{r''}} J(\mathbf{r'} - \mathbf{r''}) + 2S \sum_{\mathbf{r}\neq 0} J(\mathbf{r}) \right] |\Phi_k\rangle = \sum_{\mathbf{r}\neq 0} \left(-\frac{1}{2}S^2 + 2S \right) J(\mathbf{r})|\Phi_k\rangle$$

(6.63)

The first two terms of the Hamiltonian concern the products of step-up and step-down operators

$$\hat{S}^-(\mathbf{r'})\hat{S}^+(\mathbf{r''})\hat{S}^-(\mathbf{r})|\Phi_0\rangle = 2S\delta_{\mathbf{r''r}}\hat{S}^-(\mathbf{r'})|\Phi_0\rangle$$
$$\hat{S}^+(\mathbf{r'})\hat{S}^-(\mathbf{r''})\hat{S}^-(\mathbf{r})|\Phi_0\rangle = 2S\delta_{\mathbf{r'r}}\hat{S}^-(\mathbf{r''})|\Phi_0\rangle$$

(6.64)

Half the sum of these two terms gives

$$\frac{1}{2} \left[\hat{S}^-(\mathbf{r'})\hat{S}^+(\mathbf{r''})\hat{S}^-(\mathbf{r}) + \hat{S}^+(\mathbf{r'})\hat{S}^-(\mathbf{r''})\hat{S}^-(\mathbf{r}) \right] |\Phi_0\rangle = 2S\delta_{\mathbf{r'r}}\hat{S}^-(\mathbf{r'})|\Phi_0\rangle$$ (6.65)

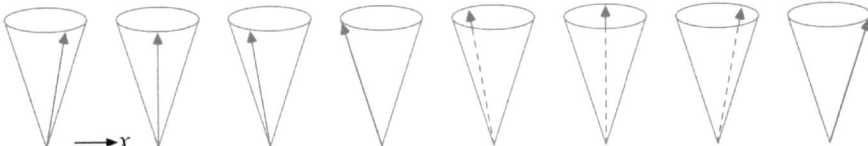

Fig. 6.17 Propagating along x of a spin wave in a one-dimensional model. The projection on the z-axis is constant ($M_{S,max} - 1$), $\langle \hat{S}_x \rangle$ and $\langle \hat{S}_y \rangle$ change from site to site

which allows us to evaluate the first terms of the Heisenberg Hamiltonian

$$
-\sum_{\mathbf{r'} < \mathbf{r''}} J(\mathbf{r'} - \mathbf{r''}) \sum_r e^{i\mathbf{k}\cdot\mathbf{r}} 2S\delta_{\mathbf{rr''}} \hat{S}^-(\mathbf{r'})|\Phi_0\rangle
$$

$$
= -2S \sum_{\mathbf{r}\neq 0} J(\mathbf{r}) e^{i\mathbf{k}\cdot\mathbf{r}} \sum_{\mathbf{r'}} e^{i\mathbf{k}\cdot\mathbf{r'}} \hat{S}^-(\mathbf{r'})|\Phi_0\rangle = -2S \sum_{\mathbf{r}\neq 0} J(\mathbf{r}) e^{i\mathbf{k}\cdot\mathbf{r}}|\Phi_k\rangle \quad (6.66)
$$

Finally, the sum of all three terms gives the eigenvalue of $|\Phi_k\rangle$

$$
E_k = \sum_{\mathbf{r}\neq 0} \left(-\frac{1}{2}S^2 + 2S - 2Se^{i\mathbf{k}\cdot\mathbf{r}} \right) J(\mathbf{r}) = E_0 + 2S \sum_{\mathbf{r}\neq 0} (1 - e^{i\mathbf{k}\cdot\mathbf{r}})J(\mathbf{r}) \quad (6.67)
$$

which is always higher than the ground state energy, except for $|\Phi_{k=0}\rangle$, which is degenerate with $|\Phi_0\rangle$. Figure 6.17 represents how the spin moment of $|\Phi_k\rangle$ propagates along the x-axis in a one-dimensional model. The total spin moment on each site is equal to S and the projection on the z-axis (the principal magnetic axis) is also constant, $M_{S,max} - 1$. The variation lies in the projection on the other two magnetic axes, which is easily demonstrated by calculating the expectation value of $\hat{S}_x(\mathbf{r})\hat{S}_x(\mathbf{r'}) + \hat{S}_y(\mathbf{r})\hat{S}_y(\mathbf{r'})$ of $|\Phi_k\rangle$, which measures the correlation of the non-z-components of the spin moments separated by $\mathbf{r}$ and $\mathbf{r'}$.

After the usual substitution of $\hat{S}_x$ and $\hat{S}_y$ by the appropriate linear combinations of $\hat{S}^-$ and $\hat{S}^+$

$$
\langle\Phi_k|\hat{S}_x(\mathbf{r})\hat{S}_x(\mathbf{r'}) + \hat{S}_y(\mathbf{r})\hat{S}_y(\mathbf{r'})|\Phi_k\rangle = \langle\Phi_k|\frac{1}{2}[\hat{S}^+(\mathbf{r'})\hat{S}^-(\mathbf{r''}) + \hat{S}^-(\mathbf{r'})\hat{S}^+(\mathbf{r''})]|\Phi_k\rangle
$$
$$(6.68)$$

we evaluate the correlation function term by term

$$
\frac{1}{2}\hat{S}^-(\mathbf{r'})\hat{S}^+(\mathbf{r''})|\Phi_k\rangle = \frac{1}{2}\sum_r e^{i\mathbf{k}\cdot\mathbf{r}} \hat{S}^-(\mathbf{r'})\hat{S}^+(\mathbf{r''})\hat{S}^-(\mathbf{r})|\Phi_0\rangle
$$
$$
= S \sum_r e^{i\mathbf{k}\cdot\mathbf{r}} \delta_{\mathbf{rr''}} \hat{S}^-(\mathbf{r})|\Phi_0\rangle = Se^{i\mathbf{k}\cdot\mathbf{r''}} \hat{S}^-(\mathbf{r'})|\Phi_0\rangle \quad (6.69)
$$

$$\langle \Phi_k | \frac{1}{2} \hat{S}^+(\mathbf{r}')\hat{S}^-(\mathbf{r}'') = \frac{1}{2}\hat{S}^-(\mathbf{r}'')\hat{S}^+(\mathbf{r}') \sum_{\mathbf{r}} e^{-i\mathbf{k}\cdot\mathbf{r}}\hat{S}^-|\Phi_0\rangle$$

$$= S \sum_{\mathbf{r}} e^{-i\mathbf{k}\cdot\mathbf{r}}\delta_{\mathbf{r}\mathbf{r}'}\hat{S}^-(\mathbf{r}'')|\Phi_0\rangle = Se^{-i\mathbf{k}\cdot\mathbf{r}'}\hat{S}^-(\mathbf{r}'')|\Phi_0\rangle \quad (6.70)$$

The sum of these two terms gives

$$\langle \Phi_k|\hat{S}_x(\mathbf{r})\hat{S}_x(\mathbf{r}') + \hat{S}_y(\mathbf{r})\hat{S}_y(\mathbf{r}')|\Phi_k\rangle = S(e^{-i\mathbf{k}\cdot\mathbf{r}'} + e^{i\mathbf{k}\cdot\mathbf{r}''}) = 2S\cos\left((\mathbf{r}'-\mathbf{r}'')\cdot\mathbf{k}\right)$$
$$(6.71)$$

showing that the orientation of the projection of the spin moment on the plane perpendicular to the principal magnetic axis varies as a cosine that depends on the separation of the spins and the lattice vector $\mathbf{k}$, exactly as the spin wave shown in Fig. 6.17.

Antiferromagnetic lattices: The description of an 'infinite' lattice with antiferromagnetic interactions is much more complicated and in fact there is no exact ground state solution for such case. The first necessary simplification towards an (approximate) description is to limit the interactions to nearest neighbours. Imagine a two-dimensional regular lattice of magnetic centers. Taking into account only nearest neighbour interactions all spins align in an anti-parallel manner. However, considering antiferromagnetic next-nearest neighbour interactions as well, the spins cannot follow the preferred alignment for centers beyond the nearest neighbours as illustrated in Fig. 6.18. This is sometimes denoted *spin frustration*. In fact, competing interactions can give rise to very interesting magnetic phenomena, and Problem 6.4 describes one of these. In the simplest case of an isolated 1D chain with only nearest neighbour antiferromagnetic interactions, an exact solution can be obtained using the Bethe ansatz.

The main problem to rigorously describe the antiferromagnetic lattice—even with the restriction of nearest neighbour interactions only—lies in the fact that the hypothetical ground state eigenfunction of the Heisenberg Hamiltonian is intrinsically multideterminantal. With increasing number of magnetic centers the

Fig. 6.18 Two-dimensional lattice of magnetic centers with antiferromagnetic nearest neighbour interactions. Next nearest neighbour antiferromagnetic interactions cannot be sustained. Note that the representation of alternating *up* and *down* spins is a simplification that is only valid for the Ising Hamiltonian

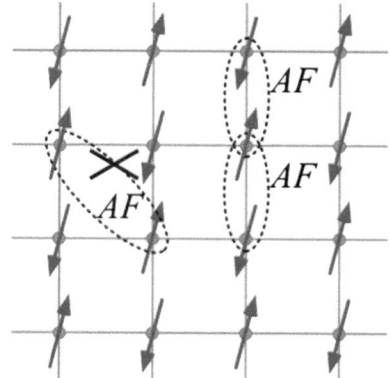

number of determinants needed to describe the antiferromagnetic state also grows. For $N = 2$, we have $\Phi_{AF} = (|\phi_1\bar{\phi}_2| - |\bar{\phi}_1\phi_2|)/\sqrt{2}$; for $N = 3$, the eigenfunction is a sum of three determinants: $\Phi_{AF} = (|2\phi_1\phi_2\bar{\phi}_3| - |\phi_1\bar{\phi}_2\phi_3| - |\bar{\phi}_1\phi_2\phi_3|)/\sqrt{6}$; and for $N = 4$, we already need a linear combination of six determinants (see Eq. 1.51). It is easy to imagine that when we consider a crystal with in principle an infinite number of magnetic sites, the wave function cannot be written down anymore.

Intuitively one could consider the state with alternating α and β spins, as drawn in Fig. 6.18, as a good representation of the ground state in an antiferromagnetic lattice. However, it is quite easy to show that this so-called Neél state is not an eigenfunction of the Heisenberg Hamiltonian and that its energy expectation value is only an upper bound to the ground state energy. Using the definition of the Heisenberg Hamiltonian given in Eq. 6.50 with $j = i + 1$ and applying periodic boundaries as mentioned in Sect. 3.3, we calculate the energy expectation value of the Neél state $\Phi_0 = |S_1, -S_2, S_3, \ldots, S_i, -S_j, \ldots - S_N|$. The action of the different products of spin operators on this function is

$$
\begin{aligned}
\hat{S}^+(i)\hat{S}^-(j)\Phi_0 &= 0 \\
\hat{S}^-(i)\hat{S}^+(j)\Phi_0 &= |S_1, -S_2, S_3, \ldots, S_i - 1, -S_j + 1, \ldots, -S_N| \\
\hat{S}_z(i)\hat{S}_z(j)\Phi_0 &= S^2\Phi
\end{aligned}
\tag{6.72}
$$

This shows that Φ_0 is not an eigenfunction of the Heisenberg Hamiltonian and that the products of spin-up and spin-down operators give both zero contribution to the energy expectation value, which becomes

$$
E(\Phi_0) = \frac{1}{2}NzS^2J
\tag{6.73}
$$

where N is the number of sites and z is the number of nearest neighbours of each magnetic center. To show that this is not the state with the lowest energy, we now generate a new spin configuration with the same total M_S value by applying the $\hat{S}^+(k)\hat{S}^-(l) + \hat{S}^-(k)\hat{S}^+(l)$ operator to the Neél state. States with different M_S values do not interact with Φ_0 and cannot lower the energy of Φ_0.

$$
\begin{aligned}
\Phi_1 &= (\hat{S}^+(k)\hat{S}^-(l) + \hat{S}^-(k)\hat{S}^+(l))|S_1, -S_2, S_3, \ldots S_i, -S_j, \ldots - S_N| \\
&= |S_1, -S_2, \ldots, S_k - 1, -S_l + 1, \ldots, -S_N|
\end{aligned}
\tag{6.74}
$$

Again, we have a state that is not an eigenfunction of the Heisenberg Hamiltonian, which is easily seen by applying the products of spin-up and spin-down operators. The energy expectation value is

$$
E(\Phi_1) = J\left(\frac{1}{2}zNS^2 - z + 1\right)
\tag{6.75}
$$

More importantly, the interaction matrix element of Φ_0 and Φ_1 is not equal to zero. Going term by term:

$$\langle \ldots S_i, -S_j, S_k, -S_l \ldots | -J \sum_{\langle i,j \rangle} \frac{1}{2} \hat{S}^+(i)\hat{S}^-(j) | \ldots, S_i, -S_j, S_k - 1, -S_l + 1, \ldots \rangle = -\frac{1}{2}J$$

$$\langle \ldots S_i, -S_j, S_k, -S_l \ldots | -J \sum_{\langle i,j \rangle} \frac{1}{2} \hat{S}^-(i)\hat{S}^+(j) | \ldots, S_i, -S_j, S_k - 1, -S_l + 1, \ldots \rangle = 0$$

$$\langle \ldots S_i, -S_j, S_k, -S_l \ldots | -J \sum_{\langle i,j \rangle} \hat{S}_z(i)\hat{S}_z(j) | \ldots, S_i, -S_j, S_k - 1, -S_l + 1, \ldots \rangle = 0 \quad (6.76)$$

where $\langle i,j \rangle$ symbolizes the sum over $i > j$ restricted to nearest neighbours. This non-zero matrix element means that the diagonalization of the 2×2 matrix spanned by Φ_0 and Φ_1 results in two new states, one of them with lower energy than Φ_0, showing that the Néel state is not the ground state of the antiferromagnetic lattice.

6.7 (a) Write down the wave function of the Néel state (Φ_0) for a system with 8 magnetic sites with $S = 1/2$ in its explicit form using the $\alpha(i)$ and $\beta(i)$ spin functions. (b) Calculate the energy expectation value of the Heisenberg Hamiltonian and compare to the outcome of Eq. 6.73. (c) Apply the $\hat{S}^+(3)\hat{S}^-(4) + \hat{S}^-(3)\hat{S}^+(4)$ operator on Φ_0 and calculate the expectation value of the so obtained wave function (Φ_1). (d) Calculate $\langle \Phi_0 | \hat{H} | \Phi_1 \rangle$.

Spin wave theory of antiferromagnets is a powerful method to study the ground state in these cases but goes beyond the scope of the book, the interested reader is referred to the monographs of Yosida [7] and Blundell [8].

Problems

6.1 Doublet ground state for mixed valence: Determine the magnitude of J in terms of t for which the model system defined in Fig. 6.4 has a doublet ground state.

6.2 Exchange interaction with s-orbital on the bridge: Consider the system depicted in Fig. 6.14 with a bridging ligand that has a s-orbital as outermost occupied valence orbital. Rationalize the antiferromagnetic coupling for this system.

6.3 Expectation value of a non-Néel state: Calculate the expectation value of $\Phi_1 = \hat{S}^+(i)\hat{S}^-(i+1)\Phi_0$ of the Heisenberg Hamiltonian with nearest neighbour interactions only for the following two cases: (a) Φ_0 is the Néel state of a one-dimensional chain with $N = 8$; (b) Φ_0 is the Néel state of a 4×4 lattice. Both systems have periodic boundaries and $S > \frac{1}{2}$.

6.4 Helical spin order: Consider a one dimensional spin chain with sizeable first (J_1) and second (J_2) neighbour interactions. In the mean-field approximation, the energy of the system is given by

$$E = -NS^2\big(J_1 \cos(\theta) + J_2 \cos(2\theta)\big)$$

In most cases the spins will align either parallel or anti-parallel, depending on the sign of J_1, but for certain ratios of J_1/J_2 spin arrangements can be observed with non-collinear spin moments. Such spin configurations are supposed to play an important role in the ferroelectric properties of magnetic materials. (a) Check that the energy expression is identical to the energy of the Neél state (Eq. 6.73) when $J_2 = 0$ and $J_1 < 0$. (b) Is there any possibility for a non-collinear alignment when $J_2 > 0$? (c) Find the three values of θ for which the energy is minimized and classify them as antiferromagnetic, ferromagnetic or non-collinear solutions. (d) Calculate the angle between two neighbouring sites with $J_2 = -0.3|J_1|$ and $J_1 = \pm 1$.

References

1. M.B. Robin, P. Day, Adv. Inorg. Chem. Radiochem. **10**, 247 (1967)
2. C. Zener, Phys. Rev. **82**(3), 403 (1951)
3. M.B. Salamon, M. Jaime, Rev. Mod. Phys. **73**, 583 (2001)
4. J.J. Girerd, V. Papaefthymiou, K.K. Surerus, E. Münck, Pure Appl. Chem. **61**, 805 (1989)
5. J.J. Borrás-Almenar, J.M. Clemente-Juan, E. Coronado, A.V. Palii, B.S. Tsukerblatt, in *Magnetism: Molecules to Materials*, ed. by J.S. Miller, M. Drillon (Wiley-VCH, 2001), pp. 155–210, chap. 5
6. P.W. Anderson, H. Hasegawa, Phys. Rev. **100**(2), 675 (1955)
7. K. Yosida, *Theory of Magnetism* (Springer, Berlin, 1996)
8. S. Blundell, *Magnetism in Condensed Matter* (Oxford University Press, Oxford, 2001)
9. W.J. Caspers, *Spin Systems* (World Scientific, Singapore, 1989)
10. J. Stöhr, H.C. Siegmann, *Magnetism: From Fundamentals to Nanoscale Dynamics* (Springer, Berlin, 2006)
11. Z. Barandiarán, L. Seijo, J. Chem. Phys. **89**, 5739 (1988)
12. K. Doll, M. Dolg, P. Fulde, H. Stoll, Phys. Rev. B **52**, 4842 (1995)
13. B. Paulus, Phys. Rep. **428**, 1 (2006)
14. V.I. Anisimov, J. Zaanen, O.K. Andersen, Phys. Rev. B **44**(3), 943 (1991)
15. L. Hedin, Phys. Rev. **139**, A796 (1965)
16. P. Fulde, H. Stoll, Found. Phys. **30**, 2049 (2000)
17. A. Stoyanova, C. Sousa, C. de Graaf, R. Broer, Int. J. Quantum Chem. **106**, 2444 (2006)
18. C. Pisani, M. Schütz, S. Casassa, D. Usvyat, L. Maschio, M. Lorenz, A. Erba, Phys. Chem. Chem. Phys. **14**, 7615 (2012)
19. E. Koch, in *Correlated Electrons: From Models to Materials*, ed. by E. Pavarini, E. Koch, M. Jarrell (Forschungszentrum Jülich GmbH Institute for Advanced Simulation, Jülich, 2012), chap. 7
20. E. Kaxiras, *Atomic and Electronic Structure of Solids* (Cambridge University Press, Cambridge, 2003)

Appendix A
Effect of the $\hat{l}$ Operator and the Matrix Elements of the p and d Orbitals

ψ	$\hat{l}_x|\psi\rangle$	$\hat{l}_y|\psi\rangle$	$\hat{l}_z|\psi\rangle$
p_x	0	$-ip_z$	ip_y
p_y	ip_z	0	$-ip_x$
p_z	$-ip_y$	ip_x	0
d_{z^2}	$-i\sqrt{3}d_{yz}$	$i\sqrt{3}d_{xz}$	0
$d_{x^2-y^2}$	$-id_{yz}$	$-id_{xz}$	$2id_{xy}$
d_{xy}	id_{xz}	$-id_{yz}$	$-2id_{x^2-y^2}$
d_{yz}	$id_{x^2-y^2}+i\sqrt{3}d_{z^2}$	id_{xy}	$-id_{xz}$
d_{xz}	$-id_{xy}$	$-i\sqrt{3}d_{z^2}+id_{x^2-y^2}$	id_{yz}

$\langle\hat{l}\rangle$	p_x	p_y	p_z
p_x	0	$-i$	1
p_y	i	0	1
p_z	1	1	0

$\langle\hat{l}\rangle$	d_{z^2}	$d_{x^2-y^2}$	d_{xy}	d_{yz}	d_{xz}
d_{z^2}	0	0	0	$i\sqrt{3}$	$-i\sqrt{3}$
$d_{x^2-y^2}$	0	0	$-2i$	i	i
d_{xy}	0	$2i$	0	i	$-i$
d_{yz}	$-i\sqrt{3}$	$-i$	$-i$	0	i
d_{xz}	$i\sqrt{3}$	$-i$	i	$-i$	0

© Springer International Publishing Switzerland 2016
C. Graaf and R. Broer, *Magnetic Interactions in Molecules and Solids*,
Theoretical Chemistry and Computational Modelling,
DOI 10.1007/978-3-319-22951-5

Appendix B

Effect of the $\hat{S}$ Operator and the Matrix Elements for $\frac{1}{2} \leq S \leq \frac{5}{2}$

$\lvert S, M_S\rangle$	$\hat{S}_x\lvert S, M_S\rangle$	$\hat{S}_y\lvert S, M_S\rangle$	$\hat{S}_z\lvert S, M_S\rangle$
$\lvert\frac{1}{2},\frac{1}{2}\rangle$	$\frac{1}{2}\lvert\frac{1}{2},-\frac{1}{2}\rangle$	$\frac{i}{2}\lvert\frac{1}{2},-\frac{1}{2}\rangle$	$\frac{1}{2}\lvert\frac{1}{2},\frac{1}{2}\rangle$
$\lvert\frac{1}{2},-\frac{1}{2}\rangle$	$\frac{1}{2}\lvert\frac{1}{2},\frac{1}{2}\rangle$	$\frac{-i}{2}\lvert\frac{1}{2},\frac{1}{2}\rangle$	$-\frac{1}{2}\lvert\frac{1}{2},-\frac{1}{2}\rangle$
$\lvert 1,1\rangle$	$\frac{\sqrt{2}}{2}\lvert 1,0\rangle$	$\frac{-i\sqrt{2}}{2}\lvert 1,0\rangle$	$\lvert 1,1\rangle$
$\lvert 1,0\rangle$	$\frac{\sqrt{2}}{2}\left(\lvert 1,1\rangle+\lvert 1,-1\rangle\right)$	$\frac{i\sqrt{2}}{2}\left(\lvert 1,1\rangle-\lvert 1,-1\rangle\right)$	0
$\lvert 1,-1\rangle$	$\frac{\sqrt{2}}{2}\lvert 1,0\rangle$	$\frac{i\sqrt{2}}{2}\lvert 1,0\rangle$	$-\lvert 1,-1\rangle$
$\lvert\frac{3}{2},\frac{3}{2}\rangle$	$\frac{\sqrt{3}}{2}\lvert\frac{3}{2},\frac{1}{2}\rangle$	$\frac{-i\sqrt{3}}{2}\lvert\frac{3}{2},\frac{1}{2}\rangle$	$\frac{3}{2}\lvert\frac{3}{2},\frac{3}{2}\rangle$
$\lvert\frac{3}{2},\frac{1}{2}\rangle$	$\frac{\sqrt{3}}{2}\lvert\frac{3}{2},\frac{3}{2}\rangle+\lvert\frac{3}{2},-\frac{1}{2}\rangle$	$\frac{i\sqrt{3}}{2}\lvert\frac{3}{2},\frac{3}{2}\rangle-i\lvert\frac{3}{2},-\frac{1}{2}\rangle$	$\frac{1}{2}\lvert\frac{3}{2},\frac{1}{2}\rangle$
$\lvert\frac{3}{2},-\frac{1}{2}\rangle$	$\lvert\frac{3}{2},\frac{1}{2}\rangle+\frac{\sqrt{3}}{2}\lvert\frac{3}{2},-\frac{3}{2}\rangle$	$i\lvert\frac{3}{2},\frac{1}{2}\rangle-\frac{i\sqrt{3}}{2}\lvert\frac{3}{2},-\frac{3}{2}\rangle$	$-\frac{1}{2}\lvert\frac{3}{2},-\frac{1}{2}\rangle$
$\lvert\frac{3}{2},-\frac{3}{2}\rangle$	$\frac{\sqrt{3}}{2}\lvert\frac{3}{2},-\frac{1}{2}\rangle$	$\frac{i\sqrt{3}}{2}\lvert\frac{3}{2},-\frac{1}{2}\rangle$	$-\frac{3}{2}\lvert\frac{3}{2},-\frac{3}{2}\rangle$
$\lvert 2,2\rangle$	$\lvert 2,1\rangle$	$-i\lvert 2,1\rangle$	$2\lvert 2,2\rangle$
$\lvert 2,1\rangle$	$\lvert 2,2\rangle+\frac{\sqrt{6}}{2}\lvert 2,0\rangle$	$i\lvert 2,2\rangle-\frac{i\sqrt{6}}{2}\lvert 2,0\rangle$	$\lvert 2,1\rangle$
$\lvert 2,0\rangle$	$\frac{\sqrt{6}}{2}\left(\lvert 2,1\rangle+\lvert 2,-1\rangle\right)$	$\frac{i\sqrt{6}}{2}\left(\lvert 2,1\rangle-\lvert 2,-1\rangle\right)$	0
$\lvert 2,-1\rangle$	$\frac{\sqrt{6}}{2}\lvert 2,0\rangle+\lvert 2,-2\rangle$	$\frac{i\sqrt{6}}{2}\lvert 2,0\rangle-i\lvert 2,-2\rangle$	$-\lvert 2,-1\rangle$
$\lvert 2,-2\rangle$	$\lvert 2,-1\rangle$	$-i\lvert 2,-1\rangle$	$-2\lvert 2,-2\rangle$

© Springer International Publishing Switzerland 2016
C. Graaf and R. Broer, *Magnetic Interactions in Molecules and Solids*,
Theoretical Chemistry and Computational Modelling,
DOI 10.1007/978-3-319-22951-5

$\lvert S, M_S\rangle$	$\hat{S}_x \lvert S, M_S\rangle$	$\hat{S}_y \lvert S, M_S\rangle$	$\hat{S}_z \lvert S, M_S\rangle$
$\lvert \frac{5}{2}, \frac{5}{2}\rangle$	$\frac{\sqrt{5}}{2}\lvert \frac{5}{2}, \frac{3}{2}\rangle$	$\frac{-i\sqrt{5}}{2}\lvert \frac{5}{2}, \frac{3}{2}\rangle$	$\frac{5}{2}\lvert \frac{5}{2}, \frac{5}{2}\rangle$
$\lvert \frac{5}{2}, \frac{3}{2}\rangle$	$\frac{\sqrt{5}}{2}\lvert \frac{5}{2}, \frac{5}{2}\rangle + \lvert \frac{5}{2}, \frac{1}{2}\rangle$	$\frac{i\sqrt{5}}{2}\lvert \frac{5}{2}, \frac{5}{2}\rangle - i\lvert \frac{5}{2}, \frac{1}{2}\rangle$	$\frac{3}{2}\lvert \frac{5}{2}, \frac{3}{2}\rangle$
$\lvert \frac{5}{2}, \frac{1}{2}\rangle$	$\sqrt{2}\lvert \frac{5}{2}, \frac{3}{2}\rangle + \frac{3}{2}\lvert \frac{5}{2}, -\frac{1}{2}\rangle$	$i\sqrt{2}\lvert \frac{5}{2}, \frac{3}{2}\rangle - \frac{3i}{2}\lvert \frac{5}{2}, -\frac{1}{2}\rangle$	$\frac{1}{2}\lvert \frac{5}{2}, \frac{1}{2}\rangle$
$\lvert \frac{5}{2}, -\frac{1}{2}\rangle$	$\frac{3}{2}\lvert \frac{5}{2}, \frac{1}{2}\rangle + \sqrt{2}\lvert \frac{5}{2}, -\frac{3}{2}\rangle$	$\frac{3i}{2}\lvert \frac{5}{2}, \frac{1}{2}\rangle - i\sqrt{2}\lvert \frac{5}{2}, -\frac{3}{2}\rangle$	$-\frac{1}{2}\lvert \frac{5}{2}, -\frac{1}{2}\rangle$
$\lvert \frac{5}{2}, -\frac{3}{2}\rangle$	$\lvert \frac{5}{2}, -\frac{1}{2}\rangle + \frac{\sqrt{5}}{2}\lvert \frac{5}{2}, -\frac{5}{2}\rangle$	$i\lvert \frac{5}{2}, -\frac{1}{2}\rangle - \frac{i\sqrt{5}}{2}\lvert \frac{5}{2}, -\frac{5}{2}\rangle$	$-\frac{3}{2}\lvert \frac{5}{2}, -\frac{3}{2}\rangle$
$\lvert \frac{5}{2}, -\frac{5}{2}\rangle$	$\frac{\sqrt{5}}{2}\lvert \frac{5}{2}, -\frac{3}{2}\rangle$	$\frac{i\sqrt{5}}{2}\lvert \frac{5}{2}, -\frac{3}{2}\rangle$	$-\frac{5}{2}\lvert \frac{5}{2}, -\frac{5}{2}\rangle$

Appendix C
Matrix Representation of the ZFS Model Hamiltonian

$\hat{S}\overline{\overline{D}}\hat{S}$ in an arbitrary axis frame. The simpler form of the Hamiltonian that applies when the system is oriented along the magnetic axis frame is easily derived by putting all D_{ij} to zero for $i \neq j$, making the trace equal to zero and substituting $D_{33} - \frac{1}{2}(D_{11} + D_{22})$ by D and $\frac{1}{2}(D_{11} - D_{22})$ by E.

$S = 1$	$|1, 1\rangle$	$|1, 0\rangle$	$|1, -1\rangle$
$\langle 1, 1|$	$\frac{1}{2}(D_{11} + D_{22} + D_{33})$	$-\frac{\sqrt{2}}{2}(D_{13} + iD_{23})$	$\frac{1}{2}(D_{11} - D_{22} + 2iD_{12})$
$\langle 1, 0|$	$-\frac{\sqrt{2}}{2}(D_{13} - iD_{23})$	$D_{11} + D_{22}$	$\frac{\sqrt{2}}{2}(D_{13} + iD_{23})$
$\langle 1, -1|$	$\frac{1}{2}(D_{11} - D_{22} - 2iD_{12})$	$\frac{\sqrt{2}}{2}(D_{13} - iD_{23})$	$\frac{1}{2}(D_{11} + D_{22} + D_{33})$

$S = \frac{3}{2}$	$|\frac{3}{2}, \frac{3}{2}\rangle$	$|\frac{3}{2}, \frac{1}{2}\rangle$	$|\frac{3}{2}, -\frac{1}{2}\rangle$	$|\frac{3}{2}, -\frac{3}{2}\rangle$
$\langle\frac{3}{2}, \frac{3}{2}|$	$\frac{3}{4}(D_{11} + D_{22} + 3D_{33})$	$-\sqrt{3}(D_{13} + iD_{23})$	$\frac{\sqrt{3}}{2}(D_{11} - D_{22} + 2iD_{12})$	0
$\langle\frac{3}{2}, \frac{1}{2}|$	$-\sqrt{3}(D_{13} - iD_{23})$	$\frac{1}{4}[7(D_{11} + D_{22}) + D_{33}]$	0	$\frac{\sqrt{3}}{2}(D_{11} - D_{22} + 2iD_{12})$
$\langle\frac{3}{2}, -\frac{1}{2}|$	$\frac{\sqrt{3}}{2}(D_{11} - D_{22} - 2iD_{12})$	0	$\frac{1}{4}[7(D_{11} + D_{22}) + D_{33}]$	$\sqrt{3}(D_{13} + iD_{23})$
$\langle\frac{3}{2}, -\frac{3}{2}|$	0	$\frac{\sqrt{3}}{2}(D_{11} - D_{22} - 2iD_{12})$	$\sqrt{3}(D_{13} - iD_{23})$	$\frac{3}{4}(D_{11} + D_{22} + 3D_{33})$

$S = 2$	$|2, 2\rangle$	$|2, 1\rangle$	$|2, 0\rangle$
$\langle 2, 2|$	$D_{11} + D_{22} + 4D_{33}$	$3(D_{13} + iD_{23})$	$\frac{\sqrt{6}}{2}(D_{11} - D_{22} + 2iD_{12})$
$\langle 2, 1|$	$3(D_{13} - iD_{23})$	$\frac{5}{2}(D_{11} + D_{22}) + D_{33}$	$-\frac{\sqrt{6}}{2}(D_{13} + iD_{23})$
$\langle 2, 0|$	$\frac{\sqrt{6}}{2}(D_{11} - D_{22} - 2iD_{12})$	$-\frac{\sqrt{6}}{2}(D_{13} - iD_{23})$	$3(D_{11} + D_{22})$
$\langle 2, -1|$	0	$\frac{3}{2}(D_{11} - D_{22} - 2iD_{12})$	$\frac{\sqrt{6}}{2}(D_{13} - iD_{23})$
$\langle 2, -2|$	0	0	$\frac{\sqrt{6}}{2}(D_{11} - D_{22} - 2iD_{12})$

© Springer International Publishing Switzerland 2016
C. Graaf and R. Broer, *Magnetic Interactions in Molecules and Solids*,
Theoretical Chemistry and Computational Modelling,
DOI 10.1007/978-3-319-22951-5

cont.	$\lvert 2, -1\rangle$	$\lvert 2, -2\rangle$
$\langle 2, 2\rvert$	0	0
$\langle 2, 1\rvert$	$\frac{3}{2}(D_{11} - D_{22} + 2iD_{12})$	0
$\langle 2, 0\rvert$	$\frac{\sqrt{6}}{2}(D_{13} + iD_{23})$	$\frac{\sqrt{6}}{2}(D_{11} - D_{22} + 2iD_{12})$
$\langle 2, -1\rvert$	$5/2(D_{11} + D_{22}) + D_{33}$	$-3(D_{13} + iD_{23})$
$\langle 2, -2\rvert$	$-3(D_{13} - iD_{23})$	$D_{11} + D_{22} + 4D_{33}$

$S = \frac{5}{2}$	$\lvert \frac{5}{2}, \frac{5}{2}\rangle$	$\lvert \frac{5}{2}, \frac{3}{2}\rangle$	$\lvert \frac{5}{2}, \frac{1}{2}\rangle$
$\langle \frac{5}{2}, \frac{5}{2}\rvert$	$\frac{5}{4}(D_{11} + D_{22}) + \frac{25}{4}D_{33}$	$2\sqrt{5}(D_{13} - iD_{23})$	$\frac{\sqrt{10}}{2}(D_{11} - D_{22} - 2iD_{12})$
$\langle \frac{5}{2}, \frac{3}{2}\rvert$	$2\sqrt{5}(D_{13} + iD_{23})$	$\frac{13}{4}(D_{11} + D_{22}) + \frac{9}{4}D_{33}$	$2\sqrt{2}(D_{13} - iD_{23})$
$\langle \frac{5}{2}, \frac{1}{2}\rvert$	$\frac{\sqrt{10}}{2}(D_{11} - D_{22} + 2iD_{12})$	$2\sqrt{2}(D_{13} + iD_{23})$	$\frac{17}{4}(D_{11} + D_{22}) + \frac{1}{4}D_{33}$
$\langle \frac{5}{2}, -\frac{1}{2}\rvert$	0	$\frac{3}{2}\sqrt{2}(D_{11} - D_{22} + 2iD_{13})$	0
$\langle \frac{5}{2}, -\frac{3}{2}\rvert$	0	0	$\frac{3}{2}\sqrt{2}(D_{11} - D_{22} + 2iD_{13})$
$\langle \frac{5}{2}, -\frac{5}{2}\rvert$	0	0	0

cont.	$\lvert \frac{5}{2}, -\frac{1}{2}\rangle$	$\lvert \frac{5}{2}, -\frac{3}{2}\rangle$	$\lvert \frac{5}{2}, -\frac{5}{2}\rangle$
$\langle \frac{5}{2}, \frac{5}{2}\rvert$	0	0	0
$\langle \frac{5}{2}, \frac{3}{2}\rvert$	$\frac{3}{2}\sqrt{2}(D_{11} - D_{22} - 2iD_{13})$	0	0
$\langle \frac{5}{2}, \frac{1}{2}\rvert$	0	$\frac{3}{2}\sqrt{2}(D_{11} - D_{22} - 2iD_{13})$	0
$\langle \frac{5}{2}, -\frac{1}{2}\rvert$	$\frac{17}{4}(D_{11} + D_{22}) + \frac{1}{4}D_{33}$	$-2\sqrt{2}(D_{13} - iD_{23})$	$\frac{\sqrt{10}}{2}(D_{11} - D_{22} - 2iD_{12})$
$\langle \frac{5}{2}, -\frac{3}{2}\rvert$	$-2\sqrt{2}(D_{13} + iD_{23})$	$\frac{13}{4}(D_{11} + D_{22}) + \frac{9}{4}D_{33}$	$-2\sqrt{5}(D_{13} - iD_{23})$
$\langle \frac{5}{2}, -\frac{5}{2}\rvert$	$\frac{\sqrt{10}}{2}(D_{11} - D_{22} + 2iD_{12})$	$-2\sqrt{5}(D_{13} + iD_{23})$	$\frac{5}{4}(D_{11} + D_{22}) + \frac{25}{4}D_{33}$

Appendix D
Analytical Expressions for $\chi(T)$

1. The Bonner-Fisher expression for a $S = 1/2$ uniform Heisenberg chain:

$$\chi(T) = \frac{N_A \mu_B^2 g_e^2}{kT} \frac{A + Bx + Cx^2}{1 + Dx + Ex^2 + Fx^3}$$

with $x = J/2kT$ and $A = 0.25$; $B = 0.14995$; $C = 0.30094$; $D = 1.9862$; $E = 0.68854$; $F = 6.0626$ [1].

2. The alternate $S = \frac{1}{2}$ Heisenberg chain with $J = J_1 = \alpha J_2$:

$$\chi(T) = \frac{N_A \mu_B^2 g_e^2}{kT} \frac{A + Bx + Cx^2}{1 + Dx + Ex^2 + Fx^3}$$

for $0 < \alpha < 0.4$ the values of A–F are [2]

$$A = 0.25$$
$$B = -0.12587 + 0.22752\alpha$$
$$C = 0.019111 - 0.13307\alpha + 0.509\alpha^2 - 1.3167\alpha^3 + 1.0081\alpha^4$$
$$D = 0.100772 + 1.4192\alpha$$
$$E = -0.0028521 - 0.42346\alpha + 2.1953\alpha^2 - 0.82412\alpha^3$$
$$F = 0.37754 - 0.067022\alpha + 5.9805\alpha^2 - 2.1678\alpha^3 + 15.838\alpha^4$$

and for $0.4 < \alpha < 1$ the values of A–F are

$$A = 0.25$$
$$B = -0.13695 + 0.26387\alpha$$
$$C = 0.017025 - 0.12668\alpha + 0.49113\alpha^2 - 1.1977\alpha^3 + 0.87257\alpha^4$$
$$D = 0.070509 + 1.3042\alpha$$

© Springer International Publishing Switzerland 2016
C. Graaf and R. Broer, *Magnetic Interactions in Molecules and Solids*,
Theoretical Chemistry and Computational Modelling,
DOI 10.1007/978-3-319-22951-5

$$E = -0.0035767 - 0.40837\alpha + 3.4862\alpha^2 - 0.73888\alpha^3$$
$$F = 0.36184 - 0.065528\alpha + 6.65875\alpha^2 - 20.945\alpha^3 + 15.425\alpha^4$$

3. The uniform two-dimensional $S = 1/2$ Heisenberg lattice [3]

$$\chi(T) = \frac{N_A \mu_B^2 g_e^2}{kT} \sum_{n=1}^{5} \frac{a_n J/kT}{b_n J/kT}$$

n	A_n	B_n
1	0.998586	−1.84279
2	1.28534	1.14141
3	0.656313	−0.704192
4	0.235862	−0.189044
5	0.277527	−0.277545

4. The generalized expression for a Heisenberg 2D lattice [4]

$$\chi(T) = \frac{N_A g_e^2 \mu_B}{3kT} \frac{Y_1 + Y_2}{(1 - u_1 v_1)(1 - u_2 v_2)}$$

with

$$Y_1 = (1 + u_1 v_1)(1 + u_2 v_2) + (u_1 + v_1)(u_2 + v_2)$$

$$Y_2 = (u_1 + v_1)(1 + u_2 v_2) + (u_2 + v_2)(1 + u_1 v_1)$$

$$u_i = \coth\left(\frac{S(S+1)J_{i-}}{kT} - \frac{kT}{S(S+1)J_{i-}}\right)$$
$$v_i = \coth\left(\frac{S(S+1)J_{i+}}{kT} - \frac{kT}{S(S+1)J_{i+}}\right)$$

5. Hexagonal 2D lattice [5, 6] (Fig. D.1)

$$\chi(T) = \frac{N_A g_e^2 \mu_B}{3kT} \frac{(1 + u_1 u_2)^2 \left(1 + u_2^2\right) + 2u_2 \left(1 + u_1 u_2\right)^2 + u_1 \left(1 - u_2^2\right)^2}{\left(1 - u_1^2 u_2^2\right)\left(1 - u_2^2\right)}$$

with

$$u_i = \coth\left(\frac{S(S+1)J_i}{kT}\right) - \frac{kT}{S(S+1)J_i}$$

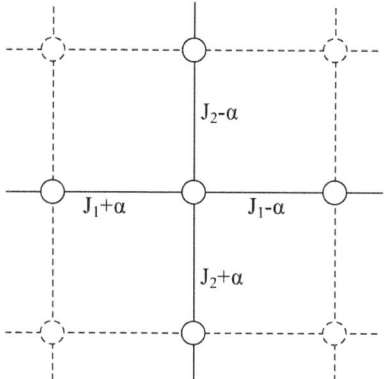

Fig. D.1 Definition of $J_{i\pm} = J_i \pm \alpha$ used in the generalized expression for the Heisenberg 2D lattice

References

1. W.E. Estes, D.P. Gavel, W.E. Hatfield, D.J. Hodgson, Inorg. Chem. **17**, 1415 (1978)
2. J.W. Hall, W.E. Marsh, R.R. Weller, W.E. Hatfield, Inorg. Chem. **20**, 1033 (1981)
3. F.M. Woodward, A.S. Albrecht, C.M. Wynn, C.P. Landee, M.M. Turnbull, Phys. Rev. B **65**, 144412 (2002)
4. J. Curély, J. Rouch, Phys. B **254**, 298 (1998)
5. J. Curély, F. Lloret, M. Julve, Phys. Rev. B **58**, 11465 (1998)
6. I. Negodaev, C. de Graaf, R. Caballol, J. Phys. Chem. A **114**, 7553 (2010)

Appendix E
Solutions

Exercises and Problems of Chap. 1

Exercise 1.1 $\Psi(1, 2, 3) = |\phi_a(1)\phi_b(2)\phi_c(3)| = \frac{1}{\sqrt{6}}(\phi_a(1)\phi_b(2)\phi_c(3) - \phi_a(1)\phi_c(2)$
$\phi_b(3) - \phi_b(1)\phi_a(2)\phi_c(3) + \phi_b(1)\phi_c(2)\phi_a(3) + \phi_c(1)\phi_a(2)\phi_b(3) - \phi_c(1)\phi_b(2)\phi_a(3))$
$\Psi(2, 1, 3) = \frac{1}{\sqrt{6}}(\phi_b(1)\phi_a(2)\phi_c(3) - \phi_c(1)\phi_a(2)\phi_b(3) - \phi_a(1)\phi_b(2)\phi_c(3) + \phi_c(1)\phi_b$
$(2)\phi_a(3) + \phi_a(1)\phi_c(2)\phi_b(3) - \phi_b(1)\phi_c(2)\phi_a(3)) = \frac{1}{\sqrt{6}}(-\phi_a(1)\phi_b(2)\phi_c(3) + \phi_a(1)\phi_c$
$(2)\phi_b(3) + \phi_b(1)\phi_a(2)\phi_c(3) - \phi_b(1)\phi_c(2)\phi_a(3) - \phi_c(1)\phi_a(2)\phi_b(3) + \phi_c(1)\phi_b$
$(2)\phi_a(3)) = -\Psi(1, 2, 3)$ Assume $\phi_a = \phi_b$, then $\Psi(1, 2, 3) = \frac{1}{\sqrt{6}}(\phi_a(1)\phi_a(2)\phi_c$
$(3) - \phi_a(1)\phi_c(2)\phi_a(3) - \phi_a(1)\phi_a(2)\phi_c(3) + \phi_a(1)\phi_c(2)\phi_a(3) + \phi_c(1)\phi_a(2)\phi_a(3) -$
$\phi_c(1)\phi_a(2)\phi_a(3)) = 0$.

Exercise 1.2 $\hat{A}(1, 2) = \frac{1}{\sqrt{2}}(1 - \hat{P}_{12}); \hat{A}\varphi_1\varphi_2 = \frac{1}{\sqrt{2}}(\varphi_1\varphi_2 - \varphi_2\varphi_1) \Rightarrow \sqrt{N!}\hat{A}\varphi_1\varphi_2 =$
$\varphi_1\varphi_2 - \varphi_2\varphi_1; \hat{A}\hat{A}\varphi_1\varphi_2 = \frac{1}{\sqrt{2}}(1 - \hat{P}_{12})\frac{1}{\sqrt{2}}(\varphi_1\varphi_2 - \varphi_2\varphi_1) = \frac{1}{2}(\varphi_1\varphi_2 - \varphi_2\varphi_1 - \varphi_2\varphi_1 +$
$\varphi_1\varphi_2) = \varphi_1\varphi_2 - \varphi_2\varphi_1$.

Exercise 1.3 For a given S, M_S runs from S to $-S$ in steps of 1. Hence, the degeneracy is $2S + 1$.

Exercise 1.4 Substituting $s = 1/2$ and $m_s = \pm 1/2$ in the normalization factor of $\hat{s}^+$ gives $\sqrt{1/2(1/2 + 1) - 1/2(1/2 + 1)} = 0$ for α and $\sqrt{1/2(1/2 + 1) - -1/2(-1/2 + 1)}$ $= \sqrt{3/4 + 1/4} = 1$ for β. **(b)** $\hat{s}^2 = 1/4(\hat{s}^+ + \hat{s}^-)(\hat{s}^+ + \hat{s}^-) - 1/4(\hat{s}^+ + \hat{s}^-)(\hat{s}^+ +$ $\hat{s}^-) + \hat{s}_z^2 = 1/4(\hat{s}^+\hat{s}^+ + \hat{s}^+\hat{s}^- + \hat{s}^-\hat{s}^+ + \hat{s}^-\hat{s}^-) - 1/4(\hat{s}^+\hat{s}^+ - \hat{s}^+\hat{s}^- - \hat{s}^-\hat{s}^+ + \hat{s}^-\hat{s}^-) + \hat{s}_z^2$ (remember $\hat{s}^+$ and $\hat{s}^-$ do no commute) $= 1/2(\hat{s}^+\hat{s}^- + \hat{s}^-\hat{s}^+) + \hat{s}_z^2 = 1/2(\hat{s}^+\hat{s}^- +$ $\hat{s}^-\hat{s}^+) + (1/2)\hat{s}^+\hat{s}^- - (1/2)\hat{s}^+\hat{s}^- + \hat{s}_z^2 = \hat{s}^+\hat{s}^- - 1/2[\hat{s}^+, \hat{s}^-] + \hat{s}_z^2 = \hat{s}^+\hat{s}^- - \hat{s}_z + \hat{s}_z^2$. **(c)** $\hat{s}^+\hat{s}^-\alpha = \alpha$, $-\hat{s}_z\alpha = -1/2\alpha$, $\hat{s}_z^2\alpha = 1/4\alpha$. Combining the three terms gives $(1 - 1/2 + 1/4)\alpha$; the expectation value is 3/4. $\hat{s}^+\hat{s}^-\beta = 0$, $-\hat{s}_z\beta = 1/2$, $\hat{s}_z^2\beta = 1/4\beta$. Combining the terms, gives the expectation value $(0 + 1/2 + 1/4) = 3/4$ for β.

© Springer International Publishing Switzerland 2016
C. Graaf and R. Broer, *Magnetic Interactions in Molecules and Solids*,
Theoretical Chemistry and Computational Modelling,
DOI 10.1007/978-3-319-22951-5

Exercise 1.5 $\hat{S}^2|\varphi_1\overline{\varphi}_2| = |\varphi_1\varphi_2|(\hat{S}^+\hat{S}^- - \hat{S}_z + \hat{S}_z^2)\alpha\beta = |\varphi_1\varphi_2|(\hat{S}^+\hat{S}^-\alpha\beta - \hat{S}_z\alpha\beta + \hat{S}_z^2\alpha\beta)$. With $\hat{S} = \hat{s}(1) + \hat{s}(2)$ we arrive at $|\varphi_1\varphi_2|((\alpha\beta + \beta\alpha) - (\frac{1}{2}\alpha\beta - \frac{1}{2}\alpha\beta) + \hat{s}_z(\frac{1}{2}\alpha\beta - \frac{1}{2}\alpha\beta))$. Since the last two terms are zero, we get $|\varphi_1\varphi_2|(\alpha\beta + \beta\alpha) = |\varphi_1\overline{\varphi}_2| + |\overline{\varphi}_1\varphi_2|$, which is the same result as obtained in Eq. 1.27 where the determinant was fully expanded; $\hat{S}^2(|\varphi_1\overline{\varphi}_2| + |\overline{\varphi}_1\varphi_2|)/\sqrt{2} = (1/\sqrt{2})|\varphi_1\varphi_2|\hat{S}^2(\alpha\beta + \beta\alpha) = (1/\sqrt{2})|\varphi_1\varphi_2|((\alpha\beta + \beta\alpha) + (\beta\alpha + \alpha\beta)) = (2/\sqrt{2})(|\varphi_1\overline{\varphi}_2| + |\overline{\varphi}_1\varphi_2|)$; The expansion of Φ_2 leads to $(1/2)(\varphi_1\overline{\varphi}_2 - \overline{\varphi}_2\varphi_1 + \overline{\varphi}_1\varphi_2 - \varphi_2\overline{\varphi}_1) = (1/2)(\varphi_1\varphi_2 - \varphi_2\varphi_1)(\alpha\beta + \beta\alpha)$. $\hat{S}^2\Phi_2 = (1/2)(\varphi_1\varphi_2 - \varphi_2\varphi_1)\hat{S}^2(\alpha\beta + \beta\alpha) = (1/2)(\varphi_1\varphi_2 - \varphi_2\varphi_1) \cdot 2(\alpha\beta + \beta\alpha) = \varphi_1\overline{\varphi}_2 - \overline{\varphi}_2\varphi_1 + \overline{\varphi}_1\varphi_2 - \varphi_2\overline{\varphi}_1 = \sqrt{2}(|\varphi_1\overline{\varphi}_2| + |\overline{\varphi}_2\varphi_1|)$.

Exercise 1.6 (a) Rewriting Eq. 1.23 gives $|S, M_S + 1\rangle = \hat{S}^+|S, M_S\rangle/\sqrt{S(S+1) - M_S(M_S + 1)} = \hat{S}^+(|a\overline{b}| + |\overline{a}b|)/(\sqrt{2}(1(1+1) - 0(0+1))) = (|ab| + |ab|)/2 = |ab|$. Similar for the $M_S - 1$ component: $\hat{S}^-(|a\overline{b}| + |\overline{a}b|)/(\sqrt{2}(1(1+1) - 0(0-1))) = (|\overline{ab}| + |\overline{ab}|)/2 = |\overline{ab}|$. (b) $|S\rangle = \hat{P}_0|ab| = (\hat{S}^2 - 2)|ab| = |a\overline{b}| + |\overline{a}b| - 2|\overline{a}b| = |a\overline{b}| - |\overline{a}b|$; $|T\rangle = \hat{P}_1|ab| = (\hat{S}^2 - 0)|\overline{a}b| = |\overline{a}b| + |a\overline{b}|$.

Exercise 1.7 $|3/2, 3/2\rangle = \hat{S}^+(|\overline{a}bc| + |a\overline{b}c| + |ab\overline{c}|)/(\sqrt{3}N)$ with $N = \sqrt{3/2(3/2 + 1) - 1(2(1/2 + 1)} = \sqrt{3} \Rightarrow |3/2, 3/2\rangle = (|abc| + |abc| + |abc|)/3 = |abc|$; $|3/2, -1/2\rangle = (|\overline{a}\overline{b}c| + |a\overline{b}c| + |\overline{a}bc|)/(\sqrt{3}N)$ with $N = \sqrt{3/2(3/2 + 1) - 1/2(1/2 - 1)} = 2 \Rightarrow |3/2, -1/2\rangle = (|a\overline{b}\overline{c}| + |\overline{a}b\overline{c}| + |\overline{a}\overline{b}c| + |a\overline{b}\overline{c}| + |\overline{a}b\overline{c}| + |\overline{a}\overline{b}c|)/2\sqrt{3} = (|a\overline{b}\overline{c}| + |\overline{a}b\overline{c}| + |\overline{a}\overline{b}c|)/\sqrt{3}$; $|3/2, -3/2\rangle = \hat{S}^-(|a\overline{b}\overline{c}| + |\overline{a}b\overline{c}| + |\overline{a}\overline{b}c|)/(\sqrt{3}N)$ with $N = \sqrt{3/2(3/2 + 1) - -1/2(-1/2 - 1)} = \sqrt{3} \Rightarrow |3/2, -3/2\rangle = (|\overline{a}\overline{b}\overline{c}| + |\overline{a}\overline{b}\overline{c}| + |\overline{a}\overline{b}\overline{c}|)/3 = |\overline{a}\overline{b}\overline{c}|$.

Exercise 1.8 (a) $\langle\Psi_A|\Psi_B\rangle = (1/\sqrt{12})\langle a\overline{b}c - \overline{a}bc|2a\overline{b}c - ab\overline{c} - \overline{a}bc\rangle = (1/\sqrt{12})(-\langle ab\overline{c}|a\overline{b}c\rangle + \langle\overline{a}bc|\overline{a}bc\rangle) = 0$. (b) $\hat{S}^+\hat{S}^-|a\overline{b}\overline{c}| = \hat{S}^+(|\overline{a}b\overline{c}| + |\overline{a}\overline{b}c|) = |a\overline{b}\overline{c}| + |\overline{a}bc| + |a\overline{b}\overline{c}| + |a\overline{b}c|$; $\hat{S}^+\hat{S}^-|\overline{a}bc| = \hat{S}^+(|\overline{a}\overline{b}c| + |\overline{a}b\overline{c}|) = |a\overline{b}c| + |\overline{a}bc| + |a\overline{b}c| + |\overline{a}bc|$; $\hat{S}_z(|a\overline{b}\overline{c}| - |\overline{a}bc|) = (1/2 + 1/2 - 1/2)|a\overline{b}\overline{c}| - (-1/2 + 1/2 + 1/2)|\overline{a}bc| \Rightarrow \hat{S}_z^2(|a\overline{b}\overline{c}| - |\overline{a}bc|) = (1/4)(|a\overline{b}\overline{c}| - |\overline{a}bc|)$. Collecting all the terms gives $\hat{S}^2(|a\overline{b}\overline{c}| - |\overline{a}bc|)/\sqrt{2} = (1 - 1/2 + 1/4)(|a\overline{b}\overline{c}| - |\overline{a}bc|)/\sqrt{2}$, and hence, $S(S + 1) = 3/4, S = 1/2$.

Exercise 1.9 (a)

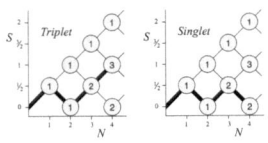

(b) Starting with $\Psi(1, 1/2, \pm 1/2)$, the application of the formula in Eq. 1.43 gives $\Psi(2, 0, 0) = (-\sqrt{0 - 0 + 1}\beta\alpha + \sqrt{0 + 0 + 1}\alpha\beta)/\sqrt{2 \cdot 0 + 2} = (\alpha\beta - \beta\alpha)/\sqrt{2}$.

Exercise 1.10 (a) $\hat{S}^+(\alpha\alpha\beta\beta - \beta\beta\alpha\alpha) = (\alpha\alpha\alpha\beta + \alpha\alpha\beta\alpha - \alpha\beta\alpha\alpha - \beta\alpha\alpha\alpha)$, which is equivalent to the function of Eq. 1.50b, except for the normalization factor. (b) Starting with $(\alpha\beta - \beta\alpha)/\sqrt{2}$, the singlet coupling for electron 1 and 2, Eq. 1.53

has to be applied. The only term with a non-zero prefactor is the second one $\Rightarrow$
$\Psi(4, 1, 0) = \left(\sqrt{2(1 + 0)(1 - 0)}((\alpha\beta - \beta\alpha)/\sqrt{2})((\alpha\beta + \beta\alpha)/\sqrt{2})\right)/\sqrt{2(2 - 1)} =$
$(1/2)(\alpha\beta\alpha\beta + \alpha\beta\beta\alpha - \beta\alpha\alpha\beta - \beta\alpha\beta\alpha)$.

Exercise 1.11 $\langle(\alpha\alpha\beta + \alpha\beta\alpha + \beta\alpha\alpha)/\sqrt{3}|\alpha\beta\alpha\rangle = 1/\sqrt{3} = 0.577\ldots$; $\langle(\alpha\alpha\beta - \beta\alpha\alpha)\sqrt{2}|\alpha\beta\alpha\rangle = 0$; $\langle(2\alpha\beta\alpha - \alpha\alpha\beta - \beta\alpha\alpha)/\sqrt{6}|\alpha\beta\alpha\rangle = 0.816\ldots$

Exercise 1.12 $\hat{H}^{(0)}\psi_0^{(3)} + \hat{V}\psi_0^{(2)} = E_0^{(0)}\psi_0^{(3)} + E_0^{(1)}\psi_0^{(2)} + E_0^{(2)}\psi_0^{(1)} + E_0^{(3)}\psi_0^{(0)}$.

Problem 1.1 Singlet: $(|a\bar{b}| - |\bar{a}b|)/\sqrt{2}$ or $(|a\bar{b}| + |b\bar{a}|)/\sqrt{2}$ maintaining the spatial or spin part, respectively. Triplet: $(|a\bar{b}| + |\bar{a}b|)/\sqrt{2}$ or $(|a\bar{b}| - |b\bar{a}|)/\sqrt{2}$ maintaining the spatial or spin part, respectively.

Problem 1.2 (a) Coulomb integral between the charge distributions $\phi_a\phi_a$ and $\phi_b\phi_b$, both on the same atom, and hence, relatively large integral. (b) Exchange integral, relatively large. (c) Exchange integral of medium size. The permutation leads to a zero integral because of the orthogonality of the spin part. (d) Neither Coulomb, nor exchange. Other small integral. (e) Exchange integral, equal to c. (f) Coulomb integral, medium.

Problem 1.3 $E^{(1)} = \int\limits_{-\infty}^{\infty} \psi^{(0)}\hat{V}\psi^{(0)} = \int\limits_{-\infty}^{0} \psi^{(0)}\hat{V}\psi^{(0)}dx + \int\limits_{0}^{a} \psi^{(0)}\hat{V}\psi^{(0)}dx +$
$\int\limits_{a}^{b} \psi^{(0)}\hat{V}\psi^{(0)}dx + \int\limits_{b}^{L} \psi^{(0)}\hat{V}\psi^{(0)}dx + \int\limits_{L}^{\infty} \psi^{(0)}\hat{V}\psi^{(0)}dx$, with $a = \frac{1}{2}L - \frac{1}{2}\gamma$ and $b = \frac{1}{2}L + \frac{1}{2}\gamma$. The first and last integrals are zero because $\psi^{(0)}$ is zero outside the box ($V = \infty$), the second and fourth integral are zero because $V = 0$ in these intervals. Remains the third integral. With $\hat{V} = V_0$, the correction for the ground state ($n = 1$) reads $E_0^{(1)} = \frac{2V_0}{L}\int\limits_{a}^{b} \sin^2\frac{\pi x}{L}dx$. Making use of the assumption that $\psi^{(0)}$ is constant in this interval, the integrand reduces to $\sin^2\frac{\pi L}{2L} = 1$ and the integral equals γ. Then, $E_0^{(1)} = \frac{2V_0\gamma}{L}$. For the first excited state ($n = 2$), the integral is equal zero ($\psi_1^{(0)} = 0$ for $x = \frac{1}{2}L$), and hence, $E_1^{(1)} = 0$. The second excited state ($n = 3$) has the same correction as the ground state.

Problem 1.4 (a) $\langle\Phi_i|\Phi_j\rangle = \delta_{ij} \Rightarrow N_k$ (the norm of the projections on the model space) $= \sum_i c_i^2(k)$ with $i = 2, 3, 4$. $N_1 = 0.769$, $N_2 = 0.277$, $N_3 = 0.928$, $N_4 = 0.784$, $N_5 = 0.242$. (b) Ψ_1, Ψ_3 and Ψ_4 have to be used to construct $\hat{H}^{eff}$. Normalized projections $\tilde{\Psi}_k = \sum_i \tilde{c}_i(k)$ with $\tilde{c}(1) = \{0.3651, 0.1826, 0.9129\}$, $\tilde{c}(3) = \{-0.1444, 0.9828, -0.1151\}$, $\tilde{c}(4) = \{-0.8732, 0.0493, 0.4849\}$. Orthogonalization of $\tilde{\Psi}_3$ by $\tilde{c}_i'(3) = \tilde{c}_i(3) - \langle\tilde{\Psi}_1|\tilde{\Psi}_3\rangle\tilde{c}_i(3)$ and subsequent normalization gives $\tilde{c}_i'(3) = \{-0.1523, 0.9791, -0.1349\}$. Orthogonalization of $\tilde{\Psi}_4$ by $\tilde{c}_i'(4) = \tilde{c}_i'(4) - \langle\tilde{\Psi}_1|\tilde{\Psi}_4\rangle\tilde{c}_i(1) - \langle\tilde{\Psi}_3|\tilde{\Psi}_4\rangle\tilde{c}_i(3)$. After normalization, we get $\tilde{c}_i^{\perp}(4) =$

$\{-0.9184, -0.0898, 0.3853\}$. (c) $\langle \Phi_k | \hat{H}^{eff} | \Phi_l \rangle$ $=$ $\sum_i \tilde{c}_i^\perp(k) \tilde{c}_i^\perp(l) E_i$

$$\hat{H}^{eff} = \begin{array}{ccc} \Phi_2 & \Phi_3 & \Phi_4 \\ -0.379595 & & \\ -0.003368 & -0.403011 & \\ -0.047489 & -0.018051 & -0.477394 \end{array}$$

Diagonalization gives $-0.5, -0.4$ and -0.36 as eigenvalues. (d) $\mu = \langle \Phi_1 | \hat{H}^{eff} | \Phi_2 \rangle = -0.003368$, $\gamma = \langle \Phi_1 | \hat{H}^{eff} | \Phi_3 \rangle = -0.047489$. $(\gamma - 4\mu)/2 = -0.017009 \approx \langle \Phi_2 | \hat{H}^{eff} | \Phi_3 \rangle$.

Exercises and Problems of Chap. 2

Exercise 2.1 $R = 13.6056925$ eV

H-$2p^1$	Ca^{19+}$-3p^1$	Ca^{19+}$-3d^1$	U^{91+}$-2p^1$	U^{91+}$-6d^1$	U^{91+}$-5f^1$
0.03 meV	1.43 eV	0.29 eV	2163.81 eV	16.03 eV	9.89 eV

Exercise 2.2 The ground state has $S = \frac{3}{2}$ (Hund's rule) and the electrons in the three different p orbitals (p_{-1}, p_0, p_1), hence $L = 0$. $J = L + S = \frac{3}{2}$. The term symbol is $^4S_{\frac{3}{2}}$.

Exercise 2.3 $p_y = (+i/\sqrt{2})(Y_{1,1} + Y_{1,-1})$; $\hat{l}_z p_y = (+i/\sqrt{2})(Y_{1,1} - Y_{1,-1}) = -ip_x$. From this follows that $\langle p_x | \hat{l}_z | p_y \rangle = -i$. $\hat{l}_z p_x = ip_y \Rightarrow \langle p_y | \hat{l}_z | p_x \rangle = i = -\langle p_x | \hat{l}_z | p_y \rangle$.

Exercise 2.4 *tetrahedral*: ground state $^3A_2$. The direct product $A_2 \times A_2$ does not contain the irreducible representation that describes the transformation of the rotation operator, hence no orbital momentum is expected. *octahedral*: ground state $^3T_{1g}$, the rotation operator transforms as T_1, which is contained in the $T_{1g} \times T_{1g}$ product, hence non-zero orbital momentum. C_{2v}: $^1A_1$, the direct product (A_1) does not contain the irrep of the rotation operator, no orbital momentum.

Exercise 2.5

$$\left(\hat{S}_x \ \hat{S}_y \ \hat{S}_z \right) \begin{pmatrix} D_{xx} & 0 & 0 \\ 0 & D_{yy} & 0 \\ 0 & 0 & D_{zz} \end{pmatrix} \begin{pmatrix} \hat{S}_x \\ \hat{S}_y \\ \hat{S}_z \end{pmatrix} = D_{xx}\hat{S}_x^2 + D_{yy} + \hat{S}_y^2 D_{zz}\hat{S}_z^2$$

Trace of the matrix: $\lambda = D_{xx} + D_{yy} + D_{zz}$. Then the traceless tensor becomes

$$\overline{\overline{\Delta}} = \begin{pmatrix} D_{xx} - \frac{1}{3}\lambda & 0 & 0 \\ 0 & D_{yy} - \frac{1}{3}\lambda & 0 \\ 0 & 0 & D_{zz} - \frac{1}{3}\lambda \end{pmatrix}$$

$\hat{S}\overline{\overline{\Delta}}\hat{S} = D_{xx}\hat{S}_x^2 + D_{yy}\hat{S}_y^2 + D_{zz}\hat{S}_z^2 - \frac{1}{3}\lambda\hat{S}^2 = D_{xx}\hat{S}_x^2 + D_{yy}\hat{S}_y^2 + D_{zz}\hat{S}_z^2 - \frac{1}{3}D_{xx}\hat{S}_x^2 -$
$\frac{1}{3}D_{xx}\hat{S}_y^2 - \frac{1}{3}D_{xx}\hat{S}_z^2 - \frac{1}{3}D_{yy}\hat{S}_x^2 - \frac{1}{3}D_{yy}\hat{S}_y^2 - \frac{1}{3}D_{yy}\hat{S}_z^2 - \frac{1}{3}D_{zz}\hat{S}_x^2 - \frac{1}{3}D_{zz}\hat{S}_y^2 - \frac{1}{3}D_{zz}\hat{S}_z^2 =$
$\frac{2}{3}D_{xx}\hat{S}_x^2 - \frac{1}{3}D_{xx}\hat{S}_y^2 - \frac{1}{3}D_{xx}\hat{S}_z^2 - \frac{1}{3}D_{yy}\hat{S}_x^2 + \frac{2}{3}D_{yy}\hat{S}_y^2 - \frac{1}{3}D_{yy}\hat{S}_z^2 - \frac{1}{3}D_{zz}\hat{S}_x^2 - \frac{1}{3}D_{zz}\hat{S}_y^2 +$
$\frac{2}{3}D_{zz}\hat{S}_z^2$ to be compared with $D(\hat{S}_z^2 - \frac{1}{3}\hat{S}^2) + E(\hat{S}_x^2 - \hat{S}_y^2) = D\hat{S}_z^2 - \frac{1}{3}D\hat{S}_x^2 - \frac{1}{3}D\hat{S}_y^2 -$
$\frac{1}{3}D\hat{S}_z^2 + E\hat{S}_x^2 - E\hat{S}_y^2$. After substituting the definitions of D and E: $D_{zz}\hat{S}_z^2 - \frac{1}{2}D_{xx}\hat{S}_z^2 -$
$\frac{1}{2}D_{xx}\hat{S}_z^2 - \frac{1}{3}D_{zz}\hat{S}_x^2 - \frac{1}{3}D_{zz}\hat{S}_y^2 - \frac{1}{3}D_{zz}\hat{S}_z^2 + \frac{1}{6}D_{xx}\hat{S}_x^2 + \frac{1}{6}D_{xx}\hat{S}_y^2 + \frac{1}{6}D_{xx}\hat{S}_z^2 + \frac{1}{6}D_{yy}\hat{S}_x^2 +$
$\frac{1}{6}D_{yy}\hat{S}_y^2 + \frac{1}{6}D_{yy}\hat{S}_z^2 + \frac{1}{2}D_{xx}\hat{S}_x^2 - \frac{1}{2}D_{yy}\hat{S}_x^2 - \frac{1}{2}D_{xx}\hat{S}_y^2 + \frac{1}{2}D_{yy}\hat{S}_y^2 = \frac{2}{3}D_{xx}\hat{S}_x^2 - \frac{1}{3}D_{xx}\hat{S}_y^2 -$
$\frac{1}{3}D_{xx}\hat{S}_z^2 - \frac{1}{3}D_{yy}\hat{S}_x^2 + \frac{2}{3}D_{yy}\hat{S}_y^2 - \frac{1}{3}D_{yy}\hat{S}_z^2 - \frac{1}{3}D_{zz}\hat{S}_x^2 - \frac{1}{3}D_{zz}\hat{S}_y^2 + \frac{2}{3}D_{zz}\hat{S}_z^2$, equal to what
is obtained from $\hat{S}\overline{\overline{\Delta}}\hat{S}$.

Exercise 2.6 $\hat{S}_z^2|\frac{1}{2}, \frac{1}{2}\rangle = \frac{1}{4}|\frac{1}{2}, \frac{1}{2}\rangle; \hat{S}_z^2|\frac{1}{2}, -\frac{1}{2}\rangle = \frac{1}{4}|\frac{1}{2}, -\frac{1}{2}\rangle \Rightarrow D(\hat{S}_z^2 - \frac{1}{3}\hat{S}^2)|\frac{1}{2}, \pm\frac{1}{2}\rangle$
$= D[\frac{1}{4}|\frac{1}{2}, \pm\frac{1}{2}\rangle - \frac{1}{3}(\frac{1}{2}(\frac{1}{2} + 1))|\frac{1}{2}, \pm\frac{1}{2}\rangle] = D(\frac{1}{4} - \frac{1}{4})|\frac{1}{2}, \pm\frac{1}{2}\rangle = 0; \hat{S}_x^2|\frac{1}{2}, \pm\frac{1}{2}\rangle =$
$\frac{1}{4}(\hat{S}^+\hat{S}^+ + \hat{S}^-\hat{S}^- + \hat{S}^+\hat{S}^- + \hat{S}^-\hat{S}^+)|\frac{1}{2}, \pm\frac{1}{2}\rangle = |\frac{1}{2}, \pm\frac{1}{2}\rangle, \hat{S}_y^2|\frac{1}{2}, \pm\frac{1}{2}\rangle = -\frac{1}{4}(\hat{S}^+\hat{S}^+ +$
$\hat{S}^-\hat{S}^- - \hat{S}^+\hat{S}^- - \hat{S}^-\hat{S}^+) = |\frac{1}{2}, \pm\frac{1}{2}\rangle \Rightarrow E(\hat{S}_x^2 - \hat{S}_y^2)|\frac{1}{2}, \pm\frac{1}{2}\rangle = E(\frac{1}{4} - \frac{1}{4})|\frac{1}{2}, \pm\frac{1}{2}\rangle = 0.$
This shows that both diagonal and off-diagonal matrix elements are zero. This means
that the ZFS model Hamiltonian cannot remove the degeneracy.

Exercise 2.7 (a) Taking $E^{(0)}$ as zero of energy, the exponent in the denominator
becomes 1, limiting the sum over the $2S + 1$ M_S-sublevels of the ground state, the
denominator simplifies to $2S + 1$. Quintet: $\sum_{M_S} M_S^2 = [(-2)^2 + (-1)^2 + 0 +$
$1^2 + 2^2] = 10; S(S + 1)(2S + 1)/3 = 2(2 + 1)(2 \cdot 2 + 1)/3 = 10.$ (b) $C =$
$N_A(\mu_B g_e)^2 S(S+1)/3k$ with $N_A \mu_B^2/3k \approx 1/8 \Rightarrow C = g_e^2 S(S+1)/8$. Taking $g_e = 2$,
$C = S(S + 1)/2; 3/8 (0.375), 1, 15/8 (1.875), 3, 35/8 (4.375), 6, 63/8 (7.875).$

Exercise 2.8 Equation 2.44 gives $\langle S_z \rangle = (-S(S+1)/3k_BT) \cdot (\mu_B g_e H - nJ\langle S_z \rangle) \Rightarrow$
$\langle S_z \rangle - (-S(S+1)/3k_BT)nJ\langle S_z \rangle = (-S(S+1)/3k_BT)\mu_B g_e H \Rightarrow \langle S_z \rangle(3k_BT - S(S+$
$1)nJ) = -S(S + 1)\mu_B g_e H$, which directly leads to Eq. 2.45.

Exercise 2.9 First for α: $\langle \psi_i^{(0)}|\hat{L} \cdot \hat{S}|\psi_0^{(0)}\rangle = \langle \psi_i^{(0)}|\hat{L}_z \cdot \hat{S}_z|\psi_0^{(0)}\rangle + \frac{1}{2}\langle \psi_i^{(0)}|\hat{L}^+ \cdot \hat{S}^-|$
$\psi_0^{(0)}\rangle + \frac{1}{2}\langle \psi_i^{(0)}|\hat{L}^+ \cdot \hat{S}^-|\psi_0^{(0)}\rangle = \langle \psi_i^{(0)}|\hat{L}_z|\psi_0^{(0)}\rangle\langle\alpha|\hat{S}_z|\alpha\rangle + \frac{1}{2}\langle \psi_i^{(0)}|\hat{L}^+|\psi_0^{(0)}\rangle$
$\langle\alpha|\hat{S}^-|\alpha\rangle + \frac{1}{2}\langle \psi_i^{(0)}|\hat{L}^-|\psi_0^{(0)}\rangle\langle\alpha|\hat{S}^+|\alpha\rangle = \frac{1}{2}\langle \psi_i^{(0)}|\hat{L}_z|\psi_0^{(0)}\rangle$. Now for β: $\langle \overline{\psi}_i^{(0)}|\hat{L} \cdot \hat{S}|$
$\psi_0^{(0)}\rangle = \langle \overline{\psi}_i^{(0)}|\hat{L}_z \cdot \hat{S}_z|\psi_0^{(0)}\rangle + \frac{1}{2}\langle \overline{\psi}_i^{(0)}|\hat{L}^+ \cdot \hat{S}^-|\psi_0^{(0)}\rangle + \frac{1}{2}\langle \overline{\psi}_i^{(0)}|\hat{L}^+ \cdot \hat{S}^-|\psi_0^{(0)}\rangle =$
$\langle \psi_i^{(0)}|\hat{L}_z|\psi_0^{(0)}\rangle\langle\beta|\hat{S}_z|\alpha\rangle + \frac{1}{2}\langle \psi_i^{(0)}|\hat{L}^+|\psi_0^{(0)}\rangle\langle\beta|\hat{S}^-|\alpha\rangle + \frac{1}{2}\langle \psi_i^{(0)}|\hat{L}^-|\psi_0^{(0)}\rangle\langle\beta|\hat{S}^+|\alpha\rangle =$
$\frac{1}{2}\langle \psi_i^{(0)}|\hat{L}^+|\psi_0^{(0)}\rangle$.

Exercise 2.10 $\hat{l}^+|l, m_l\rangle = \sqrt{l(l + 1) - m_l(m_l + 1)}|l, m_l + 1\rangle; \hat{l}^+p_0 = \sqrt{2}p_+$ and
$\hat{l}^-p_0 = \sqrt{2}p_-$ From this $\hat{l}_x|p_0\rangle = \frac{1}{2}\sqrt{2}(p_+ + p_-)$. Then $\langle p_-|\hat{l}_x|p_0\rangle = \frac{\sqrt{2}}{2}$ and

$\langle p_+|\hat{l}_x|p_0\rangle = \frac{\sqrt{2}}{2}\cdot\hat{l}_x p_+ = \frac{1}{2}(\hat{l}^+ + \hat{l}^-)p_+ = \frac{1}{2}(0 + \sqrt{2}p_0) = \frac{\sqrt{2}}{2}p_0$ and analogous for $\hat{l}_x p_- = \frac{\sqrt{2}}{2}p_0$. Then $\langle p_0|\hat{l}_x|p_+\rangle = \langle p_0|\hat{l}_x|p_-\rangle = \frac{\sqrt{2}}{2}$.

Exercise 2.11 The eigenvalues of the matrix $\begin{pmatrix} a & b & 0 \\ b & -2a & b \\ 0 & b & a \end{pmatrix}$ are $E_1 = a$ and $E_{2,3} = -\frac{1}{2}a \pm \sqrt{9a^2 + 8b^2}/2$. Shifting the diagonal elements by +2a, the matrix becomes $\begin{pmatrix} 3a & b & 0 \\ b & 0 & b \\ 0 & b & 3a \end{pmatrix}$ with eigenvalues $E'_1 = 3a$ and $E'_{2,3} = (3a/2) \pm \sqrt{9a^2 + 8b^2}/2$. The constant difference of 2a between E_i and E'_i is the same quantity by which the diagonal matrix elements have been shifted.

Problem 2.1 (1) see Appendix C. (2) $\langle\widetilde{\Psi}'_1|\widetilde{\Psi}'_1\rangle = 0.97211913$; $\langle\widetilde{\Psi}'_1|\widetilde{\Psi}'_2\rangle = 0.00000457 + 0.00001946i = \langle\widetilde{\Psi}'_2|\widetilde{\Psi}'_1\rangle^*$; $\langle\widetilde{\Psi}'_1|\widetilde{\Psi}'_3\rangle = 0.00004069 - 0.00000414i = \langle\widetilde{\Psi}'_3|\widetilde{\Psi}'_1\rangle^*$; $\langle\widetilde{\Psi}'_2|\widetilde{\Psi}'_2\rangle = 0.97480639$; $\langle\widetilde{\Psi}'_2|\widetilde{\Psi}'_3\rangle = 0.00010103 - 0.00078090i = \langle\widetilde{\Psi}'_3|\widetilde{\Psi}'_2\rangle^*$; $\langle\widetilde{\Psi}'_2|\widetilde{\Psi}'_2\rangle = 0.98401064$. (3) $\langle 1, 1|\hat{H}^{\mathit{eff}}|1, 1\rangle = 6.5359$; $\langle 1, 1|\hat{H}^{\mathit{eff}}|1, 0\rangle = \langle 1, 0|\hat{H}^{\mathit{eff}}|1, 1\rangle^* = -4.6142 - 2.2944i$; $\langle 1, 1|\hat{H}^{\mathit{eff}}|1, -1\rangle = \langle 1, -1|\hat{H}^{\mathit{eff}}|1, 1\rangle^* = 5.3478 - 0.7801i$; $\langle 1, 0|\hat{H}^{\mathit{eff}}|1, 0\rangle = 35.9344$; $\langle 1, 0|\hat{H}^{\mathit{eff}}|1, -1\rangle = \langle 1, -1|\hat{H}^{\mathit{eff}}|1, 0\rangle^* = 4.6142 + 2.2944i$; $\langle 1, -1|\hat{H}^{\mathit{eff}}|1, -1\rangle = 6.5359$. (4) $D_{11} = 23.315016$; $D_{12} = D_{21} = 0.780050$; $D_{13} = D_{31} = -6.525412$; $D_{22} = 12.619358$; $D_{23} = D_{32} = 3.244771$; $D_{33} = -11.431250$. After diagonalization: $D_{11} = 24.503124$; $D_{22} = 13.048843$; $D_{33} = -13.048843$. Using Eq. 2.16, $D = -31.8$ cm^{-1} and $E = 5.8$ cm^{-1}. Using the energies (Eq. 2.22): $D = \frac{1}{2}(11.54 + 0) - 37.55 = -31.8$ cm^{-1} and $E = \frac{1}{2}(11.54 - 0) = 5.8$ cm^{-1}.

Problem 2.2 (1) see inset Fig. 2.1 (2) Since there is only one energy difference, only one effective anisotropy parameter can be derived, under the assumption $E = 0$, this parameter can be considered to be D. Unless the orientation of the magnetic axes frame is known, the sign of D cannot be determined. (3) see Appendix C. (4) $\langle\frac{3}{2}, \frac{3}{2}|\hat{H}^{\mathit{eff}}|\frac{3}{2}, \frac{3}{2}\rangle = \langle\frac{3}{2}, -\frac{3}{2}|\hat{H}^{\mathit{eff}}|\frac{3}{2}, -\frac{3}{2}\rangle = 3.634038$; $\langle\frac{3}{2}, \frac{3}{2}|\hat{H}^{\mathit{eff}}|\frac{3}{2}, \frac{1}{2}\rangle = \langle\frac{3}{2}, \frac{1}{2}|\hat{H}^{\mathit{eff}}|\frac{3}{2}, \frac{3}{2}\rangle^* = 0.173864 + 9.819256i$; $\langle\frac{3}{2}, \frac{3}{2}|\hat{H}^{\mathit{eff}}|\frac{3}{2}, -\frac{1}{2}\rangle = \langle\frac{3}{2}, -\frac{1}{2}|\hat{H}^{\mathit{eff}}|\frac{3}{2}, \frac{3}{2}\rangle^* = 2.856287 + 0.412321i$; $\langle\frac{3}{2}, \frac{3}{2}|\hat{H}^{\mathit{eff}}|\frac{3}{2}, -\frac{3}{2}\rangle = \langle\frac{3}{2}, -\frac{3}{2}|\hat{H}^{\mathit{eff}}|\frac{3}{2}, \frac{3}{2}\rangle = 0$; $\langle\frac{3}{2}, \frac{1}{2}|\hat{H}^{\mathit{eff}}|\frac{3}{2}, \frac{1}{2}\rangle = \langle\frac{3}{2}, -\frac{1}{2}|\hat{H}^{\mathit{eff}}|\frac{3}{2}, -\frac{1}{2}\rangle = 28.831949$; $\langle\frac{3}{2}, \frac{1}{2}|\hat{H}^{\mathit{eff}}|\frac{3}{2}, -\frac{1}{2}\rangle = \langle\frac{3}{2}, -\frac{1}{2}|\hat{H}^{\mathit{eff}}|\frac{3}{2}, \frac{1}{2}\rangle = 0$; $\langle\frac{3}{2}, \frac{1}{2}|\hat{H}^{\mathit{eff}}|\frac{3}{2}, -\frac{3}{2}\rangle = \langle\frac{3}{2}, -\frac{3}{2}|\hat{H}^{\mathit{eff}}|\frac{3}{2}, \frac{1}{2}\rangle^* = 2.856287 + 0.412321i$; $\langle\frac{3}{2}, -\frac{1}{2}|\hat{H}^{\mathit{eff}}|\frac{3}{2}, -\frac{3}{2}\rangle = \langle\frac{3}{2}, -\frac{3}{2}|\hat{H}^{\mathit{eff}}|\frac{3}{2}, -\frac{1}{2}\rangle^* = -0.173864 - 9.819255i$. $D_{11} = 10.177529$; $D_{12} = D_{21} = -0.238054$; $D_{13} = D_{31} = 0.100381$; $D_{22} = 6.879372$; $D_{23} = D_{32} = -5.669150$; $D_{33} = -4.070506$. Diagonalization leads to $D_{11} = 10.246963$; $D_{22} = 9.216256$; $D_{33} = -6.476824$. From this $D = -16.21$ cm^{-1} and $E = 0.51$ cm^{-1}. Negative D indicates that the wave function of the lowest level is dominated by $M_S = \pm\frac{3}{2}$ contributions (see $\widetilde{\Psi}'_{1,2}$), hence, the molecule exhibits easy-axis magnetism.

Problem 2.3 (1) Ti^{III}: $[1s^2\ 2s^2\ 2p^6\ 3s^2\ 3p^6\ 3d^1]$, there is no ZFS for $S = \frac{1}{2}$. (2) In the first place, the projected wave functions have to be expressed in terms of the $3d$ orbitals by substituting the expressions of ϕ_i in the multideterminantal wave functions $\widetilde{\Psi}'_i$. This gives

	$\widetilde{\Psi}'_1$	$\widetilde{\Psi}'_2$	$\widetilde{\Psi}'_3$	$\widetilde{\Psi}'_4$	$\widetilde{\Psi}'_5$
$3d_{z^2}$	0.0024	0.0166	−0.3142	0.1158	−0.9433
$3d_{x^2-y^2}$	0.2569	0.1497	−0.2999	0.8799	0.2101
$3d_{xy}$	0.2249	−0.2178	−0.8409	−0.3731	0.2300
$3d_{yz}$	0.3072	−0.9071	0.2313	0.1587	−0.0679
$3d_{xz}$	−0.8883	−0.3271	−0.2250	0.2190	0.0923

The next step (cf. Eqs. 2.55 and 2.59) is the calculation of $\langle\widetilde{\Psi}'_i|\hat{L}_z|\widetilde{\Psi}'_1\rangle$ and $\langle\widetilde{\Psi}'_1|\hat{L}_z|\widetilde{\Psi}'_i\rangle$ with $i = 2, 3, 4, 5$, $i = 1$ is the ground state. The same has to be done for $\hat{L}_x$.

	$\widetilde{\Psi}'_2$	$\widetilde{\Psi}'_3$	$\widetilde{\Psi}'_4$	$\widetilde{\Psi}'_5$
$\langle i|\hat{L}_z|1\rangle$	0.7270184	−0.4334652	−0.7957402	0.0556362
$\langle 1|\hat{L}_z|i\rangle$	−0.7270184	0.4334652	0.7957402	−0.0556362
$\langle i|\hat{L}_x|1\rangle$	0.3192372	0.5594345	0.1304799	−0.7317820
$\langle 1|\hat{L}_x|i\rangle$	−0.3192372	−0.5594345	−0.1304799	0.7317820

From this, $g_{zz} = g_e − 0.150$ and $g_{xx} = g_e + 0.088$. The deviation in z is significantly larger than in x confirming the axial anisotropy.

Exercises and Problems of Chap. 3

Exercise 3.1 $\langle\Phi^{11}(0,0)|\hat{H}|\Phi^{22}(0,0)\rangle = \langle\phi_1\bar{\phi}_1|\hat{H}|\phi_2\bar{\phi}_2\rangle = \langle\phi_1\bar{\phi}_1|1/r_{12}|\phi_2\bar{\phi}_2\rangle - \langle\phi_1\bar{\phi}_1|1/r_{12}|\bar{\phi}_2\phi_2\rangle = \langle\phi_1\phi_1|1/r_{12}|\phi_2\phi_2\rangle - 0 = K_{12}$

Exercise 3.2 $\psi_a = (\phi_a + \phi_b)/\sqrt{2} = (1/\sqrt{2})((\chi_a + \chi_b)/\sqrt{2(1+S)} + (\chi_a - \chi_b)/\sqrt{2(1-S)}) = (1/\sqrt{2})(\chi_a[1/\sqrt{2(1+S)}+1/\sqrt{2(1-S)}]+\chi_b[1/\sqrt{2(1+S)}-1/\sqrt{2(1-S)}])$. For $S = 0.2$, the coefficient of χ_b is -0.1026 and for $S = 0.003$, the coefficient reduces to -0.0015.

Exercise 3.3 $(|\psi_a\bar{\psi}_b| + |\psi_b\bar{\psi}_a|) \overset{?}{=} (|\phi_1\bar{\phi}_1| - |\phi_2\bar{\phi}_2|)$: Substitute $\psi_{a,b} = (\phi_1 \pm \phi_2)/\sqrt{2}$: $\frac{1}{2}\{|(\phi_1 + \phi_2)(\bar{\phi}_1 - \bar{\phi}_2)| + |(\phi_1 - \phi_2)(\bar{\phi}_1 + \bar{\phi}_2)|\} = \frac{1}{2}\{|\phi_1\bar{\phi}_1| - |\phi_1\bar{\phi}_2| + |\phi_2\bar{\phi}_1| - |\phi_2\bar{\phi}_2| + |\phi_1\bar{\phi}_1| + |\phi_1\bar{\phi}_2| - |\phi_2\bar{\phi}_1| - |\phi_2\bar{\phi}_2|\} = (|\phi_1\bar{\phi}_1| - |\phi_2\bar{\phi}_2|)$. $(|\psi_a\bar{\psi}_b| - |\psi_b\bar{\psi}_a|) \overset{?}{=} (|\phi_1\bar{\phi}_2| - |\phi_2\bar{\phi}_1|)$: Substitute $\psi_{a,b} = (\phi_1 \pm \phi_2)/\sqrt{2}$: $\frac{1}{2}\{|(\phi_1 + \phi_2)(\bar{\phi}_1 - \bar{\phi}_2)| - |(\phi_1 - \phi_2)(\bar{\phi}_1 + \bar{\phi}_2)|\} = \frac{1}{2}\{|\phi_1\bar{\phi}_1| - |\phi_2\bar{\phi}_2| - |\phi_1\bar{\phi}_2| + |\phi_2\bar{\phi}_1| - |\phi_1\bar{\phi}_1| + |\phi_2\bar{\phi}_2| - |\phi_1\bar{\phi}_2| + |\phi_2\bar{\phi}_1|\} = -(|\phi_1\bar{\phi}_2| - |\phi_2\bar{\phi}_1|)$ (the sign is not relevant)

Exercise 3.4 Using a and b as shorthand notation for ψ_a and ψ_b, $\langle\Psi^{\mathrm{cov}}(0,0)|\hat{H}|\Psi^{\mathrm{cov}}(0,0)\rangle = \frac{1}{2}(\langle a\bar{b}|\hat{H}|a\bar{b}\rangle + \langle a\bar{b}|\hat{H}|b\bar{a}\rangle + \langle b\bar{a}|\hat{H}|a\bar{b}\rangle + \langle b\bar{a}|\hat{H}|b\bar{a}\rangle) =$ $\frac{1}{2}(2\langle a|\hat{h}|a\rangle + 2\langle b|\hat{h}|b\rangle + \langle a\bar{b}|\frac{1-\hat{P}_{12}}{r_{12}}|a\bar{b}\rangle + \langle a\bar{b}|\frac{1-\hat{P}_{12}}{r_{12}}|b\bar{a}\rangle + \langle b\bar{a}|\frac{1-\hat{P}_{12}}{r_{12}}|a\bar{b}\rangle +$ $\langle b\bar{a}|\frac{1-\hat{P}_{12}}{r_{12}}|b\bar{a}\rangle) = h_{aa} + h_{bb} + J_{ab} + K_{ab}; \langle\Psi^{\mathrm{cov}}(1,0)|\hat{H}|\Psi^{\mathrm{cov}}(1,0)\rangle = \frac{1}{2}(\langle a\bar{b}|\hat{H}|a\bar{b}\rangle -$ $\langle a\bar{b}|\hat{H}|b\bar{a}\rangle - \langle b\bar{a}|\hat{H}|a\bar{b}\rangle + \langle b\bar{a}|\hat{H}|b\bar{a}\rangle) = \frac{1}{2}(2\langle a|\hat{h}|a\rangle + 2\langle b|\hat{h}|b\rangle + \langle a\bar{b}|\frac{1-\hat{P}_{12}}{r_{12}}|a\bar{b}\rangle -$ $\langle a\bar{b}|\frac{1-\hat{P}_{12}}{r_{12}}|b\bar{a}\rangle - \langle b\bar{a}|\frac{1-\hat{P}_{12}}{r_{12}}|a\bar{b}\rangle + \langle b\bar{a}|\frac{1-\hat{P}_{12}}{r_{12}}|b\bar{a}\rangle) = h_{aa} + h_{bb} + J_{ab} - K_{ab}$. The energy difference is $2K_{ab}$.

Exercise 3.5 $|\phi_a\bar{\phi}_b| = \frac{\psi_a + \nu\psi_b}{\sqrt{1+\nu^2}} \cdot \frac{\overline{\psi}_b + \nu\overline{\psi}_a}{\sqrt{1+\nu^2}} = (\psi_a\overline{\psi}_b + \nu\psi_b\overline{\psi}_b + \nu\psi_a\overline{\psi}_a + \nu^2\psi_b\overline{\psi}_a)/$ $(1+\nu^2); |\phi_b\bar{\phi}_a| = \frac{\psi_b + \nu\psi_a}{\sqrt{1+\nu^2}} \cdot \frac{\overline{\psi}_a + \nu\overline{\psi}_b}{\sqrt{1+\nu^2}} = (\psi_b\overline{\psi}_a + \nu\psi_b\overline{\psi}_b + \nu\psi_a\overline{\psi}_a + \nu^2\psi_a\overline{\psi}_b)/(1+$ $\nu^2). |\phi_a\bar{\phi}_b| + |\phi_b\bar{\phi}_a| = \frac{1}{1+\nu^2}((1+\nu^2)(\psi_a\overline{\psi}_b + \psi_b\overline{\psi}_a) + 2\nu(\psi_a\overline{\psi}_a + \psi_b\overline{\psi}_b)) =$ $\psi_a\overline{\psi}_b + \psi_b\overline{\psi}_a + S_{ab}(\psi_a\overline{\psi}_a + \psi_b\overline{\psi}_b)$. Multiplying with $1/\sqrt{2+2S^2}$ leads to $(\psi_a\overline{\psi}_b + \psi_b\overline{\psi}_a)/\sqrt{2}$ for $S_{ab} = 0$.

Exercise 3.6 Maximum spin $S_{max} = S_1 + S_2 = 2S; E(S_{max}) = -\frac{1}{2}J(2S(2S+1) - 2S(S+1)) = -JS^2$; Minimum spin $S_{min} = S_1 - S_2 = 0; E(S_{min}) = -\frac{1}{2}J(0 - 2S(S+1)) = JS(S+1)$.

Exercise 3.7

3	21	1	6	3	11	18	3	4
$7/2$	28	1	7	3	13	$49/2$	$7/2$	4

Exercise 3.8 $S_{min} \leqslant S \leqslant S_{max}, S = 0, 1 \Rightarrow 1(1+1)(2 \cdot 1 + 1)\exp(J \cdot 1(1 + 1)/2kT)/(1 + 3\exp(J \cdot 1(1+1)/2kT)) = (6\exp(J/kT))(1 + 3\exp(J/kT)) = 6/(3 + \exp(-J/kT))$. Multiplying with $N_A\mu_B^2 g_e^2/3kT$ one arrives at $2N_A\mu_B^2 g_e^2/kT$ $(3 + \exp(-J/kT))$.

Exercise 3.9 $(\hat{s}_1^+ + \hat{s}_2^+)(\hat{s}_3^- + \hat{s}_4^-)\alpha\alpha\beta\beta = (\hat{s}_1^+ + \hat{s}_2^+)(0+0) = 0; (\hat{s}_1^+ + \hat{s}_2^+)(\hat{s}_3^- + \hat{s}_4^-)\beta\beta\alpha\alpha = (\hat{s}_1^+ + \hat{s}_2^+)(\beta\beta\beta\alpha + \beta\beta\alpha\beta) = \alpha\beta\beta\alpha + \alpha\beta\alpha\beta + \beta\alpha\beta\alpha + \beta\alpha\alpha\beta; (\hat{s}_1^+ + \hat{s}_2^+)(\hat{s}_3^- + \hat{s}_4^-)\alpha\beta\alpha\beta = (\hat{s}_1^+ + \hat{s}_2^+)(\alpha\beta\beta\beta + 0) = \alpha\alpha\beta\beta; (\hat{s}_1^+ + \hat{s}_2^+)(\hat{s}_3^- + \hat{s}_4^-)\alpha\beta\beta\alpha = (\hat{s}_1^+ + \hat{s}_2^+)(0 + \alpha\beta\beta\beta) = \alpha\alpha\beta\beta; (\hat{s}_1^+ + \hat{s}_2^+)(\hat{s}_3^- + \hat{s}_4^-)\beta\alpha\beta\alpha = (\hat{s}_1^+ + \hat{s}_2^+)(0 + \beta\alpha\beta\beta) = \alpha\alpha\beta\beta; (\hat{s}_1^+ + \hat{s}_2^+)(\hat{s}_3^- + \hat{s}_4^-)\beta\alpha\alpha\beta = (\hat{s}_1^+ + \hat{s}_2^+)(\beta\alpha\beta\beta + 0) = \alpha\alpha\beta\beta; (\hat{s}_1^- + \hat{s}_2^-)(\hat{s}_3^+ + \hat{s}_4^+)\alpha\alpha\beta\beta = (\hat{s}_1^- + \hat{s}_2^-)(\alpha\alpha\alpha\beta + \alpha\alpha\beta\alpha) = \beta\alpha\alpha\beta + \beta\alpha\beta\alpha + \alpha\beta\alpha\beta + \alpha\beta\beta\alpha; (\hat{s}_1^- + \hat{s}_2^-)(\hat{s}_3^+ + \hat{s}_4^+)\beta\beta\alpha\alpha = 0; (\hat{s}_1^- + \hat{s}_2^-)(\hat{s}_3^+ + \hat{s}_4^+)\alpha\beta\alpha\beta = (\hat{s}_1^- + \hat{s}_2^-)(0 + \alpha\beta\alpha\alpha) = \beta\beta\alpha\alpha; (\hat{s}_1^- + \hat{s}_2^-)(\hat{s}_3^+ + \hat{s}_4^+)\alpha\beta\beta\alpha = (\hat{s}_1^- + \hat{s}_2^-)(\alpha\beta\alpha\alpha + 0) = \beta\beta\alpha\alpha; (\hat{s}_1^- + \hat{s}_2^-)(\hat{s}_3^+ + \hat{s}_4^+)\beta\alpha\beta\alpha = (\hat{s}_1^- + \hat{s}_2^-)(\beta\alpha\alpha\alpha + 0) = \beta\beta\alpha\alpha; (\hat{s}_1^- + \hat{s}_2^-)(\hat{s}_3^+ + \hat{s}_4^+)\beta\alpha\alpha\beta = (\hat{s}_1^- + \hat{s}_2^-)(0 + \beta\alpha\alpha\alpha) = \beta\beta\alpha\alpha; (\hat{s}_{z,1} + \hat{s}_{z,2})(\hat{s}_{z,3} + \hat{s}_{z,4})\alpha\alpha\beta\beta = (\hat{s}_{z,1} + \hat{s}_{z,2})(-\frac{1}{2} - \frac{1}{2})\alpha\alpha\beta\beta = (\frac{1}{2} + \frac{1}{2})(-\frac{1}{2} - \frac{1}{2})\alpha\alpha\beta\beta = -\alpha\alpha\beta\beta; (\hat{s}_{z,1} + \hat{s}_{z,2})(\hat{s}_{z,3} + \hat{s}_{z,4})\beta\beta\alpha\alpha = (-\frac{1}{2} - \frac{1}{2})(\frac{1}{2} + \frac{1}{2})\beta\beta\alpha\alpha = -\beta\beta\alpha\alpha$. The action of $(\hat{s}_{z,1} + \hat{s}_{z,2})(\hat{s}_{z,3} + \hat{s}_{z,4})$ on all other determinants gives zero. Triplet: $\hat{H}(\alpha\alpha\beta\beta - \beta\beta\alpha\alpha)/\sqrt{2} = -J(0 - \frac{1}{2}(\alpha\beta\beta\alpha + \alpha\beta\alpha\beta + \beta\alpha\beta\alpha + \beta\alpha\alpha\beta) + \frac{1}{2}(\beta\alpha\alpha\beta + \beta\alpha\beta\alpha + \alpha\beta\alpha\beta + \alpha\beta\beta\alpha) - 0 - \alpha\alpha\beta\beta + \beta\beta\alpha\alpha)/\sqrt{2} = -J(-\alpha\alpha\beta\beta + \beta\beta\alpha\alpha)/\sqrt{2} \Rightarrow$ eigenvalues of the triplet is J. Singlet:

$\hat{H}(2\alpha\alpha\beta\beta + 2\beta\beta\alpha\alpha - \alpha\beta\alpha\beta - \alpha\beta\beta\alpha - \beta\alpha\alpha\beta - \beta\alpha\beta\alpha)/2\sqrt{3} = -J \cdot (1/2\sqrt{3})[\frac{1}{2}(2 \cdot 0 + 2\alpha\beta\beta\alpha + 2\alpha\beta\alpha\beta + 2\beta\alpha\beta\alpha + 2\beta\alpha\alpha\beta - \alpha\alpha\beta\beta - \alpha\alpha\beta\beta - \alpha\alpha\beta\beta - \alpha\alpha\beta\beta + 2\alpha\beta\beta\alpha + 2\alpha\beta\alpha\beta + 2\beta\alpha\beta\alpha + 2\beta\alpha\alpha\beta + 2 \cdot 0 - \beta\beta\alpha\alpha - \beta\beta\alpha\alpha - \beta\beta\alpha\alpha - \beta\beta\alpha\alpha) - 2\alpha\alpha\beta\beta - 2\beta\beta\alpha\alpha] = -J \cdot (1/2\sqrt{3})[2\alpha\beta\beta\alpha + 2\alpha\beta\alpha\beta + 2\beta\alpha\beta\alpha + 2\beta\alpha\alpha\beta - 4\alpha\alpha\beta\beta - 4\beta\beta\alpha\alpha] \Rightarrow$ eigenvalue of singlet is $2J$.

Exercise 3.10 Two-electrons: $s_1 = \frac{1}{2}, s_2 = \frac{1}{2}, S_a = 0, 1$. Three electrons: $S_a = 0, 1, s_3 = \frac{1}{2}, S_b = \frac{1}{2}, \frac{1}{2}, \frac{3}{2}$. Four electrons: $S_b = \frac{1}{2}, \frac{1}{2}, \frac{3}{2}, s_4 = \frac{1}{2}, S_{tot} = 0, 1, 0, 1, 1, 2$.

Exercise 3.11 (i) $E(T_{2,3}) - E(Q) = J = -129.7$ meV; $E(S_2) - E(Q) = J = -142.4$ meV; $[E(T_1) - E(Q)]/2 = -130.0$ meV; $[E(S_1) - E(Q)]/3 = J = -117.1$ meV. (ii) $[E(T_{2,3}) - E(S_2)] \times 4 = J_r = 50.7$ meV; $E(T_{2,3}) - E(Q) = J = -129.7$ meV; $E(T_1) - E_Q = 2J - \frac{1}{2}J_r \Rightarrow J = \frac{1}{2}\Delta E + \frac{1}{4}J_r = -117.4$ meV; $E(S_1) - E_Q = 3J + \frac{3}{4}J_r \Rightarrow J = \frac{1}{3}\Delta E - \frac{1}{4}J_r = -129.7$ meV. (iii) $E(S1) - \frac{3}{2}(E(T_1) - E_Q) = \frac{3}{2}J_r \Rightarrow J_r = 25.9$ meV; $E(T_1) - E_Q = 2J - \frac{1}{2}J_r \Rightarrow J = \frac{1}{2}(\Delta E + \frac{1}{2}J_r) = -123.5$ meV; $E(T_{2,3}) - E_Q = J + J_3 \Rightarrow J_3 = \Delta E - J = -6.2$ meV. These three parameters exactly fit the energy difference $E(S2) - E(Q) = -142.4$ meV $= J + 2J_3 - \frac{1}{4}J_r = -123.5 - 2 \times 6.2 - \frac{1}{4} \times 25.9$.

Exercise 3.12

$$\hat{H} = \begin{pmatrix} \hat{S}_x(1) & \hat{S}_y(1) & \hat{S}_z(1) \end{pmatrix} \begin{pmatrix} A_{xx} & A_{xy} & A_{xz} \\ A_{yx} & A_{yy} & A_{yz} \\ A_{zx} & A_{zy} & A_{zz} \end{pmatrix} \begin{pmatrix} \hat{S}_x(2) \\ \hat{S}_y(2) \\ \hat{S}_z(2) \end{pmatrix}$$

$$= \begin{pmatrix} A_{xx}\hat{S}_x(1) + A_{yx}\hat{S}_y(1) + A_{zx}\hat{S}_z(1), & A_{xy}\hat{S}_x(1) + A_{yy}\hat{S}_y(1) + A_{zy}\hat{S}_z(1), & A_{xz}\hat{S}_x(1) + A_{yz}\hat{S}_y(1) + A_{zz}\hat{S}_z(1) \end{pmatrix} \begin{pmatrix} \hat{S}_x(2) \\ \hat{S}_y(2) \\ \hat{S}_z(2) \end{pmatrix} = A_{xx}\hat{S}_x(1)\hat{S}_x(2) + A_{yx}\hat{S}_y(1)\hat{S}_x(2) + A_{zx}\hat{S}_z(1)\hat{S}_x(2) + A_{xy}\hat{S}_x(1)\hat{S}_y(2) + A_{yy}\hat{S}_y(1)\hat{S}_y(2) + A_{zy}\hat{S}_z(1)\hat{S}_y(2) + A_{xz}\hat{S}_x(1)\hat{S}_z(2) + A_{yz}\hat{S}_y(1)\hat{S}_z(2) + A_{zz}\hat{S}_z(1)\hat{S}_z(2).$$

Exercise 3.13

$$\begin{pmatrix} 1 & 0 & 0 & 0 \\ 0 & \frac{1}{\sqrt{2}} & \frac{1}{\sqrt{2}} & 0 \\ 0 & 0 & 0 & 1 \\ 0 & \frac{1}{\sqrt{2}} & -\frac{1}{\sqrt{2}} & 0 \end{pmatrix} \begin{pmatrix} H_{11} & H_{12} & H_{13} & H_{14} \\ H_{21} & H_{22} & H_{23} & H_{24} \\ H_{31} & H_{32} & H_{33} & H_{34} \\ H_{41} & H_{42} & H_{43} & H_{44} \end{pmatrix} \begin{pmatrix} 1 & 0 & 0 & 0 \\ 0 & \frac{1}{\sqrt{2}} & 0 & \frac{1}{\sqrt{2}} \\ 0 & \frac{1}{\sqrt{2}} & 0 & -\frac{1}{\sqrt{2}} \\ 0 & 0 & 1 & 0 \end{pmatrix}$$

$$= \begin{pmatrix} H_{11} & H_{12} & H_{13} & H_{14} \\ \frac{H_{21}+H_{31}}{\sqrt{2}} & \frac{H_{22}+H_{32}}{\sqrt{2}} & \frac{H_{23}+H_{33}}{\sqrt{2}} & \frac{H_{24}+H_{34}}{\sqrt{2}} \\ H_{41} & H_{42} & H_{43} & H_{44} \\ \frac{H_{21}-H_{31}}{\sqrt{2}} & \frac{H_{22}-H_{32}}{\sqrt{2}} & \frac{H_{23}-H_{33}}{\sqrt{2}} & \frac{H_{24}-H_{34}}{\sqrt{2}} \end{pmatrix} \begin{pmatrix} 1 & 0 & 0 & 0 \\ 0 & \frac{1}{\sqrt{2}} & 0 & -\frac{1}{\sqrt{2}} \\ 0 & \frac{1}{\sqrt{2}} & 0 & \frac{1}{\sqrt{2}} \\ 0 & 0 & 1 & 0 \end{pmatrix}$$

$$
= \begin{pmatrix}
H_{11} & \frac{H_{12}+H_{13}}{\sqrt{2}} & H_{14} & \frac{H_{12}-H_{13}}{\sqrt{2}} \\
\frac{H_{21}+H_{31}}{\sqrt{2}} & \frac{H_{22}+H_{32}+H_{23}+H_{33}}{2} & \frac{H_{24}+H_{34}}{\sqrt{2}} & \frac{H_{22}+H_{32}-H_{23}-H_{33}}{2} \\
H_{41} & \frac{H_{42}+H_{43}}{\sqrt{2}} & H_{44} & \frac{H_{42}-H_{43}}{\sqrt{2}} \\
\frac{H_{21}-H_{31}}{\sqrt{2}} & \frac{H_{22}-H_{32}+H_{23}-H_{33}}{2} & \frac{H_{24}-H_{34}}{\sqrt{2}} & \frac{H_{22}-H_{32}-H_{23}+H_{33}}{2}
\end{pmatrix}
$$

Substituting the definition of H_{ij} of the matrix in the uncoupled basis gives the representation in the coupled basis. As example: $\frac{H_{12}+H_{13}}{\sqrt{2}} = \frac{1}{\sqrt{2}}(\frac{1}{4}D_{xz} - \frac{1}{4}iD_{yz} + \frac{1}{4}D_{xz} - \frac{1}{4}iD_{yz}) = \frac{1}{2\sqrt{2}}(D_{xz} - iD_{yz}) = \langle T^+|\hat{H}|T^0\rangle$.

Exercise 3.14 From $\frac{1}{2}A_{ij} = D_{ij} - \frac{1}{2}A_{ji}$ and $-\frac{1}{2}A_{ji} = d_{ij} - \frac{1}{2}A_{ij}$ follows $\frac{1}{2}A_{ij} = D_{ij} + d_{ij} - \frac{1}{2}A_{ij} \Rightarrow A_{ij} = D_{ij} + d_{ij}$. Substituting this in the expression for A_{ji} gives $-\frac{1}{2}A_{ji} = d_{ij} - \frac{1}{2}D_{ij} - \frac{1}{2}d_{ij} \Rightarrow A_{ji} = D_{ij} - d_{ij} = D_{ji} + d_{ji}$.

Problem 3.1 $\Psi(0,0) = N'(|\phi_a\overline{\phi}_b| + |\phi_b\overline{\phi}_a|)$ with $\phi_a = N(\psi_a + \nu\psi_b)$, $\phi_b = N(\psi_b + \nu\psi_a)$, $N = 1/\sqrt{1+\nu^2}$ and $\langle\psi_a|\psi_b\rangle = 0$. So $S_{ab} = 2\nu/(1+\nu^2)$, see also Eq. 3.18. Substitution gives $\Psi(0,0) = N'[|\psi_a\overline{\psi}_b| + |\psi_b\overline{\psi}_a| + (2\nu/(1+\nu^2))(|\psi_a\overline{\psi}_a|+|\psi_b\overline{\psi}_b|)]$. Since $\psi_a = (1/\sqrt{2})(\phi_1+\phi_2)$ and $\psi_b = (1/\sqrt{2})(\phi_1-\phi_2)$ (see Eq. 3.10a), we get $|\psi_a\overline{\psi}_b|+|\psi_b\overline{\psi}_a| = |\phi_1\overline{\phi}_1| - |\phi_2\overline{\phi}_2|$ and $|\psi_a\overline{\psi}_a|+|\psi_b\overline{\psi}_b| = |\phi_1\overline{\phi}_1| + |\phi_2\overline{\phi}_2|$. Now, $\Psi(0,0) = N'[|\phi_1\overline{\phi}_1| - |\phi_2\overline{\phi}_2| + S_{ab}(|\phi_1\overline{\phi}_1| + |\phi_2\overline{\phi}_2|)] = N'[(S_{ab} + 1)|\phi_1\overline{\phi}_1| + (S_{ab} - 1)|\phi_2\overline{\phi}_2|]$. Hence, $c_2/c_1 = (S_{ab} - 1)/(S_{ab} + 1)$.

Problem 3.2 (a) $|g_1\overline{g}_1| = \frac{1}{2}|(a_1 + b_1)(\overline{a}_1 + \overline{b}_2)| = \frac{1}{2}(|a_1\overline{a}_1| + |a_1\overline{b}_1| + |b_1\overline{a}_1| + |b_1\overline{b}_1|)$, 50 % neutral, 50 % ionic, eigenfunction of $\hat{S}^2$ (singlet); $|g_1g_2| = \frac{1}{2}|(a_1 + b_1)(a_2 + b_2)| = \frac{1}{2}(|a_1a_2| + |a_1b_2| + |b_1a_2| + |b_1b_2|)$, 50 % neutral, 50 % ionic, eigenfunction of $\hat{S}^2$ (triplet); $|g_1\overline{u}_1| = \frac{1}{2}|(a_1 + b_1)(\overline{a}_1 - \overline{b}_1)| = \frac{1}{2}(|a_1\overline{a}_1| - |a_1\overline{b}_1| - |b_1\overline{a}_1| + |b_1\overline{b}_1|)$, 50 % neutral, 50 % ionic, not an eigenfunction of $\hat{S}^2$. (b) $\frac{1}{\sqrt{2}}(|g_1\overline{g}_1|+|u_1\overline{u}_1|) = \frac{1}{2}(|(a_1 + b_1)(\overline{a}_1 + \overline{b}_1)|+|(a_1 - b_1)(\overline{a}_1 - \overline{b}_1)|) = \frac{1}{2\sqrt{2}}(|a_1\overline{a}_1|+ |a_1\overline{b}_1| + |b_1\overline{a}_1| + |b_1\overline{b}_1| + |a_1\overline{a}_1| - |a_1\overline{b}_1| - |b_1\overline{a}_1| + |b_1\overline{b}_1|) = \frac{1}{\sqrt{2}}(|a_1\overline{a}_1| + |b_1\overline{b}_1|)$, 100 % ionic, eigenfunction of $\hat{S}^2$ (singlet); $\frac{1}{\sqrt{2}}(|g_1\overline{g}_1| - |u_1\overline{u}_1|) = \frac{1}{2}(|(a_1 + b_1)(\overline{a}_1 + \overline{b}_1)| - |(a_1 - b_1)(\overline{a}_1 - \overline{b}_1)|) = \frac{1}{2\sqrt{2}}(|a_1\overline{a}_1| + |a_1\overline{b}_1| + |b_1\overline{a}_1| + |b_1\overline{b}_1| - |a_1\overline{a}_1| + |a_1\overline{b}_1| + |b_1\overline{a}_1| - |b_1\overline{b}_1|) = \frac{1}{\sqrt{2}}(|a_1\overline{b}_1| + |b_1\overline{a}_1|)$, 100 % covalent, eigenfunction of $\hat{S}^2$ (singlet). (c) $|g_1u_1| = \frac{1}{2}|(a_1 + b_1)(a_1 - b_1)| = \frac{1}{2}(-|a_1b_1| + |b_1a_1|) = \frac{1}{2}(|b_1a_a|+|b_1a_1|) = |b_1a_1|$, 100 % covalent, eigenfunction of $\hat{S}^2$ (triplet). $|g_1u_1v_1| = \frac{1}{2}|(a_1 + b_1)(a_1 - b_1)c_1| = \frac{1}{2}(-|a_1b_1c_1|+|b_1a_1c_1|) = \frac{1}{2}(-|a_1b_1c_1|-|a_1b_1c_1|) = -|a_1b_1c_1|$, 100 % covalent, eigenfunction of $\hat{S}^2$ (quartet). (d) For simplicity, we drop the subscript and multiply with the normalization constant at the end. $2|gu\overline{v}|-|\overline{g}uv| - |\overline{g}uv| = |(a + b)(a - b)\overline{c}| - \frac{1}{2}|(a + b)(\overline{a} - \overline{b})c| - \frac{1}{2}|(\overline{a} + \overline{b})(a - b)c| = -|ab\overline{c}| + |ba\overline{c}|-\frac{1}{2}(|a\overline{a}c|-|a\overline{b}c|+|b\overline{a}c|-|b\overline{b}c|)-\frac{1}{2}(|\overline{a}ac|-|\overline{a}bc|+|\overline{b}ac|-|\overline{b}bc|) = -2|ab\overline{c}|+|a\overline{b}c| + |\overline{a}bc|$. After multiplying with $1/\sqrt{6}$, the doublet spin eigenfunction appears, with 100 % covalent character.

Problem 3.3 $\hat{S}^+(1)\hat{S}^-(2)\alpha\alpha = 0$; $\hat{S}^-(1)\hat{S}^+(2)\alpha\alpha = 0$; $\hat{S}_z(1)\hat{S}_z(2)\alpha\alpha = \frac{1}{4}\alpha\alpha \Rightarrow$ $-J\hat{S}(1) \cdot \hat{S}(2)\Phi(T) = -\frac{1}{4}J\Phi(T)$. $\hat{S}^+(1)\hat{S}^-(2)(\alpha\beta - \beta\alpha) = 0 - \alpha\beta$; $\hat{S}^-(1)\hat{S}^+(2)(\alpha\beta - \beta\alpha) = \beta\alpha - 0$; $\hat{S}_z(1)\hat{S}_z(2)(\alpha\beta - \beta\alpha) = -\frac{1}{4}\alpha\beta + \frac{1}{4}\beta\alpha \Rightarrow \hat{S}(1) \cdot$ $\hat{S}(2)(\alpha\beta - \beta\alpha) = -\frac{1}{2}(\alpha\beta - \beta\alpha) - \frac{1}{4}(\alpha\beta - \beta\alpha) \Rightarrow -J\hat{S}(1) \cdot \hat{S}(2)\Phi(S) = \frac{3}{4}J\Phi(S)$.

Problem 3.4 (a) All determinants have two electrons with α spin and one with β spin, hence M_S of all determinants is $\frac{1}{2}$. $\Psi_3 = \frac{1}{\sqrt{3}}(|\phi_1\phi_2\overline{\phi}_3| + |\phi_1\overline{\phi}_2\phi_3| + |\overline{\phi}_1\phi_2\phi_3|)$. Separating the spatial and spin part: $\Psi_3 = \frac{1}{\sqrt{3}}|\phi_1\phi_2\phi_3|(\alpha\alpha\beta + \alpha\beta\alpha + \beta\alpha\alpha)$. The spin part is the $M_S = \frac{1}{2}$ component of the quartet spin eigenfunction, the application of $\hat{S}^+$ gives $\alpha\alpha\alpha$. (b) Ψ_1 corresponds to D_2, then $J_{12} = J_{23} = \frac{2}{3}(E(\Psi_1) - E(\Psi_3)) = -27.24$ meV and $J_{13} = -27.24 - (E(D_2) - E(D_1)) = 0.07$ meV. (c) Model space: $|\phi_1\phi_2\overline{\phi}_3|$, $|\phi_1\overline{\phi}_2\phi_3|$, $|\overline{\phi}_1\phi_2\phi_3|$, $|\tilde{\Psi}_i|^2 = 0.8973, 0.9623, 1.0002, 0.0349, 0.0922$ (the third is due to round-off errors). (d) Ψ_1, Ψ_2 and Ψ_3. Only $\langle\Psi_1|\Psi_2\rangle \neq 0$, Gram-Schmidt orthogonalization gives $c_1 = -0.4672$, $c_2 = 0.8135$, $c_3 = -0.3463$ for Ψ_1 and $c_1' = -0.6696$, $c_2' = -0.0699$, $c_3' = 0.7394$ for Ψ_2. (e) $\hat{H}_{11}^{eff} = -27.9615094$, $\hat{H}_{21}^{eff} = -0.0001464$, $\hat{H}_{22}^{eff} = -27.9608149$, $\hat{H}_{13}^{eff} = 0.0000786$, $\hat{H}_{23}^{eff} = 0.0010935$, $\hat{H}_{33}^{eff} = -27.9587333$. $J_{12} = 7.97$ meV, $J_{13} = -4.28$ meV, $J_{23} = -59.51$ meV.

Problem 3.5 (a) Dividing the eigenvalues given in Fig. 3.6 by $-J$, we obtain $\hat{S}_1\hat{S}_2 Q = 1 \cdot Q$; $\hat{S}_1\hat{S}_2 T = -1 \cdot T$; $\hat{S}_1\hat{S}_2 S = -2 \cdot S$. Applying the operator for the second time leads to $\hat{S}_1\hat{S}_2(1 \cdot Q) = 1 \cdot Q$; $\hat{S}_1\hat{S}_2(-1 \cdot T) = 1 \cdot T$; $\hat{S}_1\hat{S}_2(-2 \cdot S) = 4 \cdot S$. Multiplying with λ gives exactly the same eigenvalues as listed in Eq. 3.75.

Problem 3.6 $(E(T) - E(Q))/2 = -42.58$ meV, $E(S) - E(T) = -37.54$ meV, no regular spacing. From Eq. 3.75 follows $E(T) - E(Q) = 2J$ and $E(S) - E(T) = J + 3\lambda$. This gives $J = -42.58$ meV and $\lambda = [-37.54 - (-42.58)]/3 = 1.68$ meV.

Exercises and Problems of Chap. 4

Exercise 4.1 $\langle\Psi_S|\Psi_S\rangle = (1/(2 + 2S))\langle a\overline{b} + b\overline{a}|a\overline{b} + b\overline{a}\rangle = (1/(2 + 2S))(\langle a\overline{b}|a\overline{b}\rangle + \langle a\overline{b}|b\overline{a}\rangle + \langle b\overline{a}|a\overline{b}\rangle + \langle b\overline{a}|b\overline{a}\rangle) = (1/(2 + 2S))(1 + S + S + 1) = 1$; $\langle\Psi_T|\Psi_T\rangle = (1/(2 - 2S))\langle a\overline{b} - b\overline{a}|a\overline{b} - b\overline{a}\rangle = (1/(2 - 2S))(\langle a\overline{b}|a\overline{b}\rangle - \langle a\overline{b}|b\overline{a}\rangle - \langle b\overline{a}|a\overline{b}\rangle + \langle b\overline{a}|b\overline{a}\rangle) = (1/(2 - 2S))(1 - S - S + 1) = 1$.

Exercise 4.2 $J_{11} = \langle\phi_1\overline{\phi}_1|1/r_{12}|\phi_1\overline{\phi}_1\rangle = (1/4)\langle(\phi_a + \phi_b)(\overline{\phi}_a + \overline{\phi}_b)1/r_{12}(\phi_a + \phi_b)(\overline{\phi}_a + \overline{\phi}_b)\rangle = (1/4)(\langle\phi_a\overline{\phi}_a|1/r_{12}|\phi_a\overline{\phi}_a\rangle + \langle\phi_a\overline{\phi}_a|1/r_{12}|\phi_a\overline{\phi}_b\rangle + \langle\phi_a\overline{\phi}_a|1/r_{12}|\phi_b\overline{\phi}_a\rangle + \langle\phi_a\overline{\phi}_a|1/r_{12}|\phi_b\overline{\phi}_b\rangle + \langle\phi_a\overline{\phi}_b|1/r_{12}|\phi_a\overline{\phi}_a\rangle + \dots$ *eleven more terms*$) = (1/4)(2J_{aa} + 2J_{ab} + 4K_{ab} + 8\langle\phi_a\overline{\phi}_a|1/r_{12}|\phi_a\overline{\phi}_b\rangle)$, where we have used that the system is centrosymmetric: $J_{aa} = J_{bb}$, etc. This is equal to the expression given in Eq. 4.20.

Exercise 4.3 *Meta*: the shortest contacts are formed by the aligned carbon atoms of the benzene ring. These have opposite spin density, and hence, $\rho_i\rho_j < 0$, indicating a ferromagnetic interaction between the two units. *fully aligned*: All carbon atoms form shortest contacts and the product of atomic spin populations is positive in all cases, hence an antiferromagnetic coupling can be expected.

Exercise 4.4 The symmetry of the complex is D_{2h} and the five d-orbitals by increasing orbital energy are $3d_{xz}$ (b_{2g}), $3d_{yz}$ (b_{3g}), $3d_{xy}$ (b_{1g}), $3d_{z^2}$ (a_g), $3d_{x^2-y^2}$ (a_g), with the irreducible representation in parentheses. Cr^{III} has a d^3 electronic configuration, occupying the b_{1g}, b_{2g} and b_{3g} orbitals. Ni^{II} has a d^8 electronic configuration and the orbitals with unpaired electrons are the two a_g's. All exchange paths that connect Cr with Ni involve orbitals of different symmetries, implying zero overlap between the magnetic orbitals, and hence, ferromagnetic coupling. Replacing Cr^{III} with Mn^{II} introduces two extra unpaired electrons on site A, and hence, $5 \times 2 = 10$ exchange paths. Among the ten exchange paths, the four involving the a_g orbitals will give a (strong) antiferromagnetic contribution, which counterbalances the (weaker) ferromagnetic contribution of the other six paths. The net coupling will be antiferromagnetic.

Exercise 4.5 There are six determinants with two doubly occupied orbitals, six $M_S = 0$ determinants can be constructed with all orbitals singly occupied and the distribution with one doubly occupied orbital, two singly occupied and one empty orbital can be realized in 24 different ways. The total CAS wave function is a linear combination of 36 different determinants. Note that the use of spin symmetry reduces the expansion to 20 CSFs for $S = 0$, 15 CSFs for $S = 1$ and 1 CSF for $S = 2$.

Exercise 4.6 We have four reference determinants $|\ldots a\bar{a}|$, $|\ldots b\bar{a}|$, $|\ldots a\bar{b}|$ and $|\ldots b\bar{a}|$. There are $2k$ occupied spin orbitals which can be replaced by one of the $2l$ unoccupied spinorbitals: $2k \times l \times 4$. In addition there are also replacements involving a spin-flip, *i.e.* $\alpha \rightarrow \beta$ and $\beta \rightarrow \alpha$. This has to be compensated by spin-flip in the CAS space, which can only be done for $|\ldots a\bar{b}|$ and $|\ldots b\bar{a}|$: $2k \times l \times 2$. In total: $12 \times k \times l$.

Exercise 4.7 The CAS(2, 2) reference contains 4 $M_S = 0$ determinants; $n = 4$. There are 72 electrons in the inactive orbitals, $k = 36$. The system has $154 - 36 - 2$ virtual orbitals, $l = 116$. This results in $36^2 \times 116 \times 4 = 601344$ $2h$-$1p$ determinants, $36 \times 116^2 \times 4 = 1937664$ $1h$-$2p$ determinants, and $36^2 \times 116^2 \times 4 = 69755904$ $2h$-$2p$ determinants. The total number of determinants in the MR-CISD wave function is ≈ 72294912, from which the $2h$-$2p$ determinants constitute more than 96%.

Exercise 4.8 $\Phi_Q = |p\bar{p}b\bar{b}|$. The contribution to $\langle \Phi_I | \hat{H}^{eff} | \Phi_L \rangle$ equals: $\frac{\langle h\bar{h}a\bar{b} | \hat{V} | p\bar{p}b\bar{b} \rangle \langle p\bar{p}b\bar{b} | \hat{V} | h\bar{h}b\bar{b} \rangle}{E_L - E_Q}$. The first matrix element is zero since the determinants Φ_I and Φ_Q differ by more than two columns.

Exercise 4.9 (a) Neutral determinant $\Phi_I = |h\bar{h}a\bar{b}|$; ionic determinant $\Phi_J = |h\bar{h}a\bar{a}|$. Because both Φ_I and Φ_J have two different columns with respect to Φ_R, $\langle \Phi_I | \hat{H} | \Phi_R \rangle \langle \Phi_R | \hat{H} | \Phi_J \rangle$ is non-zero and in consequence the effective matrix element between Φ_I and Φ_J will be different from zero (see Eq. 1.86). (b) Since the matrix element $\langle h\bar{h}a\bar{b}|\hat{H}|a\bar{a}p\bar{b}\rangle$ is not equal to $\langle h\bar{h}a\bar{a}|\hat{H}|a\bar{a}p\bar{b}\rangle$ the diagonal matrix elements of Φ_I and Φ_J are shifted non-uniformly.

Exercise 4.10 First, $\hat{S}^2\alpha\alpha\beta\beta = 2\alpha\alpha\beta\beta + \alpha\beta\alpha\beta + \beta\alpha\alpha\beta + \alpha\beta\beta\alpha + \beta\alpha\beta\alpha$. With this
$\langle \phi_1\phi_2\bar{\phi}_3\bar{\phi}_4|\hat{S}^2|\phi_1\phi_2\bar{\phi}_3\bar{\phi}_4\rangle = \langle \phi_1\phi_2\bar{\phi}_3\bar{\phi}_4|2\phi_1\phi_2\bar{\phi}_3\bar{\phi}_4 - \phi_1\phi_3\bar{\phi}_2\bar{\phi}_4 - \phi_3\phi_2\bar{\phi}_1\bar{\phi}_4 -$
$\phi_1\phi_4\bar{\phi}_3\bar{\phi}_2 - \phi_4\phi_2\bar{\phi}_3\bar{\phi}_1\rangle = 2 - \langle \phi_2|\phi_3\rangle^2 - \langle \phi_1|\phi_3\rangle^2 - \langle \phi_2|\phi_4\rangle^2 - \langle \phi_1|\phi_4\rangle^2 = 2$,
taking into account that $\langle \phi_i|\phi_j\rangle = \delta_{ij}$.

Problem 4.1 In the first place, columns three and four have to be converted to Kelvins by multiplying with 315647.5. Next the values have to substituted in Eq. 4.18. Then, J is 1.8, 17.7, 17.9 -5.3 -47.2 K for $\theta = 85° \ldots 105°$. As expected, J is maximally ferromagnetic around $90°$ and becomes antiferromagnetic for larger angles. Practically the same tendency is observed when the entries in the second and third column are replaced by the average value: J to 3.0, 22.6, 20.7, -6.8 and -55.7 K.

Problem 4.2 To calculate the contribution to the total coupling of the two ligands one has to perform two separate calculations in which only one bridge is active. This can be achieved by dividing the molecule in three fragments: bridging ligand A, bridging ligand B, and the rest of the molecule with the two magnetic centers and the external ligands. Orbitals are optimized for the three fragments. In the first calculation one superposes the charge distributions of the three fragments and relaxes the orbitals in the field of the frozen charge distribution of the ligand B. J_A is calculated by calculating the energy difference of the relevant spin states. Subsequently, the orbitals are optimized in the field of the frozen charge distribution of the ligand A and J_B is calculated. Finally, J_{tot} from the calculation of the relevant spin states without restrictions on the orbital optimizations and the counter-complementarity is quantified by comparing J_{tot} to the sum of J_A and J_B.

Problem 4.3 (a) Yamaguchi: -281, -275 and -77 cm^{-1} for Cu, Ni and Mn. Noodleman: -284, -276, -77 cm^{-1}. Ruiz: -142, -184, -65 cm^{-1}. The difference between the different expressions becomes smaller for larger spin moment and will be irrelevant for polynuclear complexes typically used in single-molecule magnets. (b) Using the spin densities gives $J = -280$ cm^{-1} for Cu.

Exercises and Problems of Chap. 5

Exercise 5.1 The substitution of $g = (a + b)/\sqrt{2}$ and $u = (a - b)/\sqrt{2}$ in the expression of S_g gives

$$S_g = \frac{1}{2}\left(\lambda|(a + b)(\bar{a} + \bar{b})| - \mu|(a - b)(\bar{a} - \bar{b})|\right)$$
$$= \frac{1}{2}\left(\lambda(|a\bar{a}| + |a\bar{b}| + |b\bar{a}| + |b\bar{b}|) - \mu(|a\bar{a}| - |a\bar{b}| - |b\bar{a}| + |b\bar{b}|)\right)$$
$$= \left((\lambda - \mu)(|a\bar{a}| + |b\bar{b}|) + (\lambda + \mu)(|a\bar{b}| + |b\bar{a}|)\right)$$

When $\lambda \approx \mu$, the term with the ionic determinants tends to zero, demonstrating that S_g is dominated by neutral determinants. In case of $\lambda \gg \mu$, an approximately 50 % mixture of neutral (or covalent) and ionic determinants appears, typical of a covalent bond in closed shell molecules such as H_2.

Exercise 5.2 $\langle a\bar{a}|\hat{H}|a\bar{a}\rangle = \langle a|\hat{h}|a\rangle + \langle a|\hat{h}|a\rangle + \langle a\bar{a}|\frac{1-\hat{P}_{12}}{r_{12}}|a\bar{a}\rangle = h_{aa} + h_{aa} + J_{aa}.\langle b\bar{b}|\hat{H}|b\bar{b}\rangle = h_{bb} + h_{bb} + J_{bb}$. Only under the assumption that $h_{aa} = h_{bb}$ and $J_{aa} = J_{bb}$ one can write the relative energy of the ionic determinants as $U = J_{aa} - J_{ab}$. This is the case for centro-symmetric systems.

Exercise 5.3 Substitution of $\sqrt{U^2 + 16t_{ab}^2} \approx \sqrt{U^2} + \frac{1}{2}16t_{ab}^2/U = U + 8t_{ab}^2/U$ in Eq. 5.7 gives $(U - U + 8t_{ab}^2/U)/2 = 4t_{ab}^2/U$.

Exercise 5.4 $|a\bar{b}| \xrightarrow{t_{ba}} |a\bar{a}| \xrightarrow{t_{ab}} |b\bar{a}|$.

Exercise 5.5 $|h\bar{h}a\bar{b}| \xrightarrow{t_{ha}} |h\bar{a}a\bar{b}| \xrightarrow{t_{hb}} |a\bar{a}b\bar{b}| \xrightarrow{t_{bh}} |a\bar{h}b\bar{a}| \xrightarrow{t_{ah}} |h\bar{h}b\bar{a}|$, with the intermediate determinants at ΔE_{CT}, ΔE_{2CT}, and ΔE_{CT}, respectively. Using the expression in Eq. 5.12, we arrive at $(t_{ha} \cdot t_{hb} \cdot t_{bh} \cdot t_{ah})/(\Delta E_{CT} \cdot \Delta E_{2CT} \cdot \Delta E_{CT})$. Since there are four different pathways, the final perturbative estimate of the contribution to J reads $-8(t_{ab}^{eff})^2/(2\Delta E_{CT}\Delta E_{2CT})$, introducing an effective hopping parameter between the magnetic centers $t_{ab}^{eff} = t_{ha}t_{hb}$.

Exercise 5.6 The two electron pairs have $S_1 = 1$ and $S_2 = 1$. The total spin of these two electron pairs can in principle take the values $S_1 + S_2 = 2$ (quintet), $S_1 + S_2 - 1 = 1$ (triplet), and $S_1 - S_2 = 0$ (singlet). For a binuclear Cu^{2+} complex (and all other systems with two $S = 1/2$ spin moments), only the triplet and singlet couplings are relevant.

Exercise 5.7 Substituting $t^{eff} = -2218 \text{ cm}^{-1}$ and $\Delta E_{ST} = -362 \text{ cm}^{-1}$ in $U^{eff} = 4t^{eff}/\Delta E_{ST}$ gives a value of 13590 cm^{-1} (6.74 eV) for the effective on-site repulsion, a lowering of ~ 19 eV with respect to the bare valence-only value.

Exercise 5.8 $\widetilde{\Psi}_i$ are the projections of Ψ on the model space, $\widetilde{\Psi}_i'$ are the normalized projections, and $\widetilde{\Psi}_i'^\dagger$ the biorthonormal projections. $|\widetilde{\Psi}_1|^2 = (-0.9224)^2 + (-0.1223)^2 = 0.8658$, $|\widetilde{\Psi}_2|^2 = (-0.6626)^2 = 0.4390$, $|\widetilde{\Psi}_7|^2 = 0.4159^2 = 0.1730$, $|\widetilde{\Psi}_8|^2 = 0.1704^2 + (-0.5324)^2 = 0.3125$; $\widetilde{\Psi}_1' = -0.9913(|a\bar{b}| + |b\bar{a}|) - 0.1315(|a\bar{a}| + |b\bar{b}|)$; $\widetilde{\Psi}_2' = (|a\bar{b}| - |b\bar{a}|)/\sqrt{2}$; $\widetilde{\Psi}_7' = (|a\bar{a}| - |b\bar{b}|)/\sqrt{2}$; $\widetilde{\Psi}_8' = 0.3048(|a\bar{b}| + |b\bar{a}|) - 0.9524(|a\bar{a}| + |b\bar{b}|)$. $\langle\widetilde{\Psi}_1'|\widetilde{\Psi}_8'\rangle = (-0.9913 \times 0.3048 + -0.1315 \times -0.9524) = -0.1769$, the other overlaps are zero. $\langle\widetilde{\Psi}_1'|\widetilde{\Psi}_1'^\dagger\rangle = -0.9913 \times -0.9524 + -0.1315 \times -0.3048 = 0.9842 = \langle\widetilde{\Psi}_2'|\widetilde{\Psi}_2'^\dagger\rangle$; $\langle\widetilde{\Psi}_1'|\widetilde{\Psi}_2'^\dagger\rangle = -0.9913 \times 0.1315 + -0.1315 \times -0.9913 = 0 = \langle\widetilde{\Psi}_2'|\widetilde{\Psi}_1'^\dagger\rangle$; $\langle\widetilde{\Psi}_1'^\dagger|\widetilde{\Psi}_2'^\dagger\rangle = -0.9524 \times 0.1315 + -0.3048 \times -0.9913 = 0.1769 = \langle\widetilde{\Psi}_1'|\widetilde{\Psi}_2'\rangle$.

Exercise 5.9 Remember that a and b are normalized, orthogonal orbitals: $\langle a'|b'\rangle = 2\sin\alpha\cos\alpha = \sin(2\alpha)/2$. Overlaps for the listed values of α are 0, 0.052, 0.155, 0.5 and 0.

Exercise 5.10 Since both the interaction elements $\langle NH|\hat{H}|I\rangle$ are larger (t versus $t/\sqrt{2}$ for triplet and singlet) and the denominators smaller (2 and $4K$), one expects a stronger energy lowering for the triplet energy than for the singlet. This means that $J_{TQ} = (E_T - E_Q)/2$ will be larger (in absolute value) than $E_{ST} = E_T - E_S$.

Exercise 5.11 Assuming a square geometry with distance r_1 between neighboring magnetic sites, the hole-particle contribution to the total energy of Φ_α is $q_1 \cdot q_2/r_1 = -1/r$. For Φ_γ, the same contribution arises. In the case of Φ_β the hole ($+1$ charge) and particle (-1 charge) are at $r_2 = \sqrt{2}r_1$ and the contribution becomes $q_1 \cdot q_2/r_2 = -1/\sqrt{2}r_1$, and hence, the energies of the ionic determinants are not strictly the same.

Exercise 5.12 $|a\overline{b}c\overline{d}| \xrightarrow{t_{ad}} |\overline{b}cd\overline{d}| \xrightarrow{t_{cb}} |b\overline{b}d\overline{d}| \xrightarrow{t_{ba}} |\overline{a}bd\overline{d}| \xrightarrow{t_{dc}} |\overline{a}bc\overline{d}|$ or
$|a\overline{b}c\overline{d}| \xrightarrow{t_{ad}} |\overline{b}cd\overline{d}| \xrightarrow{t_{cb}} |b\overline{b}d\overline{d}| \xrightarrow{t_{dc}} |b\overline{b}\overline{c}d| \xrightarrow{t_{ba}} |\overline{a}b\overline{c}d|$.

Exercise 5.13 $I_\alpha = |\varphi_1\varphi_2\overline{\varphi}_2\varphi_3|$, $I_\beta = |\varphi_1\varphi_3\varphi_4\overline{\varphi}_4|$. $\langle\Phi_4|\hat{H}|I_\alpha\rangle = \langle\varphi_1\varphi_2\varphi_3\overline{\varphi}_4|\hat{H}|\varphi_1\varphi_2\overline{\varphi}_2\varphi_3\rangle = -\langle\varphi_1\varphi_2\varphi_3\overline{\varphi}_4|\hat{H}|\varphi_1\varphi_2\varphi_3\overline{\varphi}_2\rangle = -t_{24}$, $\langle\Phi_4|\hat{H}|I_\beta\rangle = -t_{24}$.

Exercise 5.14 $-J_1(\hat{S}_A\hat{S}_B + \hat{S}_C\hat{S}_D)\alpha\alpha\alpha\alpha = -J_1(\frac{1}{4}\alpha\alpha\alpha\alpha + \frac{1}{4}\alpha\alpha\alpha\alpha) = -\frac{1}{2}J_1\alpha\alpha\alpha\alpha$; $-J_2(\hat{S}_A\hat{S}_D + \hat{S}_B\hat{S}_C)\alpha\alpha\alpha\alpha = -J_2(\frac{1}{4}\alpha\alpha\alpha\alpha + \frac{1}{4}\alpha\alpha\alpha\alpha) = -\frac{1}{2}J_2\alpha\alpha\alpha\alpha$; $-J_3(\hat{S}_A\hat{S}_C + \hat{S}_B\hat{S}_D)\alpha\alpha\alpha\alpha = -J_3(\frac{1}{4}\alpha\alpha\alpha\alpha + \frac{1}{4}\alpha\alpha\alpha\alpha) = -\frac{1}{2}J_3\alpha\alpha\alpha\alpha$; $J_r(\hat{S}_A\hat{S}_B)(\hat{S}_C\hat{S}_D)\alpha\alpha\alpha\alpha = J_r\hat{S}_A\hat{S}_B\frac{1}{4}\alpha\alpha\alpha\alpha = \frac{1}{16}J_r\alpha\alpha\alpha\alpha$; $J_r(\hat{S}_A\hat{S}_D)(\hat{S}_B\hat{S}_C)\alpha\alpha\alpha\alpha = J_r\hat{S}_A\hat{S}_D\frac{1}{4}\alpha\alpha\alpha\alpha = \frac{1}{16}J_r\alpha\alpha\alpha\alpha$; $-J_r(\hat{S}_A\hat{S}_C)(\hat{S}_B\hat{S}_D)\alpha\alpha\alpha\alpha = -J_r\hat{S}_A\hat{S}_C\frac{1}{4}\alpha\alpha\alpha\alpha = -\frac{1}{16}J_r\alpha\alpha\alpha\alpha \Rightarrow \langle\alpha\alpha\alpha\alpha|\hat{H}|\alpha\alpha\alpha\alpha\rangle = -\frac{1}{2}(J_1 + J_2 + J_3) + \frac{1}{16}J_r$.

Problem 5.1 The model space is reduced to a 2×2 matrix spanned by $|a\overline{b}|$ and $|b\overline{a}|$. The matrix representation is

	$|a\overline{b}\rangle$	$|b\overline{a}\rangle$
$\langle a\overline{b}|$	0	K_{ab}
$\langle b\overline{a}|$	K_{ab}	0

The corresponding secular determinant leads to the equation $E^2 - K_{ab}^2 = 0$, which gives the eigenvalues $E_{1,2} = \pm K_{ab}$ and the eigenfunctions $\Psi_{1,2} = |a\overline{b}| \pm |b\overline{a}|$. The energy difference is $2K_{ab}$ and the ground state is the triplet, because K_{ab} is positive.

Problem 5.2 (a) $\Psi_1 = (|a\overline{b}| + |b\overline{a}|)/\sqrt{2}$; $\Psi_2 = (|a\overline{b}| - |b\overline{a}|)/\sqrt{2}$; $\Psi_3 = (|a\overline{a}| + |b\overline{b}|)/\sqrt{2}$; $\Psi_4 = (|a\overline{a}| - |b\overline{b}|)/\sqrt{2}$. (b) $\langle\Psi_1|\hat{H}|\Psi_1\rangle = \frac{1}{2}(\langle a\overline{b}|\hat{H}|a\overline{b}\rangle + \langle a\overline{b}|\hat{H}|b\overline{a}\rangle + \langle b\overline{a}|\hat{H}|a\overline{b}\rangle + \langle b\overline{a}|\hat{H}|b\overline{a}\rangle) = h_{aa} + h_{bb} + J_{ab} + K_{ab}$; $\langle\Psi_2|\hat{H}|\Psi_2\rangle = \frac{1}{2}(\langle a\overline{b}|\hat{H}|a\overline{b}\rangle - \langle a\overline{b}|\hat{H}|b\overline{a}\rangle - \langle b\overline{a}|\hat{H}|a\overline{b}\rangle + \langle b\overline{a}|\hat{H}|b\overline{a}\rangle) = h_{aa} + h_{bb} + J_{ab} - K_{ab}$; $\langle\Psi_3|\hat{H}|\Psi_3\rangle = \frac{1}{2}(\langle a\overline{a}|\hat{H}|a\overline{a}\rangle + \langle a\overline{a}|\hat{H}|b\overline{b}\rangle + \langle b\overline{b}|\hat{H}|a\overline{a}\rangle + \langle b\overline{b}|\hat{H}|b\overline{b}\rangle) = h_{aa} + h_{bb} + \frac{1}{2}(J_{aa} + J_{bb}) + K_{ab}$; $\langle\Psi_4|\hat{H}|\Psi_4\rangle = \frac{1}{2}(\langle a\overline{a}|\hat{H}|a\overline{a}\rangle - \langle a\overline{a}|\hat{H}|b\overline{b}\rangle - \langle b\overline{b}|\hat{H}|a\overline{a}\rangle + \langle b\overline{b}|\hat{H}|b\overline{b}\rangle) = h_{aa} + h_{bb} + \frac{1}{2}(J_{aa} + J_{bb}) - K_{ab}$. (c) $\Psi_1 = S_g$; $\Psi_2 = T_u$; $\Psi_3 = S_g$; $\Psi_4 = S_u$; only $\langle\Psi_1|\hat{H}|\Psi_3\rangle$ is non-zero. (d) $\langle\Psi_1|\hat{H}|\Psi_3\rangle = \frac{1}{2}(\langle a\overline{b}|\hat{H}|a\overline{a}\rangle + \langle a\overline{b}|\hat{H}|b\overline{b}\rangle + \langle b\overline{a}|\hat{H}|a\overline{a}\rangle + \langle b\overline{a}|\hat{H}|b\overline{b}\rangle) =$

$2h_{ab} + \langle ab|1/r_{12}|aa \rangle + \langle ab|1/r_{12}|bb \rangle = 2t_{ab}$. Taking $\langle T_u|\hat{H}|T_u \rangle$ as reference energy, the CAS(2, 2) matrix becomes

	$|\Psi_1\rangle$	$|\Psi_2\rangle$	$|\Psi_3\rangle$	$|\Psi_4\rangle$
$\langle\Psi_1|$	$2K_{ab}$	0	$2t_{ab}$	0
$\langle\Psi_2|$	0	0	0	0
$\langle\Psi_3|$	$2t_{ab}$	0	$2K_{ab} + U$	0
$\langle\Psi_4|$	0	0	0	U

Problem 5.3 (a) The third term $(\sin\alpha\cos\alpha)(|a\bar{a}| + |b\bar{b}|)$ is singlet spin eigenfunction with eigenvalue $S(S+1) = 0$. (b) $\Phi_{BS} = \frac{1}{2}\cos^2\alpha|a\bar{b}| + \frac{1}{2}\cos^2\alpha|a\bar{b}| + \frac{1}{2}\sin^2\alpha|b\bar{a}| + \frac{1}{2}\sin^2\alpha|b\bar{a}| + \frac{1}{2}\cos^2\alpha|b\bar{a}| - \frac{1}{2}\cos^2\alpha|b\bar{a}| + \frac{1}{2}\sin^2\alpha|a\bar{b}| - \frac{1}{2}\sin^2\alpha|a\bar{b}| + (\sin\alpha\cos\alpha)(|a\bar{a}| + |b\bar{b}|) = \frac{1}{2}(\cos^2\alpha + \sin^2\alpha)(|a\bar{b}| + |b\bar{a}|) + \frac{1}{2}(\cos^2\alpha - \sin^2\alpha)(|a\bar{b}| - |b\bar{a}|) + (\sin\alpha\cos\alpha)(|a\bar{a}| + |b\bar{b}|) = \frac{1}{2}\sqrt{2}|S_1\rangle + \frac{1}{2}\cos 2\alpha \cdot \sqrt{2}|T\rangle + (\sin\alpha\cos\alpha) \cdot \sqrt{2}|S_2\rangle$. (c) $\langle\Phi_{BS}|\hat{S}^2|\Phi_{BS}\rangle = \frac{1}{4}\cdot 2\cdot\langle S_1|\hat{S}^2|S_1\rangle + \frac{1}{4}\cos^2(2\alpha)\cdot 2\langle T|\hat{S}^2|T\rangle + (\sin\alpha\cos\alpha)^2\cdot 2\langle S_2|\hat{S}^2|S_2\rangle = 0 + \frac{1}{2}\cos^2(2\alpha)\cdot 1(1+1) + 0 = \cos^2(2\alpha)$.

Problem 5.4 Energies relative to E_T: $\langle S|\hat{H}|S\rangle = 2K_{ab}$; $\langle I_1|\hat{H}|I_1\rangle = U + 2K_{ab}$; $\langle I_2|\hat{H}|I_2\rangle = U - 2K_{ab}$ (all the Coulomb interactions are absorbed in E_{ref}). Interaction matrix elements: $\langle S|\hat{H}|I\rangle = \frac{1}{2}[\langle a\bar{b}|\hat{H}|a\bar{a}\rangle + \langle a\bar{b}|\hat{H}|b\bar{b}\rangle + \langle b\bar{a}|\hat{H}|a\bar{a}\rangle + \langle b\bar{a}|\hat{H}|b\bar{b}\rangle] = 2t$. $\langle S|\hat{H}|I_2\rangle = 0$. Second-order energy of S : $\langle S|\hat{H}|S\rangle + \langle S|\hat{H}|I_1\rangle\langle I_1|\hat{H}|S\rangle/(E_S - E_{I_1}) = 2K_{ab} + (2t^2 \cdot 2t^2)(2K_{ab} - (U + 2K_{ab})) = 2K_{ab} - 4t^2/U$.

Problem 5.5 $\lambda = B^2/3K - J^{(2)2}/4K = [(4B^2 - 3J^{(2)2})/12K = (4t_{13}^4 + 4t_{24}^4 - 8t_{13}^2 t_{24}^2) - (3t_{13}^4 + 3t_{24}^4 + 6t_{13}^2 t_{24}^2)]/12KU^2 = (t_{13}^4 + t_{24}^4 - 2t_{13}^2 t_{24}^2)/12KU^2 = ((t_{13}^2 - t_{24}^2)^2)/12KU^2$. Biquadratic exchange is maximum for maximal difference between the two t-values and approaches zero when they become equal.

Problem 5.6 Neglecting the direct exchange contribution, the perturbative estimate of $J_r = 80t^4/U^3$; for $J_{ij} = 4t^2/U \Rightarrow J_r/(J_{12}J_{23}) = (80t^4 \times U^2)/(U^3 \times 16t^4) = 5/U$. From this immediately follows that $J_r = (5J_{12}J_{23})/U = 5 \times -25.1 \times -39.5)/3100 = 1.6$ meV.

Exercises and Problems of Chap. 6

Exercise 6.1 For a centrosymmetric system $c_1 = c_2 = 1/\sqrt{2}$. Substitution in Eq. 6.9 leads to $t_{ab}^+ = (\Delta E_{12} - (1/2 - 1/2)(H_{aa} - H_{bb}))/4 \cdot 1/2 = \Delta E/2$.

Exercise 6.2 Φ_2 and Φ_5 have two α electrons on A and B, respectively. $\Phi_1 + \Phi_3$ and $\Phi_4 + \Phi_6$ are the $M_S = 0$ components of the on-site triplets. The minus combinations of these correspond to singlet coupling on the magnetic centers. Note that these functions are not directly spin eigenfunctions of the whole complex.

Exercise 6.3 The remaining zeros correspond to $\langle \Phi_1|\hat{H}|\Phi_2\rangle$, $\langle \Phi_2|\hat{H}|\Phi_5\rangle$, $\langle \Phi_3|\hat{H}|\Phi_6\rangle$ and $\langle \Phi_4|\hat{H}|\Phi_5\rangle$. In all cases the number of different columns of the *bra*-determinant and the *ket*-determinant is larger than two. These matrix elements are zero always as reflected in the Slater-Condon rules. Because of the centrosymmetric nature of the model, a_2 can be replaced by b_2 (and *vice versa*) in the integrals of the Hamiltonian.

Exercise 6.4 Contributions from the one-electron integrals and Coulomb integrals are the same in the three cases $(h_{11} + h_{22} + h_{33} + J_{12} + J_{13} + J_{23})$ and will be omitted. $\langle \varphi_1\varphi_2\varphi_3|\hat{H}|\varphi_1\varphi_2\varphi_3\rangle = -K_{12} - K_{13} - K_{23}$. $\langle \varphi_1\varphi_2\overline{\varphi}_3|\hat{H}|\varphi_1\varphi_2\overline{\varphi}_3\rangle = -K_{12}$. With $K_{13} = K_{23} = K$, the energy difference becomes $2K$, in line with Eq. 6.29. The doublet function with triplet coupling for φ_1 and φ_2 is given in Eq. 1.46. $(1/6)\langle 2\varphi_1\varphi_2\overline{\varphi}_3 - \varphi_1\overline{\varphi}_2\varphi_3 - \overline{\varphi}_1\varphi_2\varphi_3|\hat{H}|2\varphi_1\varphi_2\overline{\varphi}_3 - \varphi_1\overline{\varphi}_2\varphi_3 - \overline{\varphi}_1\varphi_2\varphi_3\rangle = (1/6)[-4K_{12} + 2K_{23} + 2K_{13} + 2K_{23} - K_{13} - K_{12} + 2K_{13} - K_{12} - K_{23}] = -K_{12} + (1/2)(K_{23} + K_{23})$. The energy difference with $|\varphi_1\varphi_2\varphi_3|$ is $3K$ as expected from Eq. 6.30.

Exercise 6.5 The expression for the second-order correction is $\sum_a \langle \Phi_I|\hat{H}|\Phi_a\rangle\langle \Phi_a|\hat{H}|\Phi_I\rangle/(E_I - E_a)$, where $I = S, T$ and a is one of the determinants other than Ψ_1 or Ψ_2 in the matrices. Only Ψ_7 and Ψ_9 have non-zero matrix elements with Ψ_1 leading to the expression given in Eq. 6.47 for the triplet. Ψ_2 interacts directly with Ψ_8 $(-t_{pd})$, Ψ_{10} $(-t_{pd})$ and Ψ_{13} $(2t_{ab})$. The application of the formula gives the second order corrected energy for the singlet.

Exercise 6.6 $\Psi_{\pm} = \frac{1}{2}(\alpha\alpha(\alpha\beta\pm\beta\alpha) + (\alpha\beta\pm\beta\alpha)\alpha\alpha)$. $-\frac{1}{2}J(\hat{s}^+(1) + \hat{s}^+(2))(\hat{s}^-(3) + \hat{s}^-(4))(\alpha\alpha\alpha\beta\pm\alpha\alpha\beta\alpha) = 0$; $-\frac{1}{2}(\hat{s}^+(1) + \hat{s}^+(2))(\hat{s}^-(3)+\hat{s}^-(4))(\alpha\beta\alpha\alpha\pm\beta\alpha\alpha\alpha) = -\frac{1}{2}J((\hat{s}^+(1) + \hat{s}^+(2))(\alpha\beta\beta\alpha + \beta\alpha\beta\alpha \pm (\alpha\beta\alpha\beta + \beta\alpha\alpha\beta)) = \mp J(\alpha\alpha\alpha\beta + \alpha\alpha\beta\alpha)$; $-\frac{1}{2}J(\hat{s}^-(1) + \hat{s}^-(2))(\hat{s}^+(3)+\hat{s}^+(4))(\alpha\alpha\alpha\beta\pm\alpha\alpha\beta\alpha) = \mp J(\hat{s}^-(1)+\hat{s}^-(2))\alpha\alpha\alpha\alpha = \mp J(\alpha\alpha\alpha\beta + \alpha\alpha\beta\alpha)$; $-\frac{1}{2}(\hat{s}^-(1) + \hat{s}^-(2))(\hat{s}^+(3) + \hat{s}^+(4))(\alpha\beta\alpha\alpha \pm \beta\alpha\alpha\alpha) = 0$; $(\hat{s}_z(1) + \hat{s}_z(2))(\hat{s}_z(3) + \hat{s}_z(4))(\alpha\alpha\alpha\beta \pm \alpha\alpha\beta\alpha) = (\frac{1}{2} + \frac{1}{2})(\frac{1}{2} - \frac{1}{2}) + (\frac{1}{2} + \frac{1}{2})(-\frac{1}{2} + \frac{1}{2})(\alpha\alpha\alpha\beta \pm \alpha\alpha\beta\alpha) = 0$ and similar for the other term of $\Psi_{\pm}$. From this: $\hat{H}\Psi_{\pm} = \mp J\Psi_{\pm}$. The eigenvalues of $\hat{S}^2$ can be determined in a similar way: $(\hat{s}^+(1) + \hat{s}^+(2) + \hat{s}^+(3) + \hat{s}^+(4))(\hat{s}^-(1) + \hat{s}^-(2) + \hat{s}^-(3) + \hat{s}^-(4))\alpha\alpha\beta\alpha = (\hat{s}^+(1) + \hat{s}^+(2) + \hat{s}^+(3) + \hat{s}^+(4))(\beta\alpha\beta\alpha + \alpha\beta\beta\alpha + 0 + \alpha\alpha\beta\beta) = 3\alpha\alpha\beta\alpha + \beta\alpha\alpha\alpha + \alpha\beta\alpha\alpha + \alpha\alpha\alpha\beta$; $(\hat{s}^+(1)+\hat{s}^+(2)+\hat{s}^+(3)+\hat{s}^+(4))(\hat{s}^-(1)+\hat{s}^-(2)+\hat{s}^-(3)+\hat{s}^-(4))\alpha\alpha\alpha\beta = 3\alpha\alpha\alpha\beta + \beta\alpha\alpha\alpha + \alpha\beta\alpha\alpha + \alpha\alpha\beta\alpha$; $(\hat{s}^+(1) + \hat{s}^+(2) + \hat{s}^+(3) + \hat{s}^+(4))(\hat{s}^-(1)+\hat{s}^-(2)+\hat{s}^-(3) + \hat{s}^-(4))\alpha\beta\alpha\alpha = 3\alpha\beta\alpha\alpha + \beta\alpha\alpha\alpha + \alpha\alpha\beta\alpha + \alpha\alpha\alpha\beta$; $(\hat{s}^+(1) + \hat{s}^+(2) + \hat{s}^+(3) + \hat{s}^+(4))(\hat{s}^-(1) + \hat{s}^-(2) + \hat{s}^-(3) + \hat{s}^-(4))\beta\alpha\alpha\alpha = 3\beta\alpha\alpha\alpha + \alpha\beta\alpha\alpha + \alpha\alpha\beta\alpha + \alpha\alpha\alpha\beta$; $(\hat{s}_z(1) + \hat{s}_z(2) + \hat{s}_z(3) + \hat{s}_z(4))\alpha\alpha\beta\alpha = (\frac{1}{2} + \frac{1}{2} - \frac{1}{2} + \frac{1}{2})\alpha\alpha\beta\alpha = \alpha\alpha\beta\alpha$ and similar for the other terms. The $\hat{S}_z^2$ operator has the same eigenvalues and cancels the effect of $\hat{S}_z$ because they appear with opposite signs in the expression of $\hat{S}^2$. From this $\hat{S}^2\Psi_+ = 6\Psi_+$ (quintet) and $\hat{S}^2\Psi_- = 2\Psi_-$ (triplet).

Exercise 6.7 (a) $\alpha(1)\beta(2)\alpha(3)\beta(4)\alpha(5)\beta(6)\alpha(7)\beta(8)$ (b) In a step-by-step procedure, the expectation value of Φ_0 is determined. The products of spin-up and spin-down operators change the wave function and hence give a zero contribution to the expectation value due to orthogonality. $-J\sum_{i=1, j=i+1}^{8} \hat{S}_z(i)\hat{S}_z(j)\alpha\beta\alpha\beta\alpha\beta\alpha\beta =$ $-J \cdot 8 \cdot \frac{1}{2} \cdot -\frac{1}{2}\alpha\beta\alpha\beta\alpha\beta\alpha\beta$ and the expectation value $= 2J$. From Eq. 6.73 one gets $\frac{1}{2} \cdot 8 \cdot 2(\frac{1}{2})^2 J = 2J$. (c) $\alpha\beta\beta\alpha\alpha\beta\alpha\beta$. The energy expectation value has again only contributions from the product of $\hat{S}_z$ operators and reads: $-J(-1/4 + 1/4 - 1/4 + 1/4 - 1/4 - 1/4 - 1/4) = J$. (d) $-\frac{1}{2}J\hat{S}^+(3)\hat{S}^-(4)$ is the only terms that can contribute to the interaction matrix element of Φ_0 and Φ_1. $-\frac{1}{2}J\hat{S}^+(3)\hat{S}^-(4)\alpha\beta\beta\alpha\alpha\beta\alpha\beta = -\frac{1}{2}\alpha\beta\alpha\beta\alpha\beta\alpha\beta$. The matrix element is $-\frac{1}{2}J$.

Problem 6.1 $E(D) = -\frac{1}{2}t + \frac{3}{2}J$; $E(Q) = -t$. The doublet is the ground state when $J < -\frac{1}{3}t$. That is, when it becomes more "antiferromagnetic" than $1/3\,t$.

Problem 6.2 Replacing the p_x and p_y orbitals on the bridge by an orbital of s symmetry activates the superexchange and semi covalent exchange between the half-filled orbitals, both favoring antiferromagnetic interaction. The semi covalent exchange between filled d-orbitals of t_{2g} character and the half-filled orbitals can no longer take place because there is zero overlap with the s function. The semi covalent exchange involving the filled $d(e_g)$ orbital gives a small ferromagnetic contribution.

Problem 6.3 (a) Only $\hat{S}_z(i)\hat{S}_z(i+1)$ gives non-zero contributions to the energy: $E(\Phi_0) = -J \cdot 8(\frac{1}{2} \cdot -\frac{1}{2}) = 2J$; $E(\Phi_1) = -J \cdot (6(\frac{1}{2} \cdot -\frac{1}{2}) + \frac{1}{2} \cdot \frac{1}{2} + (-\frac{1}{2} \cdot -\frac{1}{2})) = J$. Both in agreement with Eqs. 6.73 and 6.75. (b) $E(\Phi_0) = -J \cdot 16 \cdot (\frac{1}{2} \cdot -\frac{1}{2} + \frac{1}{2} \cdot -\frac{1}{2})J = 8J$ (each center has four neighbours, to avoid double counting only the ones with higher index (two centers) are taken into account). The eleven centers outside the shaded area contribute in the same way to the energy as in the ground state: $-J \cdot 11(\frac{1}{2} \cdot -\frac{1}{2} + \frac{1}{2} \cdot -\frac{1}{2}) = \frac{11}{2}J$. From the remaining five centers, four centers contribute with one parallel and one anti-parallel connection: $-J \cdot 4(\frac{1}{2} \cdot -\frac{1}{2} + \frac{1}{2} \cdot \frac{1}{2}) = 0$, and one centers with two ferromagnetic connections: $-J(\frac{1}{2} \cdot \frac{1}{2} + \frac{1}{2} \cdot \frac{1}{2}) = -\frac{1}{2}J$. In total, the energy becomes $5J$. Again in agreement with the general equations.

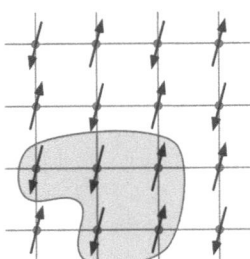

Problem 6.4 (a) When $J_2 = 0$ and $J_1 < 0$, the system corresponds to an antiferromagnetic one-dimensional chain with $\theta = 180°$, then $E = -NS^2 J$, equal to the energy expression of Eq. 6.73 with $z = 2$. (b) When $J_1 > 0$, the spins on all centers tend to align ferromagnetically, which is reinforced by a positive J_2. When $J_1 > 0$, the antiparallel alignment of the nearest neighbours results in a parallel alignment of the next-nearest neighbours (−up-down-up-down−), in line with a positive (ferromagnetic) J_2. (c) Using the trigonometric relation $\cos(2\theta) = 2\cos^2\theta + 1$, the energy

expression is first rewritten to $E = -NS^2 J_1 \cos \theta - NS^2 J_2 2 \cos^2 \theta - NS^2 J_2$. Then the energy is minimized with respect to θ: $\partial E / \partial \theta = NS^2 J_1 \sin \theta + 4NS^2 J_2 \cos \theta \sin \theta = 0 \Rightarrow \sin \theta (J_1 + 4J_2 \cos \theta) = 0$. The solutions are $\theta = 0$ (ferromagnetic), 180 (antiferromagnetic) and $\cos \theta = -J_1 / 4J_2$ (helical). (d) $J_1 = 1$, $J_2 = -0.3$; $\theta = \arccos(1/1.2) = 0.586$ rad $= 33.56°$; $J_1 = -1$, $J_2 = -0.3$; $\theta = \arccos(-1/1.2) = 2.556$ rad $= 146.44°$.

Index

A

Anderson model, 67
Anisotropic exchange, 95
   anti-symmetric, 98
   biquadratic, 101
   symmetric, 95
Anisotropic interaction, 69
Antisymmetrization operator, 2
Axial magnetic anisotropy (D), 41, 52, 98

B

Band structure, 193, 197
Bilinear exchange, 94, 159
Bilinear operator, 87, 92, 93
Biorthogonal, 29, 154
Biquadratic exchange, 88, 90, 102, 159, 166, 168, 171, 175, 193
Biquadratic exchange operator, 89
Bleaney-Bowers equation, 77
Bloch equation, 29, 72, 155
Bloch function, 193, 196, 207
Boltzmann distribution, 44
Bonner-Fisher equation, 78
Bottom-up approach, 79
Branching diagram, 13, 17
Brillouin theorem, 128, 148, 149, 151, 155, 158, 164
Broken symmetry approach, 131, 138
Broken symmetry determinant, 135, 157, 168

C

Complete active space, 121, 127, 142, 161
   CAS+S, 127, 164

CASCI, 61, 147, 174
CASPT2, *see* perturbation theory
CASSCF, 68, 128, 141
Configuration interaction, 63
Configuration state function, 61, 161, 163, 165, 174, 186
Constrained DFT, 137
Coulomb integral, 5, 31, 63, 161, 184
Coulomb interaction, 167, 191
Coulomb repulsion, 191
Counter-complementarity, 118, 138
Coupled cluster, 137
Covalent lattices, 190
Curie law, 46
Curie-Weiss law, 46

D

Density functional theory (DFT), 74, 135, 171, 174, 192
Des Cloizeaux orthogonalization, 30
Difference dedicated CI (DDCI), 123, 141, 159
   CAS+S, *see* complete active space
   DDCI1, 127
   DDCI2, 126
Dirac equation, 36
Direct exchange, 65–67, 107, 110, 143, 144, 148, 150, 155–157, 198
Doped systems, 177
Double exchange, 182, 183, 187, 211
   Anderson-Hasegawa model, 187
   Girerd-Papaefthimiou model, 186
   Zener model, 182
D tensor, 41, 52, 54

© Springer International Publishing Switzerland 2016
C. Graaf and R. Broer, *Magnetic Interactions in Molecules and Solids*,
Theoretical Chemistry and Computational Modelling,
DOI 10.1007/978-3-319-22951-5

Dzyaloshinskii-Moriya interaction, *see* anisotropic exchange

E
Easy axis magnetism, 43
Easy plane magnetism, 43
Effective Hamiltonian, 27, 28, 32, 70, 72, 73, 94, 101, 153, 166, 172, 181
Embedded cluster, 190
Embedding
   ab initio model potential, 191
   frozen density, 191
   imethod of increments, 196
   point charges, 191
   polarization, 192
Exchange-correlation, 191
Exchange integral, 5, 31, 62, 63, 65, 77, 120, 162, 184
   on-site, 161, 166, 168, 172, 185–187
   two-center, 161, 185, 202
Exchange interaction, 164, 188, 191
Exchange pathways, 107

F
Ferrimagnetism, 80
Ferroelectricity, 212
Fock operator, 25
Four-component methods, 36
Four-spin cyclic exchange, *see* ring exchange

G
Goodenough-Kanamori rules, 197, 211
Gram-Schmidt orthogonalization, 30, 155
G tensor, 48, 52, 56

H
Hartree-Fock, 21, 25, 137, 164
   Hartree product, 1, 4
   post HF, 122
   restricted open–shell, 136
   unrestricted, 135
Hay-Thibeault-Hoffmann (HTH), 64, 108, 110, 113, 144
Heisenberg Hamiltonian, 69, 71, 73, 76, 82, 89, 101, 120, 204, 206, 210
Helical spin order, 212
Hopping integral (t), 143, 145, 156, 166, 168, 175, 179, 202
Hund's rule, 14, 65, 182, 198

I
Intermolecular interactions, 80
Intramolecular interactions, 80
Intruder states, 129
Ionic crystals, 190
Ionic determinant, 124, 143, 145, 149, 162, 164–166, 181
   relaxation, 150
Ionic determinants, 64
Irreducible representation, 35, 37, 39, 61
Ising Hamiltonian, 74, 76, 193
Isotropic interactions, 69

J
Jahn-Teller distortion, 39, 199

K
Kahn-Briat, 66, 105, 107, 144
Kinetic exchange, 66, 147, 148, 150, 155, 157, 158, 165, 175
Koopmans' theorem, 127
Kramers doublet, 43, 55

L
Ladder operators, 6, 8, 42, 205, 210
Landé pattern, 70, 89, 159, 164, 165
Ligand-to-metal charge transfer (LMCT), 148, 151
LS term, 35

M
Madelung potential, 190
Magnetic susceptibility, 44, 77
Magnetoresistance, 182
Many center interactions, *see* ring exchange
McConnell's model, 110
Mermin-Wagner theorem, 85
Metal-to-ligand charge transfer (MLCT), 151
Metropolis algorithm, 85
Mixed valence systems, 177
Model Hamiltonian, 29, 41, 43, 52, 68, 74, 95, 100, 185, 188
Molecular crystals, 190
Molecular dynamics, 85
Molecular orbital diagram, 118, 119
Monte Carlo simulation, 82, 85
MR-CISD, 121
Multiconfigurational, 120
Multideterminantal, 120, 209

Multiferroic, 95
Multireference, 121, 127

N

Neél state, 210, 211
Neutral determinant, 64, 124, 144, 145, 149, 162, 165, 174, 181
Non-Hund state, 77, 160, 162, 164, 165, 186
Noodleman equation, 135

O

On-site repulsion (U), 143, 145, 156, 166, 168, 177
Open-shell singlet, 60, 120
Orbital ordering, 199
Orbitals, 1
 active, 61, 68, 121, 142, 147
 antibonding, 60, 63, 66, 110, 180
 barred, 6
 bonding, 60, 61, 63, 66, 110, 180
 degenerate, 65
 delocalized, 60, 62–64, 101
 empty, 129, 178, 179, 198–200
 energies, 26, 109, 110, 119
 filled, 178, 198, 200, 201
 gerade/ungerade, 114, 118, 142, 147, 174, 180
 half-filled, 198–201, 203
 inactive, 121, 124, 125, 178, 179
 localized, 62–66, 101, 108, 109, 111, 142
 magnetic, 60, 67, 68, 106, 107, 119, 158, 159, 177, 180
 molecular, 59, 62, 108, 109, 121
 molecular orbital theory, 1
 nonorthogonal, 66, 67, 105, 108, 133, 134
 orbital moment, 33–35, 37–40, 43, 48, 49
 orbital moment quenching, 39
 spin orbital, 6, 131, 158, 193
 state-specific, 152
 virtual, 121, 124, 125

P

Partition function, 83
Periodic boundaries, 79, 210
Periodic calculation, 193
Permutation operator, 3, 91, 106
Perturbation operator, 22, 25, 40, 127
Perturbation theory, 21, 36, 40, 48
 (un)contracted first order wave function, 130

CASPT2, 123, 127, 131, 151
Møller-Plesset, 21, 25, 127, 197
NEVPT2, 127, 129, 131, 151
quasi-degenerate, 21, 27, 124, 144–146, 148, 166, 168, 201
Rayleigh-Schrödinger, 21, 31
zeroth order Hamiltonian, 22, 25, 127, 129
Post HF, 196
Projection operator, 9, 10, 29

R

Renormalization group theory, 82
Restricted ensemble Kohn–Sham (REKS) DFT, 136
Restricted open-shell Kohn–Sham (ROKS) DFT, 136, 157
Rhombic magnetic anisotropy (E), 41, 54, 98
Ring exchange, 90, 91, 95, 159, 166, 168, 172, 175, 193
Robin and Day classification, 177
Russell-Saunders, 36, 39

S

Scalar relativistic effects, 36
Semi covalent exchange, 198–201
Shift operators, *see* ladder operators
Slater determinant, 1, 3, 7, 8, 21, 59, 65, 66, 120, 121
Slater-Condon rules, 3, 25, 62, 125, 179, 184
Spectral decomposition, 29
Spin contamination, 20, 132, 137, 174
Spin density, 111
Spin eigenfunction, 8
 by diagonalization, 10
 by projection, 9
 genealogical approach, 12
Spin-flip DFT, 137
Spin multiplicity, 35
Spin-orbit coupling, 34, 36, 40, 41, 48
Spin polarization, 132, 148, 156–158
Spin wave, 209, 211
Step-up/step-down operators, *see* ladder operators
Strong overlap limit, 133, 136
Superexchange, 67, 68, 147, 182, 197–200, 203

T

Through-bond interaction, 80
Through-space interaction, 80

Total spin operator, $\hat{S}^2$, 5–13, 16, 34, 35, 41, 42, 52, 69, 73, 101, 131, 132, 135, 136, 174
Two-electron/two-orbital, 10, 31, 59, 61, 68, 74, 120, 124, 131, 137

V
Valence bond theory, 64, 66, 105
Van Vleck equation, 45, 77

W
Wannier orbital, 196

Weak overlap limit, 133

Y
Yamaguchi's relation, 132, 134, 157

Z
Z-component of the spin operator, $\hat{S}_z$, 5–7, 9, 11, 34, 35, 41–43, 47, 49, 52, 74, 207
Zeeman effect, 43, 49, 52
Zero field splitting, 39, 52